BARRON'S

E-Z

EARTH SCIENCE

Alan D. Sills, M.A.

BARRON'S

THOUSAND OAKS LIBRARY
1401 E. Janss Road
Thousand Oaks, CA 91362

Better Grades or Your Money Back!

As a leader in educational publishing, Barron's has helped millions of students reach their academic goals. Our E-Z series of books is designed to help students master a variety of subjects. We are so confident that completing all the review material and exercises in this book will help you, that if your grades don't improve within 30 days, we will give you a full refund.

To qualify for a refund, simply return the book within 90 days of purchase and include your store receipt. Refunds will not include sales tax or postage. Offer available only to U.S. residents. Void where prohibited. Send books to **Barron's Educational Series, Inc., Attn: Customer Service** at the address on this page.

All inquiries should be addressed to:
Barron's Educational Series, Inc.
250 Wireless Boulevard
Hauppauge, New York 11788
www.barronseduc.com

Library of Congress Catalog Card No.: 2009040338

ISBN-13: 978-0-7641-4464-6
ISBN-10: 0-7641-4464-2

Library of Congress Cataloging-in-Publication Data
Sills, Alan D.
    E-Z earth science / Alan D. Sills.
       p.   cm.—(E-Z series)
    Includes bibliographical references and index.
    ISBN-13: 978-0-7641-4464-6 (alk. paper)
    ISBN-10: 0-7641-4464-2 (alk. paper)
    1. Earth sciences. I. Title.

    QE28.S54   2010
    550—dc22
                        2009040338

PRINTED IN THE UNITED STATES OF AMERICA
9 8 7 6 5 4 3 2 1

CONTENTS

Preface   viii

1 What Is Earth Systems Science?   1
What You Will Learn   1
Natural Sciences   4
    Astronomy   4
    Geology   4
    Meteorology   5
    Oceanography   8
A Related Study—Environmental Science   8
Earth's Major Systems   8
    Atmosphere   9
    Hydrosphere   9
    Cryosphere   10
    Geosphere   11
    Biosphere   11
Earth as a System   12
Studying Earth Systems Science   12
Good Starting Points for Your Search on   13
    the Internet for Images and Information
Review Exercises for Chapter 1   14
    Word Study Connection   14
    Self-Test Connection   14
    Connecting to Concepts   18
    Connecting to Life/Job Skills   18
    Answers   19

**PART I: ASTRONOMY: A STUDY OF
THE UNIVERSE AND EARTH'S
PLACE IN IT / 21**

2 Beyond Our Solar System: 23
    A Study of the Universe
What You Will Learn   23
Origin of the Universe and Its Key Features   24
    The Big Bang Theory   24
    Background Radiation   24
    Redshift—Our Expanding Universe   25
    Alternative Theories Explaining   27
        the Origin of the Universe
Tools for Studying the Universe—   27
    Remote Sensing Instruments
    Telescopes   28
    Other Remote Sensing Instruments   30
Galaxies   32
    What Is a Galaxy?   32
    How Galaxies Form   32
    Types of Galaxies   33
    Galactic Clusters   33
Stars   34
    What Is a Star?   34

A Star's Source of Energy   34
Nebulae and Stellar Formation   36
Stellar Evolution   36
Supernovae   40
Stellar Remnants   42
Hertzsprung-Russell Diagram   43
Our Sun   44
Binary Stars   44
Measuring Distances   **44**
    What Is a Light-Year?   44
    Parallax   45
    A Star's Brightness: Apparent Versus   45
        Absolute Magnitude
    Cepheid Variable Stars   46
Constellations   **46**
Related Internet Resources   **47**
    for Great Images and Information
Review Exercises for Chapter 2   **48**
    Word-Study Connection   48
    Self-Test Connection   49
    Connecting to Concepts   52
    Connecting to Life/Job Skills   52
    Answers   53

3 Our Solar System   55
What You Will Learn   **55**
Major Components of Our Solar System   **56**
    The Sun   56
    The Planets   56
    Natural Satellites   62
    Asteroid Belt   64
    Meteoroids   65
    Comets   66
How Did the Solar System Form?   **66**
    Nebular Hypothesis   66
    The Inner Planets Versus the Outer Planets   68
An Early View of Our Solar System—   **68**
    The Ptolemaic System
Contributions of the First Modern Astronomers   **70**
    Nicolaus Copernicus and the Heliocentric Theory   70
    Tycho Brahe and Stellar Parallax   70
    Johannes Kepler and the Laws of Planetary Motion   72
    Galileo Galilei and the Telescope   74
    Sir Isaac Newton and Gravity   75
Motions of the Planets   **76**
    Rotation   76
    Revolution   77
    Precession   77
Recent Discoveries/Special Topics   **79**
    Other Solar Systems   79
    Venus and Its Runaway Greenhouse Effect   80
    Large Quantities of Ice Found on Mars   80

Comet Shoemaker-Levy and   81
    Its Impact upon Jupiter
History of Human Exploration   **81**
    of the Solar System
    U.S. Contributions   81
    Soviet/Russian Contributions   83
    Future Endeavors   84
Related Internet Resources   **84**
    for Great Images and Information
Review Exercises for Chapter 3   **85**
    Word-Study Connection   85
    Self-Test Connection   86
    Connecting to Concepts   89
    Connecting to Life/Job Skills   90
    Answers   90

4 The Earth–Moon System   **93**
What You Will Learn   **93**
Earth and the Moon—A Comparison   **94**
    Major Systems and Processes   94
    Surface Features   94
    Relative Size—A Unique Relationship   96
Importance of the Moon   **98**
    Scientific Study   98
    Impact on the Earth's Oceans   98
    Lunar Motions   99
    Phases of the Moon   100
    Sidereal Versus Synodic Month   101
Lunar and Solar Eclipses   **103**
    Lunar Eclipses   103
    Solar Eclipses   103
    Lunar Climate   104
Internet Resources to Lunar Phase   **105**
    and Tide Data
Review Exercises for Chapter 4   **105**
    Word-Study Connection   105
    Self-Test Connection   106
    Connecting to Concepts   109
    Connecting to Life/Job Skills   109
    Answers   109

PART II: GEOLOGY: A STUDY OF EARTH'S SURFACE AND ITS INTERIOR / 111

5 Minerals and Rocks—Evidence   **113**
    of Continual Change
What You Will Learn   **113**
Minerals   **114**
    Basic Chemistry   114
    Properties of Minerals   117
    Mineral Groups   119
    Minerals as Important Resources   121
    Mineral or Rock?   121

Igneous Rocks   **121**
    Magma Versus Lava   121
    Classifying Igneous Rocks   123
    Common Igneous Rocks and Their Uses   125
Sedimentary Rocks   **125**
    Formation of Sedimentary Rocks   125
    Classifying Sedimentary Rocks   126
    Importance and Value of Sedimentary Rocks   128
Metamorphic Rocks   **129**
    Metamorphism   129
    Classifying Metamorphic Rocks   130
    Common Metamorphic Rocks and Their Uses   131
The Rock Cycle   **131**
Related Internet Resources for Great Images   **132**
    and Information
Review Exercises for Chapter 5   **132**
    Word-Study Connection   132
    Self-Test Connection   133
    Connecting to Concepts   136
    Connecting to Life   137
    Answers   138

6 Geologic Time and Dating Rocks   **141**
What You Will Learn   **141**
The Geologic Time Scale   **142**
    Recognizing the Need for a Time Scale   142
    Organization of the Geologic Time Scale   142
Relative Dating of Rocks   **148**
    Law of Superposition   148
    Principle of Original Horizontality   149
    Correlation of Rock Layers   149
    Cross Cutting   150
    Unconformities   151
    Fossils   151
Absolute Dating of Rocks   **152**
    Radioactive Decay   152
    Basics of Nuclear Chemistry   153
    Radiometric Dating   154
The Age of Earth   **155**
Related Internet Resources to   **157**
    Great Images and Information
Review Exercises for Chapter 6   **158**
    Word-Study Connection   158
    Self-Test Connection   159
    Connecting to Concepts   162
    Answers   163

7 Plate Tectonics—Our   **165**
    Dynamic Planet
What You Will Learn   **165**
Continental Drift   **166**
    Early Evidence Presented   166
    Recent Evidence Presented   167

Convection Cells and the Mantle 169
Relationship of the Mantle to the Crust 169
Tectonic Plates 169
Pangaea—Then Versus Now 171
Plate Boundaries and Tectonic Activity 171
Divergent Boundaries 171
Convergent Boundaries 171
Transform Fault Boundaries 175
Hot Spots and the Hawaiian Islands 176
Plate Boundaries and Earthquakes 177
What Is an Earthquake? 177
P-, *S*-, and *L*-Waves 178
Aftershocks and Foreshocks 179
Reading a Seismograph 179
Epicenter Versus Focus 179
Locating an Epicenter 180
Magnitude and Intensity 181
Special Topics 183
Earthquakes—The Key to Understanding 183
Earth's Interior
Predicting Earthquakes 184
Tsunamis 184
Geology of Our National Parks 185
Internet Resources to Study Current 185
Earthquakes and Related Information
Review Exercises for Chapter 7 186
Word-Study Connection 186
Self-Test Connection 186
Connecting to Concepts 189
Connecting to Life/Job Skills 190
Answers 191

8 Volcanoes 193
What You Will Learn 193
Volcanoes and Plate Tectonics—The Ring 194
of Fire
Volcanic Eruptions 195
Origin of Magma 195
Gases and Viscosity in Magma 195
Lava Flows 196
Pyroclastic Materials 196
Types of Volcanoes 196
Shield Volcanoes 197
Cinder Cones 197
Composite Cones 198
Lava Plateaus 198
Intrusive Features 198
Batholiths 198
Laccoliths 199
Dikes and Sills 199
Case Studies of Volcanoes 199
Mount Saint Helens 199
Kilauea 200

Internet Resources for Current Volcanic Activity 201
Review Exercises for Chapter 8 201
Word-Study Connection 201
Self-Test Connection 201
Connecting to Concepts 204
Connecting to Life/Job Skills 205
Answers 205

9 Earth's Surface 207
What You Will Learn 207
Our Dynamic Planet 208
Forces Affecting Earth's Surface: 208
Weathering and Erosion
Running Water 208
Wind 209
Animals and Plants 209
Mass Wasting 210
Glaciers 210
Types of Mountains 213
Fault-Block Mountains 213
Folded Mountains 214
Domed or Upwarped Mountains 214
Volcanic Mountains 215
Types of Deformation 215
Folds 215
Faults 215
Joints 216
Review Exercises for Chapter 9 216
Word-Study Connection 216
Self-Test Connection 217
Connecting to Concepts 220
Connecting to Life/Job Skills 220
Answers 221

PART III: OCEANOGRAPHY AND WATER ON EARTH / 223

10 Water, Water Everywhere 225
What You Will Learn 225
Distribution of Water 226
Important Properties of Water 226
Groundwater 231
Hot Springs and Geysers 232
Wells 232
Caverns and Caves 233
Streams and Rivers 234
Stream Flow 234
Drainage Basins 234
Review Exercises for Chapter 10 236
Word-Study Connection 236
Self-Test Connection 237
Connecting to Concepts 240
Connecting to Life/Job Skills 240
Answers 241

11 The World's Oceans — 243

What You Will Learn — 243
Seawater — 244
  Composition — 244
  Sources of Minerals — 244
  Resources from Seawater — 246
  The Layered Structure of the Oceans — 246
  Marine-Life Zones — 249
Mapping the Oceans — 251
  Early Discoveries — 251
  More Recent Projects — 253
Features of the Seafloor — 254
  Deep Ocean Trenches — 254
  Abyssal Plains — 255
  Seamounts — 256
  Mid-ocean Ridges — 257
  Continental Margins — 257
Review Exercises for Chapter 11 — 258
  Word-Study Connection — 258
  Self-Test Connection — 258
  Connecting to Concepts — 261
  Connecting to Life/Job Skills — 262
  Answers — 262

12 The Dynamic Oceans — 265

What You Will Learn — 265
Surface Currents — 266
  Ocean Circulation Patterns — 266
  Upwelling — 266
  Surface Ocean Currents and Climatic Patterns — 267
  El Niño and La Niña — 269
Deep Ocean Circulation — 271
Tides — 272
  Causes of Tides — 272
  Spring and Neap Tides — 273
  Tidal Patterns — 273
Waves — 275
  Causes of Waves — 275
  Characteristics of Waves — 275
  Wave Erosion — 276
Human Efforts to Restore Our Beaches — 278
  Groins — 278
  Seawalls — 279
  Dunes — 280
  Beach Restoration — 280
Related Internet Resources for Great Images and Information — 281
Review Exercises for Chapter 12 — 281
  Word-Study Connection — 281
  Self-Test Connection — 282
  Connecting to Concepts — 285
  Connecting to Life/Job Skills — 286
  Answers — 286

PART IV: ATMOSPHERIC SCIENCE— THE STUDY OF OUR ATMOSPHERE, WEATHER, AND CLIMATE / 289

13 Earth's Atmosphere — 291

What You Will Learn — 291
Composition of the Atmosphere — 292
  Major Components — 292
  Variable Components — 292
  The Early Atmosphere — 293
Structure of the Atmosphere — 294
  Troposphere — 294
  Stratosphere — 295
  Mesosphere — 296
  Thermosphere — 296
Important Earth–Sun Relationships — 296
  Radiation — 296
  Paths Taken by Incoming Solar Radiation — 298
The Dynamic Earth — 299
  Albedo and Reradiated Energy — 299
  The Natural Greenhouse Effect — 299
  Convection — 300
The Earth Rotates on a Tilted Axis — 301
  Solstices, Equinoxes, and Seasons — 301
Circulation of the Atmosphere—Global Convection Cells — 307
Review Exercises for Chapter 13 — 312
  Word-Study Connection — 312
  Self-Test Connection — 313
  Connecting to Concepts — 316
  Connecting to Life/Job Skills — 317
  Answers — 317

14 Atmospheric Moisture, Pressure, and Winds — 321

What You Will Learn — 321
Moisture — 322
  Relative Humidity — 322
  Dew Point — 325
  Measuring Dew Point and Relative Humidity — 326
  Latent Heat — 328
Clouds and Fog — 328
  Composition — 328
  Cloud Formation — 328
  Types of Clouds — 329
  Atmospheric Stability — 330
Precipitation — 330
  How Precipitation Forms — 330
  Forms of Precipitation — 331
  Measuring Precipitation — 333
Air Pressure — 334
  Measuring Air Pressure — 334
  Lows and Highs: Synoptic-Scale Weather Systems — 335

Winds 339
    Relationship Between Air Pressure and Wind 339
    Factors Affecting Winds 340
Review Exercises for Chapter 14 341
    Word-Study Connection 341
    Self-Test Connection 342
    Connecting to Concepts 345
    Connecting to Life/Job Skills 346
    Answers 347

15 Weather and Weather Systems 349
What You Will Learn 349
Cyclones and Anticyclones 350
    Air Masses and Their Characteristics 350
    Fronts 351
    Cyclogenesis—Formation of 354
      a Midlatitude Storm System
    Development of a Cyclone 355
    Precipitation Patterns Associated 355
      with a Cyclone
    Dissipation of a Cyclone 357
Other Forms of Severe Weather 357
    Thunderstorms and Lightning 357
    Tornadoes 359
Tropical Weather Systems 361
    Tropical Storms and Hurricanes 361
    Paths of Destruction 363
    Storm Surge 364
Weather Forecasting 364
    Is It Science or Art? 364
    Modern Methods of Forecasting 365
Related Internet Resources 365
    for Great Images and Information
Review Exercises for Chapter 15 366
    Word-Study Connection 366
    Self-Test Connection 367
    Connecting to Concepts 370
    Connecting to Life/Job Skills 372
    Answers 373

16 Climate and Climatic Zones 375
What You Will Learn 375
Climate Defined 376
    Global Climatic Patterns 376
    Factors Affecting Climate 377
    Importance of Vertical Motion 377
Global Climatic Zones 379
    The Intertropical Convergence Zone 379
    The Horse Latitudes 380
    The Polar Storm Track Westerlies 380
    Subsidence at the Poles 380
Climate Patterns Across the United States 381
    A Three-City Study of Climatic Data from 381
      Selected Locations
    An Analysis of the Western Seaboard 384
    The Mountain Rain Shadow Effect 386
    The Desert Southwest and Semiarid Regions 388
    Mountain Climates 389
    The Midwest 390
    The Great Lakes Effect 390
    The Northeast 390
    The Subtropical Southeast 391
Climate Changes 392
    Natural Changes 392
    Human Impact on Global Climate 392
Related Internet Resources 394
    for Great Images and Information
Review Exercises for Chapter 16 395
    Word-Study Connection 395
    Self-Test Connection 395
    Connecting to Concepts 398
    Connecting to Life/Job Skills 400
    Answers 401

Practice Exam 403
Index 425

PREFACE

E-Z Earth Science is a useful guide intended for learners who wish to advance their understanding of the natural sciences known collectively as the Earth sciences. Included in the content are topics related to the study of geology, oceanography, meteorology, and astronomy. This study guide is designed to serve as an excellent resource for high school and college students taking their first Earth science, Earth systems science, and geology courses.

E-Z Earth Science can also provide the reader who is not enrolled in a formal course with an insight into all major Earth systems science–related topics. The writing style is conversational, and all topics, some of which are technical, are presented in a scientifically accurate, yet highly understandable and interesting, fashion.

Current environmental issues are addressed at strategic points throughout the book. Also, the guide takes a "systems" approach toward the study of Earth science. Connections between major systems and interrelationships are emphasized, such as the interrelationships between events occurring within the atmosphere, hydrosphere, and biosphere. One such example is the phenomenon known as acid rain. Acid rain is produced by human activities that cause chemical changes in the atmosphere, which then results in acidification of our planet's waters. Acidification of the waters then has an impact on other life-forms. Many other environmental issues and examples of the interactions between Earth's major systems are given throughout the book.

Each chapter contains many useful features, including photographs, line drawings explaining concepts, and Internet references, inserted at key points within the chapter as well as in a general list at the end of several chapters. At the end of each chapter are a list of key vocabulary terms, a variety of questions (with complete answers) given in different formats to enhance learning, and a section called "Connecting to Life/Job Skills," which enables readers to see how the material learned relates to the current world around them.

The hundreds of illustrations provided throughout the book allow the reader to gain a visual picture of the material being discussed, thus aiding in comprehension. Finally, a 100-question test is given at the end of the book to help solidify the material learned.

It is hoped that this book will help the reader gain a greater appreciation for Earth and the natural systems that are continuously at work on it, within it, and beyond it. Students who are enrolled in an Earth science or Earth systems science course will find this book to be an invaluable resource to supplement classroom instruction and provide greater insight into many of the topics usually discussed.

What Is Earth Systems Science?

WHAT YOU WILL LEARN

This chapter focuses on an introduction to Earth science and the different Earth systems that make up the growing field. In this chapter you will learn

- important vocabulary and scientific terms relating to Earth science;
- about the different fields within Earth science, such as astronomy and meteorology;
- about the Earth's different systems, their functions, and how to research further;
- different interactions between the Earth's systems and how they affect one another.

SECTIONS IN THIS CHAPTER

- Natural Sciences
- A Related Study—Environmental Science
- Earth's Major Systems
- Earth as a System
- Studying Earth Systems Science
- Good Starting Points for Your Search on the Internet for Images and Information
- Review Exercises for Chapter 1

Earth systems science is an exciting new and emerging field. A major thrust of Earth systems science is to understand the ways in which our planet is changing and the potential consequences of these changes for life on Earth. Traditionally, an Earth scientist has studied and specialized in one of four major disciplines: geology, the study of the solid Earth; meteorology, the study of atmosphere, climate, and weather; oceanography, the study of oceans; and astronomy, the study of Earth's place in the solar system and the universe, but Earth scientists have rarely looked at the planet as a whole.

Earth systems science brings together scientists interested in each of the disciplines mentioned above and challenges them, and us, to search for links between processes observed in each of Earth's major systems. For example, an Earth systems scientist may choose to study the oceans and any changes within them for the purpose of understanding their potential impact on our planet's climate and on global climatic change.

Earth systems scientists are also responsible for developing models for tracking changes in social-risk hazards and natural hazards. Models are tools used by scientists to develop a better understanding of what they are studying. For example, they may model the most common causes of groundwater contamination and then seek to discover means by which these can be minimized. Models are often based upon mathematical formulas known to govern or control a process. With modern computers, these models can be studied as visualizations. Visualizations help scientist and layperson alike to better understand the process being studied and the interrelationships that exist between major systems.

Oil and coal exploration continues to be an important social issue as our society is heavily dependent upon fossil fuels, but these same fuels are probably responsible for fouling our environment. Earth systems scientists build models emulating this process and then search for new methods of extracting oil and coal from the planet in a more environmentally friendly way, and of burning them without adding to the air pollution already prevalent in our cities.

Satellites and other technologies have fueled this move toward a "systems" approach as they enable us to view and study our planet in ways that highlight the interdependency of Earth's major systems. In recent years, the National Aeronautics and Space Administration (NASA) has launched a series of satellites designed to study our planet. One such satellite is Terra; launched December 18, 1999, it started collecting data and images February 24, 2000. Armed with a series of instruments, Terra is capable of studying climate, sea-surface temperatures, surface vegetation, air pollution, and more. A complete description, along with recent images acquired by Terra, can be found on the Internet at *http://terra.nasa.gov/*.

Terra: Studying conditions on Earth. Courtesy NASA.

Terra and other satellites like it are helping increasing numbers of scientists to make this shift toward an Earth systems view of our planet. This change is essential because many of the issues we face as a society transcend the traditional boundaries established by the individual disciplines: geology, meteorology, oceanography, and astronomy.

For example, the issue of global warming has been discussed at length both by the media and within scientific circles. Global warming will cause changes in all major systems, and changes in one system will impact other systems. Understanding this issue requires a broad knowledge that transcends the traditional areas studied by meteorologists, oceanographers, and geologists.

The issues we face as a society involve a number of different time scales. Some changes may occur over a matter of minutes or hours; an example is the buildup of ozone over the course of a hot summer day in a metropolitan region. Change can also occur over a period of years, centuries, or even eons. As for global warming, many scientists fear that human activity is the root cause of Earth's recent warming and by the time these suspicions are confirmed, the process may not be reversible.

"Wouldn't it be nice if Earth were a little warmer?" Consider the systems that would be affected by global warming AND the changes that would take place in the interactions between these systems.

A system is a part of a whole that may be studied separately, but is necessarily still dependent upon changes in other systems. Our planet has five major systems (descriptions of each begin in a later section called "Earth's Major Systems"): the atmosphere, hydrosphere, geosphere, cryosphere, and (unique to Earth) biosphere. The kinds of connections that exist among these systems can be seen by considering humans, who are part of the biosphere, and the activities of a modern society. These activities inevitably produce pollutants, which cause changes in the atmosphere and hydrosphere (waters) that can then have an impact on other life-forms. This is just one simple example of the kinds of interdependencies that exist among the five major systems.

In this chapter, you will be introduced to the various natural sciences that together make up the Earth sciences. Although all are intimately interrelated, each discipline is often treated as a full course of study in college. As you read this guide, you will find that each natural science paints a dynamic picture of the world and universe around us, filled with imagery and fascinating stories.

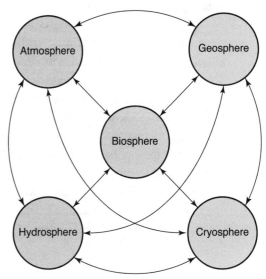

Interactions and interdependencies between and among each of the major systems.

Natural Sciences

ASTRONOMY

The science of astronomy concerns itself with the study of the universe and Earth's place within it. Astronomers, those who engage in the study of astronomy, are concerned with understanding the origins of our universe, galaxy, and solar system. A greater understanding of these entities may help us to answer questions such as "What is the probability of finding an Earth-like planet, with life as we know it, elsewhere in the universe?" and "Will the universe, which began with the 'big bang,' end with a 'big crunch'?"

Astronomers generally accept that the universe began with a cataclysmic explosion, often called the *big bang*. Evidence to support this event is discussed in Chapter 2. Since the time of the big bang, 13 billion years ago, the universe has been expanding. A significant, and as yet unanswered, question is whether this expansion will continue indefinitely or whether the expansion will cease and the universe will begin to contract, ultimately producing a "big crunch."

GEOLOGY

Geologists investigate natural processes occurring both on the surface and within the interior of Earth. Geology can be subdivided into two major time-related disciplines: physical geology and historical geology. Physical geology is the study of the materials the planet is composed of and the present-day processes that operate on the surface and within Earth's deeper layers. Understanding these processes involves study of the tectonic plates that create earthquakes, volcanic eruptions, and other interesting phenomena as these plates slowly slip and slide over the mantle.

Historical geology is the study of Earth's 4.6-billion-year history, its formation, development, and evolution. Historical geologists strive to create a time line of events that have occurred since Earth's formation. Rocks estimated to be about 3.5 billion years old are used to re-create the sequence of events that led to the formation of life on this planet and the changes that have transpired since. From these rocks, scientists have found evidence that life began more than 3 billion years ago in the oceans. When the first organisms crawled onto the land, they left behind, in the form of fossils, a record of their presence. Fossils (see Figure 1.1) are the remnants of ancient organisms. By determining the age of the rock in which a fossil is found, geologists can infer the age of the fossil. Following the rock record, a geologic timescale (see Figure 1.2) has been developed that traces the progress of life on Earth from its earliest simple forms to the wide variety of complex and in some cases highly evolved species that exist today.

Major issues of concern to geologists include finding suitable sites to drill for oil and natural gas, and predicting earthquakes and volcanic eruptions far enough in advance to minimize loss of human life and property.

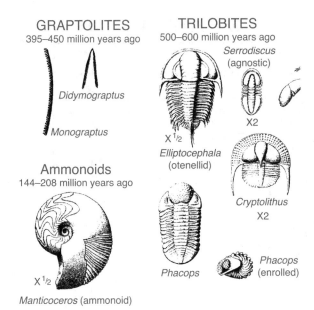

GRAPTOLITES
395–450 million years ago

Didymograptus

Monograptus

Ammonoids
144–208 million years ago

X ½

Manticoceros (ammonoid)

TRILOBITES
500–600 million years ago

Serrodiscus
(agnostic)

X2

X ½

Elliptocephala
(otenellid)

Cryptolithus
X2

Phacops

Phacops
(enrolled)

FIGURE 1.1 Illustration of various fossils, including graptolites, trilobites, and ammonoids. Each fossil is labeled with the time period that it roamed the planet. Source: The New York State Museum, Educational Leaflet #28, *Geology of New York, A Simplified Account*, Second Edition. Printed with permission of the New York State Museum, Albany, N.Y.

METEOROLOGY

On a daily basis, meteorology is the branch of the Earth sciences on which the average person is probably the most reliant. The science of meteorology is essentially the study of our atmosphere, and the primary responsibility of many meteorologists is to prepare daily weather forecasts. Meteorologists who are directly involved in the issuing of weather forecasts are practicing what is known as *synoptic meteorology*. Most forecasting meteorologists are serving in some capacity for the National Weather Service. In contrast, dynamic meteorologists study atmospheric processes in an effort to improve human understanding of the planet's atmosphere.

A scientific field closely related to meteorology is climatology. Climatologists are concerned with the study of long-term weather patterns, that is, with climate. In recent years headlines in newspapers and periodicals have cited the increasing frequency and severity of heat waves, along with record seasonal warmth, causing many to question whether Earth's climate system is changing. Extensive research on global climate change is now under way to tackle these issues.

Changes in the hydrosphere, or oceans, have also been tied to climate change. Perhaps the most familiar of these changes are the phenomena known as El Niño and its counterpart, La Niña. El Niño is a warming of the surface waters in the eastern equatorial Pacific Ocean, and La Niña is a cooling of these waters. Both anomalies can impact global atmospheric circulation and, consequently, weather and climate patterns. El Niño and La Niña are discussed in greater detail in Chapter 12.

If our planet's climate is now changing more rapidly than the changes identified over the past several hundred thousand years indicate, many ecosystems will be

GEOLOGIC HISTORY

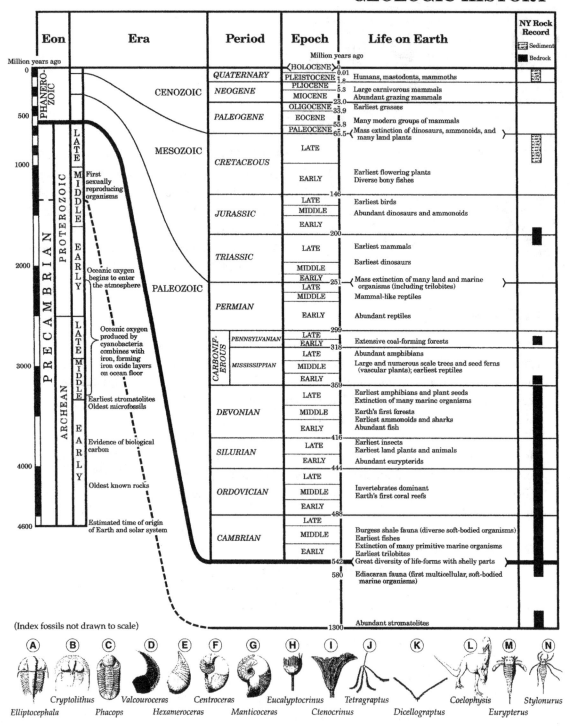

FIGURE 1.2 The geologic time scale. Source: The State Education Department, *Earth Science Reference Tables*, 2006 ed. (Albany, New York: The University of the State of New York).

OF NEW YORK STATE

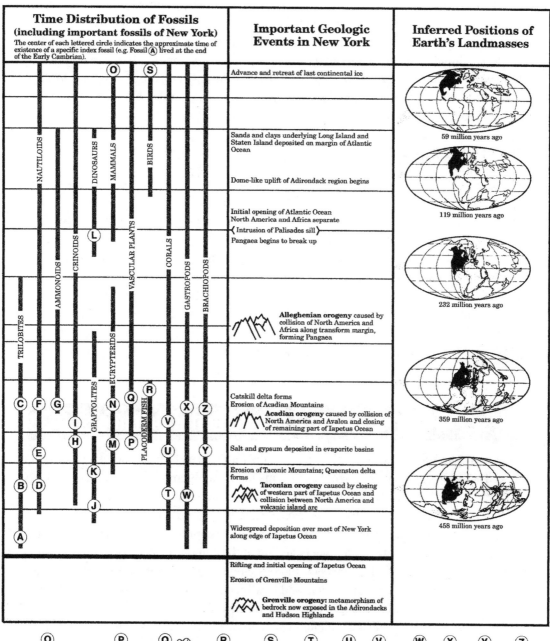

Time Distribution of Fossils (including important fossils of New York)	Important Geologic Events in New York	Inferred Positions of Earth's Landmasses

The center of each lettered circle indicates the approximate time of existence of a specific index fossil (e.g. Fossil (A) lived at the end of the Early Cambrian).

Fossil column labels: TRILOBITES, NAUTILOIDS, AMMONOIDS, CRINOIDS, DINOSAURS, MAMMALS, GRAPTOLITES, EURYPTERIDS, VASCULAR PLANTS, PLACODERM FISH, BIRDS, CORALS, GASTROPODS, BRACHIOPODS

Important Geologic Events in New York:
- Advance and retreat of last continental ice
- Sands and clays underlying Long Island and Staten Island deposited on margin of Atlantic Ocean
- Dome-like uplift of Adirondack region begins
- Initial opening of Atlantic Ocean; North America and Africa separate
- ⟨ Intrusion of Palisades sill ⟩
- Pangaea begins to break up
- **Alleghenian orogeny** caused by collision of North America and Africa along transform margin, forming Pangaea
- Catskill delta forms
- Erosion of Acadian Mountains
- **Acadian orogeny** caused by collision of North America and Avalon and closing of remaining part of Iapetus Ocean
- Salt and gypsum deposited in evaporite basins
- Erosion of Taconic Mountains; Queenston delta forms
- **Taconian orogeny** caused by closing of western part of Iapetus Ocean and collision between North America and volcanic island arc
- Widespread deposition over most of New York along edge of Iapetus Ocean
- Rifting and initial opening of Iapetus Ocean
- Erosion of Grenville Mountains
- **Grenville orogeny:** metamorphism of bedrock now exposed in the Adirondacks and Hudson Highlands

Inferred Positions of Earth's Landmasses:
- 59 million years ago
- 119 million years ago
- 232 million years ago
- 359 million years ago
- 458 million years ago

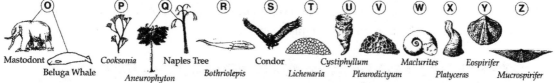

O Mastodont / Beluga Whale
P Cooksonia / Aneurophyton
Q Naples Tree
R Bothriolepis
S Condor
T Lichenaria
U Cystiphyllum / Pleurodictyum
V
W Platyceras
X Maclurites
Y Eospirifer
Z Mucrospirifer

ESC/BW/TN (2009)

impacted. Earth systems science will play a role in identifying the ecosystems that are most fragile and most likely to be affected by changes in global climate patterns. For example, the Great Plains of the United States are blessed with the best soil on the planet for growing crops that produce large amounts of food for the entire world. If global climate change results in less favorable climatic conditions for that region, how will the world be fed? Can the "bread basket" simply be shifted farther north? Most experts say no, as Canada does not have the rich soils that characterize the Great Plains.

OCEANOGRAPHY

Oceanographers study our planet's oceans. Oceanography includes study of the composition of seawater; the movement of water, including waves and tides; coastal geomorphology (changes in our coastlines); seafloor topography; and marine life. Oceanographers receive extensive training in chemistry, physics, geology, and biology, all of which involve interrelated aspects of oceans. The oceans may provide important clues about future changes in Earth's climate system. Changes in sea-surface temperatures or in the flow of ocean currents could cause rapid and profound changes in climate patterns worldwide. Earth systems scientists are actively engaged in the study of the complex relationships that exist among ocean currents, seawater temperatures, both on the surface and at depth, and the planet's overall climate system.

A Related Study–Environmental Science

Environmental science is so closely related to the Earth sciences that it deserves mention here. The study of our environment has become increasingly important in recent years, as humans have stressed Earth's natural systems. Our society now faces critical environmental issues that include global climate change, stratospheric ozone depletion, ground-level ozone accumulation, acid rain, and tropical deforestation. These problems are best studied by first developing a firm understanding of the natural sciences described above, along with the fundamentals of physics, biology, and chemistry. With a firm grasp of the Earth sciences, one can begin to understand how human activities may be causing drastic and potentially devastating changes in our natural environment.

Increasingly, social scientists, including economists, sociologists, and geographers, are being drawn into the picture as their contributions add valuable perspectives to the issues being researched by Earth systems scientists.

Earth's Major Systems

Our planet, along with each of the other known planets, is made up of natural systems. These systems can include an atmosphere, a hydrosphere, a biosphere, a cryosphere, and a geosphere. The entire nature of a planet, or the entire character of its surface environment, is controlled by the existence and nature of these major systems.

For example, photographs of the Moon reveal only the presence of a geosphere, or solid surface. The Moon lacks all other major systems; there is no atmosphere, hydrosphere (water), cryosphere (although recent discoveries reveal the presence of trace amounts of ice deep within some craters), or biosphere (life). In many respects, despite its proximity to Earth, the Moon is certainly a very different celestial body from Earth.

When more than one major system is present, as is the case on Earth, a significant feature of these systems is their interdependency, which helps to shape the characteristics of each planet. Earth is the only planet that we know of that has all five of the major systems.

> **REMEMBER**
> "Atmos" from air or gaseous terrestrial matter.

ATMOSPHERE

Earth's envelope of gases above the surface is known as the *atmosphere*. The atmosphere (see Figure 1.3) is layered and contains a mixture of life-sustaining gases, the most important of which to all animal life on this planet is oxygen. Earth is the only planet known to contain a significant amount of free (uncombined) oxygen in its atmosphere.

> **REMEMBER**
> "Hydros" from water or liquid terrestrial matter.

Earth's atmosphere is composed primarily of two gases, nitrogen (78%) and oxygen (21%). Earth's atmosphere also contains a small (1%) but important amount of trace gases, including argon, ozone, water vapor, and carbon dioxide; without these, life, as we know it on our planet, would not be possible.

Most of the other known planets also have atmospheres; however, none appears to have the life-supporting mix of oxygen and nitrogen that Earth's atmosphere provides.

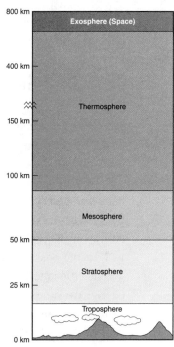

HYDROSPHERE

The *hydrosphere* is a major system that includes all the liquid water on our planet. Oceans, lakes, rivers, streams, and groundwater are all part of the hydrosphere. In 1968 we saw firsthand images from the *Apollo 8* command module as it sped toward the Moon. For the first time it was possible to see the entire disk, the entire planet Earth, against the black background of space. From this perspective, Earth appears to be a cloud-covered blue planet (see Figure 1.4) because of the abundance of water on the surface. More than 70 percent of Earth's surface is covered by water, and more than 97 percent of that water is found in the oceans. Earth is the only planet that we know to have a hydrosphere, or abundant quantities

FIGURE 1.3 The structure of Earth's atmosphere. Source: *Astronomy Explained*, Gerald North, Springer-Verlag, 1997. Used with permission.

FIGURE 1.4 Photograph of Earth from space. Source: NASA.

of water existing in the liquid state. These two major systems—a hydrosphere and an oxygen-rich atmosphere—make Earth a unique find, at least with our current knowledge of other planets within our galaxy.

CRYOSPHERE

The *cryosphere* is the system that includes all of Earth's frozen water, otherwise known as ice. The cryosphere is sometimes treated as part of the hydrosphere, and in reality there is necessarily some interaction and interdependency between these two major systems.

> **REMEMBER**
> "Cryos" from the Greek for "COLD"

In recent years, scientists have begun to explore the relationship between the cryosphere and Earth's climate system. For example, over the course of the past century, as the world's glaciers have receded, scientists have expressed concern over the potential for greater flooding in heavily populated coastal regions, as well as the possibility that the observed glacial retreat is a sign of global warming.

The cryosphere contains valuable clues that provide insights about ancient climate change over the past several hundred thousand years. By studying ice cores extracted from areas long buried by ice and glaciers, scientists have been able to identify past periods of relatively warm and cold conditions. By studying remnants of pollen literally frozen in the layers of ice, scientists can infer the climatic conditions that existed at that time.

Even the cryosphere is not devoid of life. Certain insects and plants can survive in this environment. Bacteria have been found as deep as 4 kilometers beneath the surface of glacial ice.

GEOSPHERE

Earth's solid portion is known as the *geosphere*, or *lithosphere*. The crust, our planet's outermost layer, is composed mostly of solidified rock and mineral material. It is actually quite similar in composition to the other terrestrial planets, Mercury, Venus, and Mars. Earth's Moon, although not a planet, is also structured much like the terrestrial planets, but without an iron core.

Earth's interior (see Figure 1.5), also considered to be part of the geosphere, is composed of denser material, some of which is partially molten or liquefied. The interior layers include the mantle and the outer and inner cores, which are defined by their relative densities and compositions.

BIOSPHERE

The *biosphere* is a major system that, as far as we know, is unique to Earth. The biosphere is composed of a wide variety of life-forms and is quite interdependent on all the other major systems. Various forms of life can also be found in each of the other major systems, where they act to modify and shape the system. One simple example is the connection between photosynthetic plants and the atmosphere. Plants have an impact on the composition of the atmosphere by removing carbon dioxide and producing oxygen.

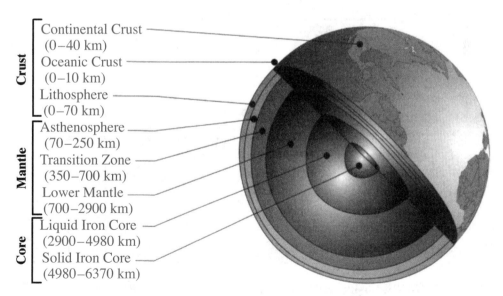

FIGURE 1.5 Cross section showing Earth's layered interior. Source: *Earth: Evolution of a Habitable World*, Jonathan I. Lunine, Cambridge University Press, 1999.

Earth as a System

As you have seen, each of the systems discussed above is quite interdependent on the other systems. The concept that "everything is connected to everything else" is an important one to embrace when studying the natural functioning of our planet.

Studying Earth Systems Science

The Earth systems science approach has changed the way we view Earth and is revolutionizing our understanding of some of the major environmental issues we face, including global warming, acid rain, and ozone depletion. These issues will be explored in later chapters.

The study of Earth systems science can be a dynamic and rewarding experience; however, it requires the careful collection and detailed analysis of data. Fortunately for us, a tremendous amount of data has already been, and continues to be, collected by scientists and students of science.

In one such initiative, former Vice President Al Gore was instrumental in getting the GLOBE Project (*www.globe.gov*) off the ground. The GLOBE Project is a prime example of the value of having students collect data. Figure 1.6 shows a map produced on the GLOBE Web site.

Initiatives such as the GLOBE Project, as well as various government organizations including the National Aeronautics and Space Administration (NASA), the

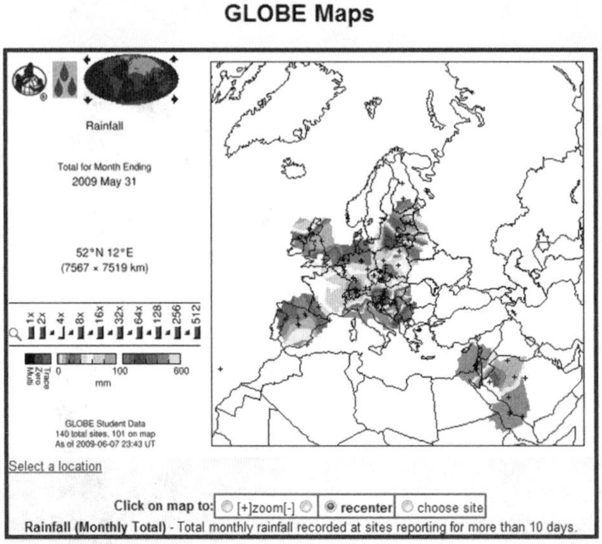

FIGURE 1.6 Illustration of a map produced on the GLOBE Web site.
Source: The GLOBE Program at *www.globe.gov*.

National Oceanic and Atmospheric Administration (NOAA), and the United States Geological Survey (USGS), have generated great collections of data, many of which are available at no charge for use and analysis by students everywhere from numerous Web sites via the Internet. This book will highlight worthwhile Web sites where data are available to support the theory and scientific understandings that you will study in each unit.

GOOD STARTING POINTS FOR YOUR SEARCH ON THE INTERNET FOR IMAGES AND INFORMATION

www.nasa.gov

This is the link to NASA's main page. It provides a good starting point for access to many images of Earth and other objects in space.

www.noaa.gov

NOAA is the agency responsible for the study of Earth's atmosphere and hydrosphere. From this site, many links exist on which you can access data, many of which are collected on a daily basis.

www.dlese.org

The Digital Library for Earth Systems Education (DLESE) provides a search engine that has been developed by Earth systems professionals and is accessible in a number of ways, including "grade-appropriate level."

www.ag.ohio-state.edu/~earthsys/

Ohio State has a good offering that addresses Earth systems education. The site provides links to courses that can be taken in Earth systems, as well as a description of the basic principles of Earth systems science.

www.cotf.edu/ete/

The "Classroom of the Future" educational team, based in Wheeling College in Wheeling, West Virginia, has been developing online "problem-based learning" modules for years. These modules are designed to help the individual student and student groups to design and complete broad, nonlinear, open-ended scientific investigations that relate to Earth systems science.

www.usgs.gov

The United States Geological Survey serves the scientific community as well as the general public with maps, imagery, data, and background information on earthquakes, volcanic activity, surface features, and much more. There is a wealth of educational resources within this expansive Web site.

REVIEW EXERCISES FOR CHAPTER 1

WORD-STUDY CONNECTION

acid rain	environmental science	outer core
astronomy	eruption	oxygen
atmosphere	evolution	ozone
big bang	galaxy	physical geology
biology	geology	physics
biosphere	geomorphology	pollutant
carbon dioxide	geosphere	seawater
chemistry	global warming	solar system
climate	historical geology	stratosphere
climatology	hydrosphere	synoptic meteorology
crust	ice	system
cryosphere	inner core	tectonic plates
deforestation	lithosphere	topography
dynamic meteorology	mantle	universe
Earth	meteorology	volcano
Earth scientists	nitrogen	water
earth systems science	ocean	weather
earthquake	oceanography	
environment	organisms	

SELF-TEST CONNECTION

PART A. Completion. *Write in the word or words that correctly complete the statement.*

1. The science that involves study of the world's oceans and other surface waters is called _____.

2. _____ is concerned with the study of ozone depletion, global warming, and other global-change-related topics.

3. _____ is a branch of meteorology that is concerned with understanding how the atmosphere works.

4. The major system that involves all forms of life is called the _____.

5. The solar system and galaxies are subjects that would be of interest to students of _____.

6. The discipline likely to be involved in establishing a time line of events that occurred on Earth millions of years ago is _____.

7. _____ is the study of long-term changes in weather and weather patterns.

8. Earth contains five major systems: an atmosphere, a hydrosphere, a geosphere, a _____, and a biosphere.

9. Earth's interior is layered; the innermost layer is known as the _____.

10. Earth's atmosphere is composed primarily of nitrogen and _____.

11. A _____ is a part of a whole that may be studied separately, but is necessarily dependent upon changes in other systems.

12. The major system that comprises all the ice on this planet is the _____.

13. Earth's crust is broken into several large sections called _____.

14. The _____ is a major system also known as the lithosphere.

15. _____ is the science concerned with forecasting daily weather changes.

PART B. Multiple Choice. *Circle the letter of the item that correctly completes the statement.*

1. Geologists are primarily concerned with the study of
   (a) Earth
   (b) the oceans
   (c) the atmosphere
   (d) life on our planet
   (e) the solar system, our galaxy, and the greater universe

2. Meteorologists are primarily concerned with the study of
   (a) Earth
   (b) the oceans
   (c) the atmosphere
   (d) life on our planet
   (e) the solar system, our galaxy, and the greater universe

3. Astronomers are primarily concerned with the study of
    (a) Earth
    (b) the oceans
    (c) the atmosphere
    (d) life on our planet
    (e) the solar system, our galaxy, and the greater universe

4. Oceanographers are likely to receive extensive training in
    (a) chemistry
    (b) physics
    (c) geology
    (d) biology
    (e) all of the above

5. Earth systems science encourages collaboration between previously isolated disciplines. For example, a climatologist may find it beneficial to collaborate with an oceanographer because
    (a) the oceans have a pronounced impact upon the atmosphere
    (b) the oceans may provide clues about future changes in Earth's climate
    (c) the oceans are affected by the Moon
    (d) both a and b
    (e) both a and c

6. Environmental scientists are researching problems that include
    (a) global climate change
    (b) stratospheric ozone depletion
    (c) ground-level ozone accumulation
    (d) acid rain
    (e) all of the above

7. Earth's major systems include
    (a) the atmosphere
    (b) the cryosphere
    (c) the Sahara Desert
    (d) the continent of Antarctica
    (e) both a and b

8. The most abundant gas in the atmosphere is
    (a) nitrogen
    (b) oxygen
    (c) hydrogen
    (d) carbon dioxide
    (e) ozone

9. The percentage of Earth's surface that is covered by water is close to
    (a) 50
    (b) 60
    (c) 70
    (d) 80
    (e) 90

10. When you study the hydrosphere, you are probably investigating
    (a) the composition of air
    (b) the oceans
    (c) lakes
    (d) both a and b
    (e) both b and c

PART C. Modified True/False. *If a statement is true, write "true" for your answer. If a statement is incorrect, change the <u>underlined</u> expression to one that will make the statement true.*

1. <u>Meteorologists</u> are primarily concerned with the study of the universe and Earth's place within it.

2. The <u>hydrosphere</u> is a major system that includes the oceans and all other liquid water on Earth.

3. The atmosphere is composed primarily of <u>hydrogen</u> and oxygen.

4. <u>Earth systems science</u> brings together geologists, meteorologists, and oceanographers to study major environmental issues here on Earth.

5. The branch of meteorology concerned with issuing daily weather forecasts is called <u>dynamic</u> meteorology.

6. One subject that a traditional <u>geologist</u> may study is the composition of seawater.

7. Global climate change is an issue faced by <u>environmental scientists</u>.

8. Climate is the study of <u>short-term</u> weather patterns.

9. Earth is composed of <u>five</u> major systems.

10. The <u>biosphere</u> is the major system that is composed of ice and frozen water.

CONNECTING TO CONCEPTS

1. How does the Earth systems science approach differ from earlier approaches used in the Earth sciences?

2. What five major systems are found on Earth?

3. Cite at least one interaction between any two of Earth's five major systems.

4. What sciences collectively make up the Earth sciences?

5. With what major systems do humans interact?

6. Some scientists are concerned about rapid and profound changes in ocean currents. One such current, the Gulf Stream, has a profound impact upon western Europe. This warm water current, originating in the Gulf of Mexico, flows northeastward along the U.S. eastern seaboard to a point just off the North Carolina coast; from there it proceeds east-northeastward and is known as the North Atlantic Drift Current. This warm water current moderates the climate of northerly locales, such as England. If this current were to cease, what changes might occur in each of the major systems discussed in this chapter?

7. Do some searching on the Internet to find at least two Web sites that allow you to access data collected by scientists. Begin to create your own list of valuable Web sites that will help you to study Earth. Many word processing programs will allow you to write a summary about each Web site and then create a "hotlink" to that Web site when you type the url (universal resource locater) or simply the Web site's address into the document.

8. List at least three environmental issues that the world is facing at this time. For each, identify the major systems that are threatened and, if possible, in what way(s). Each "global change" issue that you identify is a complex problem. Feel free to do some outside research on each, as you may choose to explore one extensively as a research topic for any Earth systems course that you may be enrolled in.

CONNECTING TO LIFE/JOB SKILLS

Earth systems science is a growing field and will be integral in working to solve many of the major environmental issues we face as a global society today. Global change is not just an "American problem"; it is an issue that all nations are facing and will need to work together to solve. In coming years, many opportunities will present themselves for those with the proper training. This training should include an extensive background in mathematics, including calculus, differential equations, multivariate mathematics, and statistics, as well as courses in physics, chemistry, biology, and the emerging field of Earth systems science.

ANSWERS
SELF-TEST CONNECTION
Part A

1. oceanography	6. historical geology	11. system
2. Environmental science	7. Climatology	12. cryosphere
3. Dynamic meteorology	8. cryosphere	13. tectonic plates
4. biosphere	9. inner core	14. geosphere
5. astronomy	10. oxygen	15. Synoptic meteorology

Part B

1. **(a)**	3. **(e)**	5. **(d)**	7. **(e)**	9. **(c)**
2. **(c)**	4. **(e)**	6. **(e)**	8. **(a)**	10. **(d)**

Part C

1. False; Astronomers	5. False; synoptic	8. False; long-term
2. True	6. False; oceanographer	9. True
3. False; nitrogen	7. True	10. False; cryosphere
4. True		

CONNECTING TO CONCEPTS

1. The Earth systems science approach involves scientists from various disciplines in the study of our planet and its natural systems. Earth systems science adopts a view that all systems are interconnected and that what occurs within one system will necessarily affect other systems. An Earth systems scientist needs to have a general background in several scientific disciplines, rather than specializing in one limited area.

2. The five major systems include an atmosphere, a hydrosphere, a lithosphere (geosphere), a cryosphere, and a biosphere.

3. Many interactions between the systems can be cited. For example, pollution that enters the atmosphere eventually is transported to the oceans (hydrosphere), land (lithosphere), ice (cryosphere), and life-forms (biosphere) via wind and precipitation.

4. The Earth sciences traditionally include geology, oceanography, meteorology, and astronomy. Recently, environmental science has been included by many scientists as environmental issues have become more important and familiar to the general public.

5. Humans interact with all of the major systems.

6. If the Gulf Stream/North Atlantic Drift system were to cease flowing, western Europe and, to a lesser extent, the southeastern United States would become significantly colder. Other changes are likely, but cannot be anticipated in advance of such an event occurring.

7. Many Web sites on which to start a search are mentioned in this guide.

8. Three environmental issues include global warming, acid rain, and ozone depletion. Each of these issues will impact all major systems on Earth. Ultimately, each threatens the biosphere and the life-forms within it. All these issues begin with the atmosphere, but changes in the atmosphere will cause changes in each of the other major systems. For example, global warming caused by increased carbon dioxide and other gases in the atmosphere may lead to melting of the cryosphere; this will cause a rise in global sea level, which will impact world populations (biosphere) that live near coasts.

Beyond Our Solar System: A Study of the Universe

WHAT YOU WILL LEARN

This chapter focuses on what lies beyond our solar system. In this chapter you will learn

- how the universe formed;
- evidence for the big bang theory;
- how scientists study the universe today;
- what galaxies are and how they form;
- what stars are and how they form and evolve over time;
- how distance is measured in the universe.

SECTIONS IN THIS CHAPTER

- Origin of the Universe and Its Key Features
- Tools for Studying the Universe—Remote Sensing Instruments
- Galaxies
- Stars
- Measuring Distances
- Review Exercises for Chapter 2

Origin of the Universe and Its Key Features

Humans have long wondered about their collective origins. Stories abound in ancient texts, including the Bible, that strive to explain how our universe began. What follows is a survey of the most widely accepted scientific theories that address this question and, in particular, of some of the evidence that supports the big bang theory.

THE BIG BANG THEORY

The big bang theory is the most widely accepted explanation of how the universe formed. Let's begin with a definition of "universe." The *universe* may be thought of as everything that has been detected and is detectable.

According to the big bang theory, the universe began as an incredibly dense, hot, primordial fireball of expanding space, energy, and matter. Before this explosion, or big bang, there was no universe as we know it, and space and time as we now know them did not exist. How long this state existed is unknown; but, according to astronomers' best estimates, about 13 to 14 billion years ago the event known as the *big bang* occurred. This explosion marks the beginning of the universe, including time, space, and the processes of nature, as we know them. Figure 2.1 shows an artist's conception of the changes that have occurred from the time of the big bang to the present.

Ordinary subatomic particles, protons, neutrons, and electrons are thought to have formed during or shortly after the big bang. Hydrogen, the simplest element, with only one proton in its nucleus, was the first element to form. After the big bang, galaxies, stars, planetary systems, and even life on our planet ultimately came into being. Much evidence can be cited to explain why scientists believe an event such as the big bang occurred many billions of years ago.

BACKGROUND RADIATION

In 1965, American physicists Arno Penzias and Robert Wilson made a startling discovery. Disproving the expectations of most scientists of

REMEMBER
The big bang theory describes the events surrounding the formation of the universe as we know it— NOT our solar system.

(a) A region of the universe during the big bang

(b) A region of the universe now

(c) The present universe as it appears from our galaxy

FIGURE 2.1 The big bang: an artist's conception of the stages. (a) During the big bang, matter was thrown out uniformly in all directions. (b) Thereafter, gravity caused the matter to coalesce into clumps—galaxies. (c) Near us we see galaxies, farther away we see young galaxies (dots), and at a great distance we see radiation (arrows) coming from the hot clouds of the Big Bang explosion. Source: *Horizons: Exploring the Universe*, Michael A. Seeds, Wadsworth, 1987.

the day, they discovered that the universe is not entirely cold, but has an ambient temperature of about 3 Kelvin (K) due to background radiation. Kelvin is the absolute temperature scale on which 0 K is absolute zero, the temperature marking the absence of all energy.

The recently developed Cosmic Background Explorer (COBE) detected, as it looked into deep space, a small amount of background radiation in all directions. The existence of background radiation agreed perfectly with predictions made by proponents of the big bang. Scientists interpret the energy observed by COBE as essentially leftover radiation from the big bang. It is amazing that an event that occurred so long ago has "embers" still glowing throughout the known universe!

REDSHIFT–OUR EXPANDING UNIVERSE

Additional evidence supporting the big bang theory is the observed expansion of the universe over time. This expansion is seen as a redshift in the light spectra emitted by stars and galaxies. To visualize the redshift, one needs to understand that stars are rich in the elements hydrogen and helium. Certain conditions in stars cause the atoms of all elements to emit energy. The energy released by each element emits a characteristic electromagnetic pattern that is analogous to the bar code found on the packaging of most products at the local supermarket (see Figure 2.2).

Since the patterns for various elements have been established on Earth, astronomers have detected these same emission patterns or spectra of elements on distant stars (see Figure 2.3). The patterns observed are the same as those seen here on Earth, except that the entire pattern is shifted toward the red end of the visible spectrum in what astronomers call the *redshift* (see Figure 2.4). The redshift occurs because of a phenome-

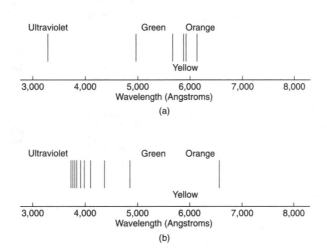

FIGURE 2.2 Bright-line spectra of (a) sodium and (b) hydrogen. The pattern of bright lines in every element's bright-line spectrum is unique; this spectral "fingerprint" can be used to identify elements. Source: *Discovering Astronomy*, Robert Chapman, W.H. Freeman, 1978.

non known as the *Doppler effect*. It is not unlike the change observed in the pitch of a police car's siren as the car approaches and then moves away. Light behaves in much the same manner as sound. When a redshift is observed, one may conclude that the object is moving away as "light waves" are being stretched by the increasing distance between observer and object. A good explanation of redshift appears at the following Web site in an article called "The Red Shift": *www.arachnoid.com/sky/redshift.html*.

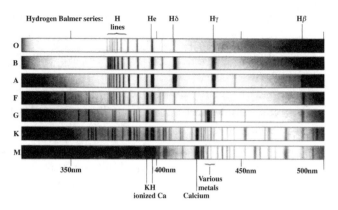

FIGURE 2.3 Representations of the spectra of stars from each of the major spectral classes. Source: *Astronomy: The Cosmic Journey*, William K. Hartman, Wadsworth, 1987.

Edwin Hubble, an American astronomer who spent much of his life studying galaxies, discovered that the more distant the galaxy, the faster it is receding from Earth. With redshift data available for numerous stars and galaxies, the rate at which the universe is expanding can be determined. Using the Hubble constant, a value determined by Hubble during his years of research, astronomers are able to determine, based upon its measured redshift, the distance of a galaxy from Earth. The farthest galaxies observed to date exceed a distance of 13 billion light-years (1 light-year equals about 600 trillion miles, or 950 trillion kilometers). Since galaxies are thought to have formed on the order of 1 or 2 billion years after the big bang, astronomers estimate the age of the universe to be approximately 13.7 billion years (Source: NASA).

From our Earth-based perspective, we note that all stars and galaxies are moving away from us. This does not mean, however, that we are the center of the universe,

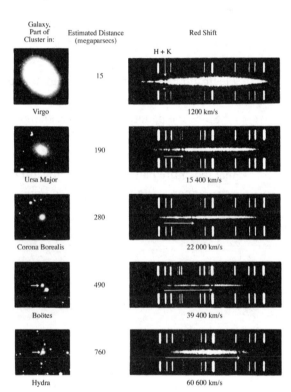

FIGURE 2.4 Redshifted spectra and corresponding velocities of several galaxies. Source: The Hale Observatories.

as everything is moving away from everything else. To visualize this situation, consider a loaf of raisin bread baking and expanding. Here the loaf represents the universe, and the raisins represent galaxies, or even individual stars. As the loaf expands, the distance between raisins increases, regardless of where a particular raisin is within the loaf.

ALTERNATIVE THEORIES EXPLAINING THE ORIGIN OF THE UNIVERSE

The big bang is not the only modern theory that attempts to explain how the universe began. The steady-state theory rivaled the big bang in scientific circles until the discovery of background radiation by COBE. Basically, this theory assumes that the universe always was and always will be. There was no beginning, and there will be no end. The steady-state theory does not, however, adequately explain the fact that the universe is expanding.

To account for the observed expansion, proponents of this alternative theory suggest that new hydrogen atoms (the type of atoms that make up most of the universe) are created continuously in empty space at a rate equal to the rate at which matter is carried away by receding galaxies. The steady-state theory, however, does not offer a reasonable explanation as to where the new hydrogen atoms come from. Although its proponents like the steady-state theory's philosophical appeal, most astronomers today reject it, arguing that, since energy (and mass) must be conserved in an isolated system (the universe), new matter (hydrogen atoms) can't be spontaneously generated.

> **REMEMBER**
> Background radiation and red shift are two of the best pieces of evidence that can be observed today that support the big bang theory.

Other theories regarding the formation of the universe exist and enjoy limited support by some astronomers today. Among these theories is the oscillating universe theory. According to this alternative theory, the universe has existed over eons of time and has oscillated between a relatively small and a relatively large size. Since the universe today is expanding, those who accept this theory believe that the universe will some day reach a maximum size and will then contract toward a minimum, only to expand again.

Tools for Studying the Universe— Remote Sensing Instruments

One may reasonably ask how astronomers study the heavens and collect data from which the theories discussed above are formulated. This section discusses the tools used and the methods employed for studying the universe. All such study is conducted by a technique known as remote sensing. *Remote sensing* is the process of studying an object or measuring its characteristics from a distance. This technique is particularly valuable when studying a large system, such as the atmosphere, for the purpose of predicting weather, or when studying objects that cannot be directly sensed by virtue of their distance or inaccessibility. In the study of astronomy, this would include just about everything!

Consider this: The Moon is the only celestial body that humans have ever sensed directly. This was done by only a few men between the years 1969 and 1972 and has not been repeated since. Thus, when you look out at night at the heavens, the only practical method for studying the various objects you can see is remote sensing.

Many remote sensing instruments have been developed to aid astronomers in studying the universe. Perhaps the most common is the telescope. The next section describes the telescope in detail and offers a short history of its development. Other remote sensing instruments are discussed in the section with that title.

TELESCOPES

Among the many instruments used to study the heavens, the most familiar is the *telescope*. The telescope was invented by Dutch spectacle-maker Hans Lippershey in 1608, and was used and popularized as an astronomical tool shortly afterward by Galileo Galilei. Galileo, as he is known historically, lived at a time when the scientific method began to replace intuitive reasoning as a way of conducting scientific research. Galileo used the telescope and the new approach toward scientific research to make many important observations that form the framework in which we view our solar system today.

Two major types of telescopes exist today. The *reflecting telescope* uses mirrors to collect light from an object. Some astronomers prefer reflecting telescopes for deep-sky observing, and they are often popular also among beginning sky observers because of their low cost. Professional astronomers frequently use giant reflecting telescopes when light-gathering ability is most important, particularly when studying the most distant, faint objects. The *refracting telescope* uses lenses to bring light from an object to a focus and is preferred for planetary observation because of its ability to produce images with excellent resolution. Refractors are also more rugged and require less maintenance than reflectors. Galileo used a refracting telescope to make his famous observations. The two types of telescopes are illustrated in Figure 2.5.

Major improvements in the telescope resulted in keener observations and better images in the years following Galileo's

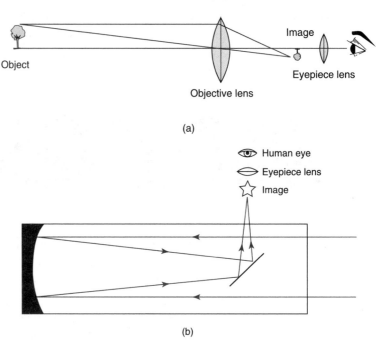

(a)

(b)

FIGURE 2.5 (a) A simple refracting telescope. (b) A Newtonian focus reflecting telescope. Source: *Discovering Astronomy*, Robert Chapman, W. H. Freeman, 1978.

work. Precision of observation was improved in the mid-1600s when three new instruments intended to improve the quality of observations made through a telescope were invented: the telescopic sight, which made alignment of a telescope possible; the micrometer, which allowed measurements to be made in the field of view of a telescope; and the pendulum clock, which enabled astronomers to time star transits accurately.

With the invention of the reflecting telescope by Newton and improvements in mirror making by William Herschel in the late 1700s, the telescope became a much more powerful tool than the instrument with about a 30-fold magnifying power used by Galileo. The key to Herschel's craftsmanship was to switch from the glass mirror used by Newton to a bright metal mirror, shaped into a parabolic curve. This enabled him to capture much more light and thus to see more deeply into space than was previously possible. In 1781, Herschel's new design, along with some luck, enabled him to discover Uranus—a previously unknown planet.

Important developments during the nineteenth century included improvements in the construction of refracting telescopes, the addition of motor-driven stands (for tracking celestial objects over time), and better technologies for engineering reflecting telescope mirrors. All of these resulted in a wealth of important discoveries, including the first observations of galaxies.

Recent improvements to telescopes include improvements to visual eyepieces as well as the use of cameras and charge-coupled devices (CCDs) to gather light and allow computer imaging of celestial objects. A CCD is an electronic detector. This silicon chip turns starlight into electric pulses for computers and computer software to interpret and display.

Today, perhaps the most impressive tool available is the Hubble Space Telescope (HST) (see Figure 2.6). In orbit above Earth's atmosphere, the HST is not hindered by the distortion that our atmosphere creates for all ground-based telescopes.

FIGURE 2.6 The Hubble Space Telescope (HST) orbits at a height of 600 kilometers (375 mi) above Earth's surface. Launched in 1990 and updated in 2002, the HST has a long list of accomplishments. (NASA)

This important new tool has enabled astronomers to see more clearly and farther into "deep space" than ever before. With the HST, distant galaxies have been discovered that may have formed within a billion years of the big bang. Much more information, as well as many images from the HST, is available at the following Web site: *http://hubble.nasa.gov/*.

OTHER REMOTE SENSING INSTRUMENTS

The *spectroscope* (see Figure 2.7) rivals the telescope in its importance to modern astronomy. The spectroscope separates starlight into its component wavelengths for

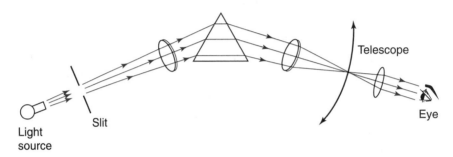

FIGURE 2.7 A simple prism spectroscope. The telescope pivots around a point centered on the prism so that the viewer can examine all visible wavelengths. Source: *Discovering Astronomy*, Robert Chapman, W. H. Freeman, 1978.

viewing. The light enters through a narrow slit and passes through a lens, which produces a beam of parallel rays of light. A prism or diffraction grating disperses this light into its component wavelengths (colors); by comparing this pattern to known patterns for elements, scientists can determine the elemental composition of a star. It was the spectroscope that enabled astronomers to determine the elemental makeup of the universe. This knowledge has helped to clarify and confirm many of the theories accepted today regarding the "workings" of the universe from its formation through stellar evolution.

Radio astronomers make use of the *radio telescope* (see Figure 2.8). This instrument is designed to sense radio waves emitted by stars and galaxies rather than light. The radio telescope enables the astronomer to "see" farther into space and to study objects too faint to be seen by optical telescopes. Radio telescopes are also useful in studying objects blocked by interstellar dust clouds in our own galaxy. The radio waves emitted by celestial objects

FIGURE 2.8 Photograph of a radio telescope, similar to those at the National Radio Astronomy Observatory located in Green Bank, West Virginia. With a 100-meter diameter, this is the world's largest fully steerable radio telescope. Source: NRAO.

are collected using a large, curved-dish antenna, which is designed to focus the energy. Computers then analyze the data and produce false color images (see explanation below).

Sensors that pick up emissions in other parts of the electromagnetic spectrum have also been developed (see Figure 2.9).

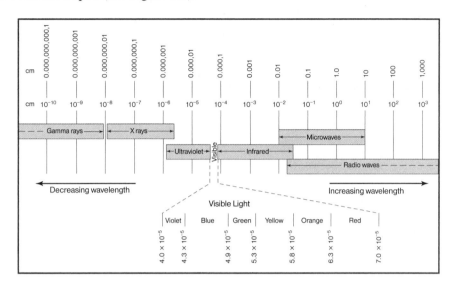

FIGURE 2.9 Electromagnetic spectrum. This diagram illustrates the regions into which electromagnetic waves are classified according to their lengths. Source: The State Education Department, *Earth Science Reference Tables*, 2001 ed. (Albany, New York: The University of the State of New York).

These high-energy sensors include infrared telescopes and telescopes that can detect ultraviolet, X-ray, and gamma ray radiation. All of the sensors can be used day and night and are relatively recent additions to the range of tools available for studying the universe. Recent additions include Chandra, launched in 1999. Chandra detects and images X-ray sources that lie within our solar system to those billions of light-years away. The results from Chandra help explore high-energy phenomena and provide insights into the universe's structure and evolution.

Another new tool, GALEX, the Galaxy Evolution Explorer, launched in 2003, is designed to study the shape, brightness, size, and distance of galaxies up to 10 billion light-years from Earth. It makes use of the electromagnetic radiation that reaches Earth from these distant objects. Once energy is detected, a "false color image" can be computer generated. Since the emissions do not fall within the visible portion of the electromagnetic radiation spectrum, colors are assigned to represent various kinds of emissions. These false color images facilitate the study of various phenomena, including emissions by stars, galaxies, and other deep-space objects.

Incidentally, the use of false color is a common practice even when studying our own planet. Meteorologists, for example, create enhanced infrared satellite images using colors to represent various cloud heights. Armed with these images, scientists can assess weather patterns with greater precision.

REMEMBER
Our study of the universe is largely conducted by remote sensing techniques.

Galaxies

WHAT IS A GALAXY?

A *galaxy* is a collection of hundreds of billions of stars, gas, and dust, all held together by the force of gravity. All of the individual stars that we earthbound observers can see at night are located within our Milky Way galaxy. Our planet and solar system are part of this same galaxy. The cloudy band of light stretches across the sky and is best seen on dark nights. This band is created by the billions of stars located closer to the center of our Milky Way galaxy than we are. We can see our galaxy from this perspective because planet Earth is located in a very "rural" section of the galaxy, far away from its center.

The average distance between stars in our galaxy is about 5 light-years, and the galaxy itself is shaped like a spiral with a diameter of over 100,000 light-years (see Figure 2.10). A light-year is simply the *distance* that light travels in a year. Since light travels at 3.0×10^8 meters per second (186,000 mi/sec), it covers about 950 million kilometers (600 trillion mi) in a year!

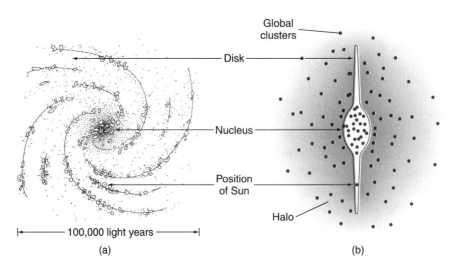

FIGURE 2.10 The Milky Way galaxy. Viewed face on (a) and edge on (b), the shapes and locations of the nucleus, disk, and halo can be seen. Note the position of the Sun, about two-thirds of the way out from the nucleus in the Orion arm of the disk. Sources: *Astronomy: The Cosmic Journey*, William K. Hartman, Wadsworth, 1987. *Horizons: Exploring the Universe*, Michael A. Seeds, Wadsworth, 1987.

HOW GALAXIES FORM

The first galaxies are thought to have formed perhaps a billion years after the big bang. One scientific model suggests that a galaxy originates as a turbulent, rotating cosmic cloud composed primarily of hydrogen and helium.

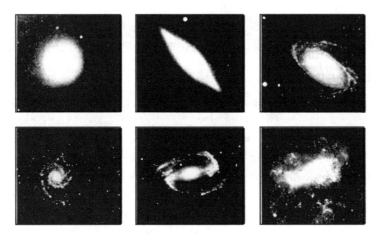

FIGURE 2.11 Galaxies are systems containing billions of stars. The universe is filled with galaxies of all shapes and sizes. Source: *A Self-Teaching Guide*, 4th Ed., Dinah L. Moché, John Wiley, 1998.

This giant cloud eventually collapses, forming one of a few simple structures when the pull of gravity exceeds any outward forces of expansion. Rotation and gravity then help to shape the galaxy into one of the familiar forms seen in many astronomy texts (see Figure 2.11).

TYPES OF GALAXIES

Galaxies are typed or classified by their shapes. There are four basic types of galaxies. The Milky Way is a large spiral galaxy. *Spiral galaxies* are disk shaped and tend to have a greater concentration of stars near their centers. A second type, the *barred spiral galaxies*, looks like regular spiral galaxies except that there is a bar-shaped concentration of material near the center with spiral arms attached to the bar. *Elliptical galaxies* are the most abundant type; about 60 percent of all galaxies have elliptical shapes. These galaxies are generally smaller than spiral galaxies and are sometimes referred to as dwarf galaxies. *Irregular galaxies* are the fourth type. The Magellanic Clouds, easily seen by the unaided eye in the Southern Hemisphere, are examples of irregular galaxies.

Interestingly, the shape of a galaxy seems to be related to the age of the stars within it. Irregular galaxies are composed primarily of young stars, and elliptical galaxies of old stars; spiral galaxies contain both young and old, with the youngest stars found in the spirals.

GALACTIC CLUSTERS

Astronomers have discovered that, just as stars cluster together to form galaxies, galaxies cluster to form galactic clusters. The force of gravity holds these clusters together, as clusters ranging from dozens to thousands of galaxies orbit each other. Our own galaxy is part of a galactic cluster known as the *local group*. The local group contains at least 28 galaxies. Finally, galactic clusters reside in huge groups called *superclusters*, which may be the largest entities in the known universe.

Stars

WHAT IS A STAR?

A *star* is a massive, self-luminous, or glowing, ball of gas that is held together largely by its own force of gravity. The gravitational tendency of a star to collapse is offset by the expansive forces associated with the ongoing nuclear fusion within. In other words, there are two competing forces in every star: gravity, which would cause the star to collapse, and nuclear fusion, a process producing large amounts of energy (described in detail in the next section), which would cause the star to expand (see Figure 2.12). The balance of opposing forces results in an entity that emits its own light and other forms of energy and retains a relatively constant size for most of its lifetime.

Two competing forces:

1: Gravity acts to try to collapse the star. This is the tendency toward collapse.

2: Nuclear fusion causes the star to attempt to expand.

During the main sequence stage, the star's size remains stable as these two competing forces balance each other.

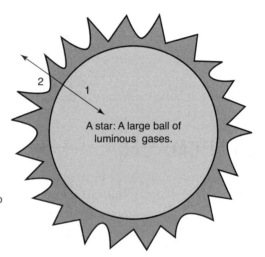

A star: A large ball of luminous gases.

FIGURE 2.12 Main-sequence, medium-sized stars remain relatively stable for billions of years as gravitational and expansive forces balance each other.

A STAR'S SOURCE OF ENERGY

The primary source of energy in a star is nuclear fusion. *Nuclear fusion* is a process whereby small nuclei fuse or combine to form larger nuclei. Intense heat and pressure within stars force positively charged atomic nuclei (often hydrogen nuclei) to overcome their natural repulsive forces and to combine, forming larger, heavier nuclei (often helium nuclei).

During the process of nuclear fusion, a small amount of matter is destroyed or converted into a tremendous amount of energy. This energy is observed as the heat and light emitted by stars. It is this same energy, emitted by our closest star, the Sun, that makes life possible here on Earth. The complete process is illustrated and discussed in Figure 2.13.

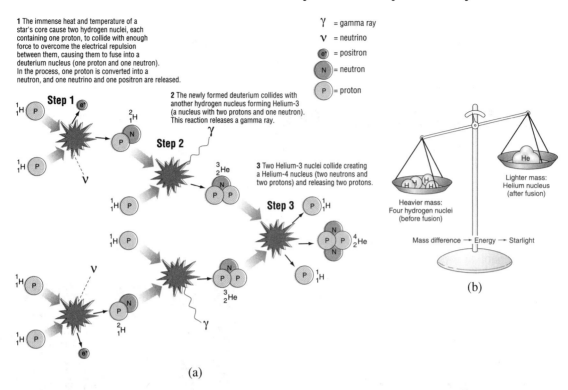

1 The immense heat and temperature of a star's core cause two hydrogen nuclei, each containing one proton, to collide with enough force to overcome the electrical repulsion between them, causing them to fuse into a deuterium nucleus (one proton and one neutron). In the process, one proton is converted into a neutron, and one neutrino and one positron are released.

γ = gamma ray
ν = neutrino
e⁺ = positron
N = neutron
P = proton

Step 1

2 The newly formed deuterium collides with another hydrogen nucleus forming Helium-3 (a nucleus with two protons and one neutron). This reaction releases a gamma ray.

Step 2

3 Two Helium-3 nuclei collide creating a Helium-4 nucleus (two neutrons and two protons) and releasing two protons.

Step 3

Heavier mass: Four hydrogen nuclei (before fusion)

Lighter mass: Helium nucleus (after fusion)

Mass difference → Energy → Starlight

(b)

(a)

FIGURE 2.13 The process of nuclear fusion in stars. (a) Step 1 illustrates the fusion of two hydrogen atoms to form deuterium (heavy hydrogen). In step 2, an additional hydrogen atom fuses with the deuterium to form a new element, helium. Step 3 illustrates the conversion of helium-3 into helium-4. (b) The formation of helium results in the destruction of a small amount of matter, which is converted into energy. Einstein's famous equation, $E = mc^2$, teaches us that when a small amount of matter is converted into energy, a large release of energy occurs as the mass of matter lost is multiplied by the speed of light squared, which is itself an incredibly large value. Source: *Astronomy: A Self-Teaching Guide*, 4th Ed., Dinah L. Moché, John Wiley, 1998.

Nuclear fusion is also the only known natural process by which heavier elements are created from lighter elements. Shortly after the big bang, according to theories, within the first 3 minutes temperatures in the still small but rapidly expanding universe exceeded a few trillion degrees Celsius. Under these conditions, matter could not exist separately from energy. Nuclei, or the cores of atoms, could not yet form.

Between 3 and 10 minutes after the big bang, however, simple nuclei were able, by a process known as *nucleosynthesis*, to form from collisions between elementary particles. Hydrogen nuclei, each with a one-proton nucleus, formed at this time, as well as some helium, with a two-proton nucleus, and a little lithium, with a three-proton nucleus.

Heavier elements such as those now found here on Earth and elsewhere in our solar system, however, are comparatively extremely rare throughout much of the rest of the universe. Since our solar system is populated with heavier nuclei, a theory explaining their formation is warranted. According to this theory, nuclear fusion in stars provides a mechanism by which heavier elements can be produced. This is a first step in accounting for the elemental composition of our solar system.

NEBULAE AND STELLAR FORMATION

A *nebula* (see Figure 2.14) is an interstellar cloud of dust and gases. Stars are born within nebulae, which are primarily composed of hydrogen. Although the density of hydrogen is very low in a nebula, the total mass is quite large—many times that of our local star, the Sun.

Occasionally, a region of a nebula will begin to concentrate hydrogen gas, causing it to condense or collapse further under the force of its own gravitational attraction. The mechanism for this initial action is not well understood; however, it is theorized that a major explosion or a nearby supernova could act as the trigger by causing a shock wave that initiates the collapse in a portion of the hydrogen gas cloud.

STELLAR EVOLUTION

The life cycle of a star varies according to its initial mass, although the early stages of evolution are quite similar for all stars. As the hydrogen gas from the nebula condenses, gravitational contraction creates great pressure, which causes the core of the developing star to heat up to temperatures of at least 10 million K (see Figure 2.16). Hydrogen "burning," or, more appropriately, nuclear fusion, begins to occur when these conditions of pressure and heat are met. The onset of nuclear fusion essentially marks the beginning of a star's life. While nuclear fusion is first beginning and a stable size is being achieved, a star is said to be in the *protostar* stage (see Figure 2.16).

Once nuclear fusion begins, a star may continue to burn hydrogen for millions of years if it's a large-mass star, or even billions of years if it's a small-mass star. In other words, the rate of burn, or the rate at which hydrogen is converted to helium and later to heavier elements, is dependent upon the star's mass. More massive stars have much

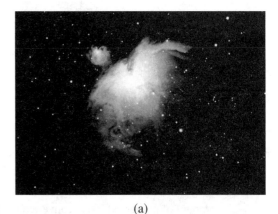

(a)

(b)

FIGURE 2.14 Nebulae. (a) Orion nebula. Source: *The Nature Company's Guides: Advanced Skywatching*, Robert D. Burnham et al., Time-Life Books, 1997. (b) Horsehead nebula. Source: N.A. Sharp/NOAO/AURA/NSF.

FIGURE 2.15 A nebula condenses into cloudlets of matter in the first stage of star formation. Source: *Astronomy Explained*, Gerald North, Springer-Verlag, 1997. Used with permission.

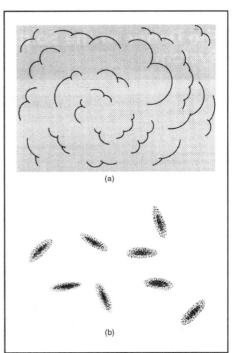

faster burn rates than less massive ones and also undergo much more spectacular "deaths."

A star spends the majority of its life span as a *main-sequence* star. An important chart called the Hertzsprung-Russell diagram (H-R diagram) is useful in studying stars. The H-R diagram can serve as a road map for tracking and understanding the changes a star will undergo during the course of its life. (See Figure 2.17, the H-R diagram, and Figure 2.18, a diagram of stellar evolution using the H-R diagram as a backdrop.) A more in-depth discussion involving the H-R diagram appears on page 38 of this chapter.

As noted earlier, the main-sequence stage of a star is highlighted by a balancing of the outward forces of gas pressure and the inward forces of gravity, resulting in a stable star of relatively constant size and brilliance. An average star spends about 90 percent of its life as a hydrogen-burning, main-sequence star. After some period of time, however, the hydrogen begins to deplete and helium becomes the dominant gas in the aging star. Rapid changes begin to occur in the star at this point.

Stars with less than one-half of the Sun's mass remain in the main-sequence stage for over 100 billion years. These "small" stars are too cool to support the fusion of anything other than hydrogen; therefore, when the hydrogen is largely depleted, nuclear fusion ceases and a hot, extremely dense *white dwarf* forms. Eventually, after billions of years, most of the remaining heat

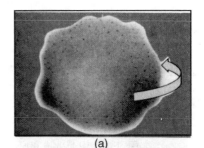

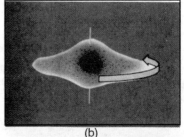

FIGURE 2.16 Three stages in the evolution of a protostar. (a) An interstellar gas cloud begins to contract because of its own gravity. (b) A central condensation forms, and the cloud rotates faster and flattens. (c) The star forms in the cloud center, surrounded by a rotating disk of gas. Source: *Astronomy: The Cosmic Journey*, William K. Hartman, Wadsworth, 1987.

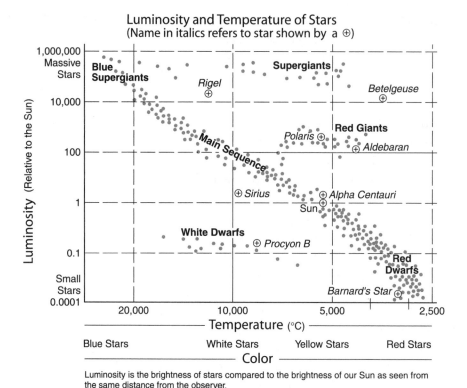

Luminosity and Temperature of Stars
(Name in italics refers to star shown by a ⊕)

Luminosity is the brightness of stars compared to the brightness of our Sun as seen from the same distance from the observer.

FIGURE 2.17 The Hertzsprung-Russell diagram. When luminosity is plotted versus temperature, stars fall into distinct regions in patterns that represent the relationship between luminosity and temperature. Source: The State Education Department, *Earth Science Reference Tables,* 2001 Edition (Albany, New York; The University of the State of New York).

radiates to space and the white dwarf ceases to give off any light, becoming a dark, cold object in space known as a *black dwarf.*

Medium-mass stars, such as the Sun, remain in their main-sequence stages for upward of 10 billion years. Toward the end of this stage, after much of the available hydrogen has been converted into helium, they have sufficient internal heat to initiate another fusion sequence. At this time, the zone where fusion occurs begins to migrate outward from the core. The hydrogen fusion, now occurring in the star's outer layers, causes the star to expand, and subsequently its surface cools. The cooler surface takes on a reddish appearance, and the star is then known as a *red giant.* It has grown to a size thousands of times larger than its size when it was stable during its main-sequence stage. The core, which had cooled below the point of being able to sustain hydrogen fusion, begins to gravitationally contract, causing it to heat up again. Eventually, the core reaches the incredible temperature of 100 million K, at which time the helium nuclei begin to fuse to form carbon in Sun-like (medium-mass) stars.

After the helium has fused into carbon, the core again collapses and the outer layers of the star are expelled. A planetary nebula, easily recognizable as a glowing, spherical

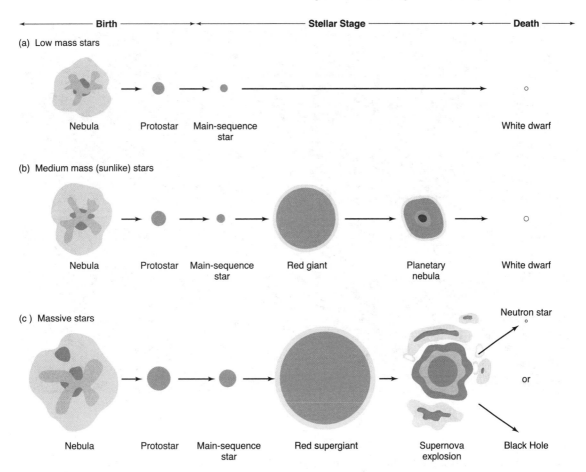

FIGURE 2.18 The evolutionary development of stars varies according to their masses. Source: *Earth Science*, 9th edition, Tarbuck and Lutgens, © 2000. Adapted by permission of Pearson Education, Inc., Upper Saddle River, N.J.

gas cloud, forms from the remains of the outer layers. The Hourglass nebula, pictured in Figure 2.19, is an example of a young planetary nebula. A color photo would reveal red, green, and blue colors. These colors would indicate the presence of ionized nitrogen (appears red), ionized hydrogen (appears green), and doubly ionized oxygen (appears blue.)

When all the nuclear fuel is consumed in a red giant, the force of gravity will take over and will squeeze the core into one of the densest objects imaginable—a white dwarf. A white dwarf is about the size of Earth! After further cooling, the white dwarf converts into a black dwarf, as previously stated.

Stars more massive than the Sun and having greater gravity to pull carbon atoms together will engage in additional nuclear fusion sequences. Oxygen, nitrogen, and eventually iron, with atomic number 26, are produced. Once the core is composed mainly of iron, fusion in the core ceases. Iron is the most stable of all elements, and therefore more energy is required to break up the nucleus of an iron atom than an atom of any other ele-

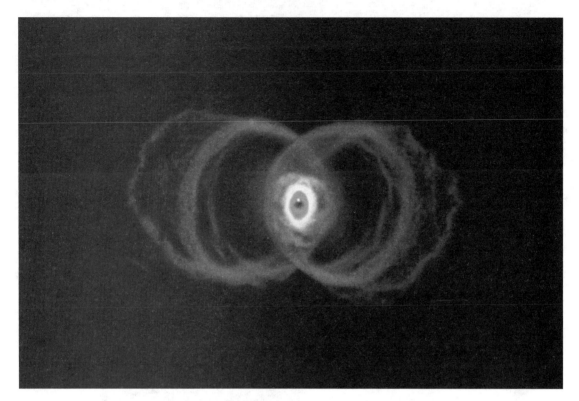

FIGURE 2.19 Hourglass nebula—an image of a young planetary nebula.

ment. To fuse iron atoms, thus creating heavier elements, requires a tremendous input of energy—more than can be produced from the events discussed thus far.

From the perspective that has been described, a star may be seen as an element factory, taking the hydrogen produced in abundance during the big bang and fusing it to produce elements as heavy as iron. A Sun-like star will spend about 10 billion years as a main-sequence star, but less than 1 billion years as a red giant.

Stars more massive than the Sun, however, are not yet done. Once the core consists mostly of iron, nuclear fusion ceases and energy is no longer radiated. The core rapidly collapses; temperatures exceed 100 billion K as iron atoms are crushed. Repulsive forces between iron nuclei overcome the force of gravity, and a shock wave recoils from the core. What is described here is an event referred to as a *supernova explosion*.

SUPERNOVAE

A *supernova* (see Figure 2.20) is perhaps one of the most spectacular events known to occur in the cosmos. Massive stars, those with more than three times the mass of the Sun, or three solar masses, follow a sequence during their lifetimes similar to that of medium-size stars, but execute it faster until they leave their main-sequence stage. A massive star may spend only a few million years in its main sequence, rapidly burning the available hydrogen fuel. These stars can be identified by their hot surfaces, which appear to be blue in color. In contrast, medium- and low-mass stars have lower surface temperatures, appearing yellow and red, respectively, during their main-sequence stages.

When the available hydrogen is nearly gone, a massive star rapidly progresses through a *red supergiant* phase during which rapid expansion is followed by an implosion caused by immense gravitational forces. The diameter of this massive star when it implodes is thought to be as little as 20 kilometers! This compression creates incredible heating; the core may reach a temperature of 1 billion K. This condition is known to produce very heavy elements, again by nuclear fusion. A supernova, in fact, is the only event known that can create the heaviest elements, those heavier than iron. The implosion creates a shock wave that moves outward from the star's core and blasts the star's outer layers into space, thus triggering the supernova event.

Our own solar system is testimony to the importance of supernovae and the elements produced when they occur. The heavier elements found throughout our solar system, and on Earth in particular, strongly suggest that our world is composed of materials produced in one or more ancient supernova explosions. Our own star, the Sun, is considered to be a second-generation star, as it contains all 92 naturally occurring elements. Although rich in hydrogen, the Sun differs from first-generation stars, which lack heavier elements.

Observations of supernovae have been recorded over the centuries. In A.D. 1054 the Chinese documented a supernova. The Crab nebula is believed to be the remnant of this event. Today it can be seen as a fuzzy spot through a small telescope. The Crab nebula contains a pulsar, which is the remains of the supernova. A *pulsar* is a rapidly spinning star that emits radiation in pulses and is a form of neutron star. The Crab nebula pulsar is

FIGURE 2.20 A supernova captured on film. These two photographs were taken before and after the supernova, which can be seen as the intensely bright star to the lower right of the galaxy in the photograph to the right. Source: *Discovering Astronomy*. Robert Chapman, W. H. Freeman, 1978.

spinning at a rate of 30 times per second. More recently, in 1987, observers in the Southern Hemisphere were treated to a supernova explosion dubbed SN 1987A. If a nearby star were to go supernova, it would outshine the Sun for a period of time!

STELLAR REMNANTS

All stars eventually consume all of their nuclear fuel and achieve one of three final states—black dwarf, neutron star, or black hole. A black dwarf, as previously noted, is often the final stage of a small or medium-mass star. The entire mass of the star is compressed into an object the size of Earth. The resulting average density is approximately 1 million times greater than that of water. To put this into perspective, if 1 liter of water weighs 1 kilogram, 1 liter of material from a black dwarf would weigh 1 million kilograms or approximately 1100 tons!

A massive star, as noted, will reach the supernova stage, after which there are two possible outcomes. A *neutron star* forms when a star originally with a mass of up to three solar masses goes supernova; then the remaining core collapses into an object so dense that electrons are forced to combine with protons, forming neutrons—hence the name neutron star. Pulsars, or neutron stars that emit pulses of strong radio waves, were first theorized and then later observed in the Crab nebula.

Black holes (see Figure 2.21) form when the most massive stars go supernova. The resulting implosion of the star's core forms an object so dense, with such an intense

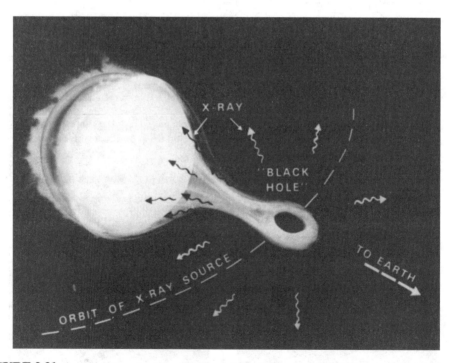

FIGURE 2.21 An artist's conception of a black hole. A stream of matter is ripped out of a star by the immense gravity of an orbiting black hole. X rays are emitted as the matter accelerates into the black hole. Source: *Astronomy: A Self-Teaching Guide*, 4th Ed., Dinah L. Moché, John Wiley, 1998.

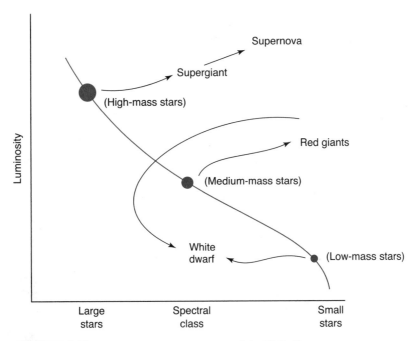

FIGURE 2.22 You can use the structure of the H-R diagram to create a flow chart showing the basic life cycle stages of low-, medium-, and high-mass stars.

gravitational field, that even light cannot escape. Black holes are naturally difficult to identify, as they are incredibly tiny and emit no light. Astronomers theorize, however, that objects swept by chance into the gravitational field of a black hole emit X rays before being engulfed.

Figure 2.22 illustrates the possible life cycle of a star.

HERTZSPRUNG-RUSSELL DIAGRAM

The *Hertzsprung-Russell* (H-R) *diagram*, named for its two discoverers, Ejnar Hertzsprung of Denmark and Henry N. Russell of the United States, helps us to see the relationship between the luminosity (apparent brightness) and surface temperature of a star. The chart actually has four axes: luminosity and absolute magnitude (actual brightness) both appear on the vertical axis, and surface temperature and spectral class are represented on the horizontal axis (refer to Figure 2.17). Take note as you study the diagram that it shows surface temperature as highest on the left and lowest on the right; this is counter to the trend on most graphs depicting temperature.

As you study the H-R diagram, you can see the "belt" of main-sequence stars, from the hottest blue stars in the upper left to the coolest red stars in the lower right. Stars out of the main sequence can be seen in the upper right as red giants and supergiants, and in the lower left as white dwarfs. Supergiants are quite bright but have cool surfaces, and white dwarfs are quite dim but have hot surfaces; hence neither group fits into the main sequence. The H-R diagram is also a valuable tool with which to study the sequence of changes a star will undergo from its formation to its death. For additional information about the H-R diagram, visit the following Web site: *www.windows.ucar.edu/tour/link=/cool_stuff/tourstars_1.html* "ExploraTour: A Peek into the Lives of Stars."

OUR SUN

Our nearest star, the Sun, is a relatively average star, both in mass and in luminosity. The Sun is truly "middle aged," as it and the rest of our solar system are approximately. 4.5–5 billion years old. Since the Sun is expected to have a 10-billion-year life span before expanding to become a red giant, it has several billion years left as a main-sequence star. As a red giant, the Sun will probably engulf the planets Mercury and Venus and render Earth a hot, lifeless ball of baked rock.

We may take comfort, however, in the fact that we have about 5 billion years left before this cataclysm occurs. Nevertheless, the Sun has been slowly increasing its energy output (luminosity) over time; and in all probability, in 1 billion years or less, well before the Sun enters its red giant stage—life as we know it here on Earth will become impossible.

> **REMEMBER**
> The life cycle of a star is determined by its mass.

BINARY STARS

About half of all the stars in our galaxy are thought to be born in pairs. These stars orbit about a common center of mass and usually remain gravitationally tied to each other for the duration of their lives. If the stars are of equal mass, the common center of mass will be a midpoint between the two; otherwise, the point about which they orbit will be closer to the larger star.

Astronomers who hope to find someday an Earth-like planet that may have life on it have generally avoided searching for any planetary systems in which binary stars orbit. They reason that any planet orbiting in such a system will be too close to one or the other binary star at times and too far at other times to allow for the kind of reasonably stable climatic conditions that have characterized Earth as it orbits the Sun.

Measuring Distances

WHAT IS A LIGHT-YEAR?

A natural question when studying the heavens is to ask how far the objects we see are from our terrestrial perspective. Astronomers have developed a unit of length or distance called the light-year. As the name suggests, a *light-year* is the distance that light travels in 1 year. To clarify this further, consider that light travels 186,000 miles per second (3.00×10^8 m/s). Here's an interesting comparison. Sunlight that reflects off the Moon reaches Earth in just over 1 second. When NASA sent the Apollo spacecraft and its teams of astronauts to the Moon, humans took almost 4 days to cover that same distance!

The Sun is considered to be "next door" to our planet even though it is 150,000,000 kilometers (93,000,000 mi) distant. This is a distance that humans have never traveled in space, yet light can cover it in just over 8 minutes! Since distances between objects in space are so vast, the light-year is the most appropriate unit in which to express these distances.

PARALLAX

While the light-year is the preferred unit when measuring the distance to an object in space, determining the actual distance is challenging as we cannot simply travel to the object in question and then check our odometer! Astronomers have developed a number of techniques that enable us to determine how far an object is from Earth. One such method, which utilizes a phenomenon known as *parallax*, enables astronomers to determine the distances to "nearby" stars.

Some stars appear to change their positions in the sky over the course of a year. Imagine that you are a child on a merry-go-round. The merry-go-round is near the ocean, and you can see a nearby boat and several distant boats. As the merry-go-round turns, the nearby boat appears to change position but the distant boats do not.

A similar condition is created as Earth revolves around the Sun. The diameter of Earth's orbit is approximately 300 million kilometers (186 million mi). This diameter creates a large enough change from our "January position" to our "June position" (half a year later) in space so that some nearby stars appear to change their positions when compared to more distant stars in the cosmic background. This change in the apparent positions of nearby objects is called the *parallax shift* (see Figure 2.23). Using geometry, astronomers can determine the distance of a star from Earth by measuring the degree of this shift.

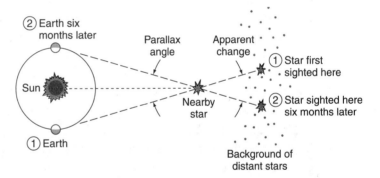

FIGURE 2.23 Stellar parallax: The apparent shift in position of a nearby star against the background of more distant stars. Stars are so distant that the parallax angle was too small to be detected by early astronomers. Source: *Astronomy: A Self-Teaching Guide*, 4th Ed., Dinah L. Moché, John Wiley, 1998.

A STAR'S BRIGHTNESS: APPARENT VERSUS ABSOLUTE MAGNITUDE

Determining stellar distance by parallax becomes impossible beyond a distance of about 200 light-years. For more distant objects, the apparent brightness method is used. The *apparent brightness* of an object is its intensity as measured by observers on Earth. Apparent brightness is dependent upon the actual brightness (absolute magnitude) of a star and its distance from Earth. In other words, an object appearing bright in our sky may be quite close and have a low absolute magnitude (a positive value), or it may be quite distant and have a high absolute magnitude (an increasingly negative value).

If we know the distance to a nearby star, we can obtain the distance to a distant star by determining the ratio of their distances. This is determined by using the ratio of their apparent brightnesses. The method works as long as the stars have the same absolute magnitude, as determined by using the H-R diagram.

NOTES:

The Universe Adventure sponsored by the Physics Division of the Lawrence Berkeley National Laboratory illustrates this concept nicely here: *www.universeadventure.org/fundamentals/popups/light-sq-magnitude.htm.*

CEPHEID VARIABLE STARS

Cepheid variable stars deserve special mention, as they have proved helpful in identifying the distances to important phenomena in the heavens, including the Magellanic Cloud (a nearby satellite galaxy of the Milky Way that can be seen from the Southern Hemisphere) and the Andromeda galaxy. Andromeda is a "nearby" galaxy at a mere distance of 2 million light-years and is visible to the unaided eye as a fuzzy patch in the sky that stretches about six times wider than a full Moon.

Cepheid variables are supergiants with much greater luminosity than the Sun. Polaris, the North Star, is a Cepheid variable star. Cepheids tend to pulse, or get brighter and dimmer, over a regular period that can be established. Early in the twentieth century, American astronomer Henrietta Leavitt established a relationship between the pulsations and the brightness of Cepheid variables. A Cepheid's *light period* is the time interval, usually between 2 and 50 days, between periods of maximum apparent brightness. Scientists have determined that the longer the light period, the greater the star's absolute magnitude (actual brightness). Once absolute magnitude is known, it can be compared to apparent magnitude to determine the star's distance from Earth.

Constellations

No chapter on stars and astronomy would be complete without at least a brief discussion of constellations. Ancient peoples, as long as 5000 years ago, in an attempt to make sense of the heavens, grouped stars according to recognizable shapes and explained them as mythological characters or great heroes. The Babylonians, Egyptians, and Mesopotamians were probably the first civilizations to organize the heavens into the patterns we now recognize as constellations.

Modern science recognizes 88 constellations (see Figure 2.24) and uses them to divide the sky into units or sectors. Each observable star is placed within one of these sectors, even though it may not be part of the actual constellation in that sector. Constellations

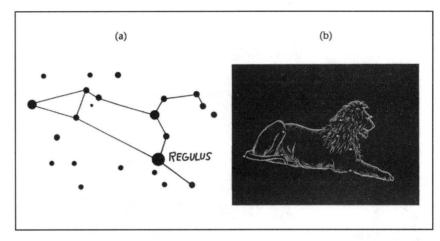

FIGURE 2.24 Constellations are imaginary patterns of stars. Source: *Astronomy: A Self-Teaching Guide*, 4th Ed., Dinah L. Moché, John Wiley, 1998.

also provide a good way to become familiar with the night sky. Specific objects can then be located using the constellations. For example, if you know that Mars is in Gemini, you will know where to look in the sky to find Mars on a particular night.

Related Internet Resources for Great Images and Information

www.pbs.org/deepspace/timeline/

The Public Broadcasting System has a Web site titled "Mysteries of Deep Space—Timeline." This site helps the student explore the universe from the big bang to the "end of the universe." It appears to be quite complete and scientifically accurate.

www.damtp.cam.ac.uk/user/gr/public/bb_history.html

Cambridge University in England has a good Web site, titled "A Brief History of the Universe," that deals with many areas of cosmology. Students can research such topics as standard, particle, and quantum cosmology.

www.pbs.org/wgbh/nova/universe/history.html

NOVA Online has a "History of the Universe" Web site. Students can view a universe time line, witness a supernova, or get a guided tour of the universe from this well-designed site.

www.enchantedlearning.com/subjects/astronomy/stars/

Enchanted Learning has created a very comprehensive site on stars. Topics covered include life cycles, galaxies, nuclear fusion, and constellations. This site also provides a number of activities to explore. In its own words, Enchanted Learning "produces children's educational web sites and games which are designed to capture the imagination while maximizing creativity, learning, and enjoyment."

REVIEW EXERCISES FOR CHAPTER 2

WORD-STUDY CONNECTION

absolute magnitude	iron	pulsar
background radiation	Kelvin	radio astronomy
big bang theory	light-year	radio waves
black dwarf	Lippershey	red giant
black hole	local group	redshift
carbon	luminosity	reflecting telescope
CCD	Magellanic Clouds	refracting telescope
celestial	main sequence	remote sensing
constellation	mass	scientific method
Crab nebula	micrometer	space
diffraction grating	Milky Way galaxy	spectroscope
electromagnetic energy	Moon	star
electron	nebula	steady-state theory
emission	neutron	subatomic particles
energy	neutron star	supercluster
galactic cluster	nuclear fusion	supergiant
galaxy	nuclei	supernova
gamma ray	nucleosynthesis	telescope
gravity	parallax	telescopic sight
helium	pendulum	ultraviolet
Hertzsprung-Russell diagram	planet	universe
Hubble constant	primordial	Uranus
Hubble space telescope	prism	wavelength
hydrogen	proton	white dwarf
interstellar	protostar	X-ray

SELF-TEST CONNECTION

PART A. Completion. *Write in the word or words that correctly complete the statement.*

1. The _____ is the actual brightness of a star.

2. _____ are shapes and forms in the night sky named by the ancients; today they present a reasonable way of locating objects in the sky.

3. When the most massive stars reach the end of their life cycles, they explode in what is known as a _____ explosion.

4. A _____ is a region in space of extremely dense matter that has so much gravity that even light cannot escape.

5. Our galaxy is located in the same vicinity as several other galaxies; together, these galaxies are known as the _____.

6. The lightest element, and the first one formed shortly after the big bang, is _____.

7. Distances are so vast in space that they are often expressed in terms of this unit: the _____.

8. Stars that are in the main portion of their life cycles are known as _____ stars.

9. _____ is the process by which stars produce heat and light.

10. The distance to relatively nearby stars can be determined by the _____ method.

11. _____ telescopes are often preferred by beginning sky observers because of their low initial cost.

12. Atoms are composed of _____.

13. A star in its formative stage is known as a _____.

14. Our own star, the Sun, is destined to become a _____ as it expands and its surface cools.

15. That Earth rotated on its axis was not proved until Foucault invented the _____.

PART B. Multiple Choice. *Circle the letter of the item that correctly completes the statement.*

1. The most widely accepted scientific theory regarding the formation of our universe is known as the
   (a) steady-state theory
   (b) big bang theory
   (c) nebular hypothesis
   (d) Hubble constant
   (e) plate tectonics theory

2. Current estimates place the age of the universe at
   (a) 15–20 million years
   (b) 5–10 billion years
   (c) 10–15 billion years
   (d) 15–20 billion years
   (e) 100 billion years

3. Ordinary subatomic particles include
   (a) protons
   (b) neutrons
   (c) electrons
   (d) deuterons
   (e) a, b, and c

4. The existence of background radiation is offered as evidence supporting the
   (a) steady-state theory
   (b) big bang theory
   (c) nebular hypothesis
   (d) Hubble constant
   (e) Both c and d

5. The redshift, first identified by Hubble, confirms that _____ objects are moving away from each other _____.
   (a) distant, slower
   (b) nearby, slower
   (c) distant, faster
   (d) nearby, faster
   (e) both b and c

6. The only celestial body (other than Earth) that we have directly sensed by landing on and physically exploring it is
   (a) Mars
   (b) the Sun
   (c) the Moon
   (d) Venus
   (e) none of the above

7. The telescope was invented by
   (a) Hans Lippershey
   (b) Galileo Galilei
   (c) Sir Isaac Newton
   (d) Edwin Hubble
   (e) William Herschel

8. Herschel is credited with making improvements in the design of the telescope, as well as the discovery of
   (a) Saturn
   (b) Uranus
   (c) Neptune
   (d) Pluto
   (e) Helium

9. Perhaps the most exciting invention of our day is the
   (a) telescope
   (b) spectroscope
   (c) CCD camera
   (d) Hubble space telescope
   (e) Saturn V rocket

10. The largest of the following is
    (a) a star
    (b) a planet
    (c) a galaxy
    (d) an asteroid
    (e) the universe

PART C. Modified True/False. *If a statement is true, write "true" for your answer. If a statement is incorrect, change the <u>underlined</u> expression to one that will make the statement true.*

1. <u>The big bang theory</u> is currently the most widely accepted scientific theory that explains the formation of our universe.

2. Our <u>supercluster</u> is a group of galaxies clustered about the Milky Way galaxy.

3. Stars convert hydrogen into <u>uranium</u> when they first begin the process of nuclear fusion.

4. Our own galaxy is called the <u>Andromeda</u> galaxy.

5. A method utilizing <u>parallax</u> is employed to determine the distances of nearby stars.

6. <u>Planets</u> formed shortly (about 1 billion years) after the big bang.

7. <u>Absolute magnitude</u> is the actual brightness of a star. A star's distance from Earth determines its apparent brightness.

8. <u>Remote sensing</u> is the act of studying something without actually touching or directly sensing it.

9. Most stars will spend the majority of their lives as <u>protostars</u>.

10. Large-diameter <u>refracting telescopes</u> are used to gather great quantities of light.

CONNECTING TO CONCEPTS

1. Compare and contrast the big bang theory and the steady-state theory.

2. What evidence exists to support the big bang theory?

3. Create a time line according to the information presented in the chapter, identifying the major events from the beginning of our universe to the appearance of humans.

4. What is remote sensing, and why is it so important to the study of astronomy?

5. Describe the four types of galaxies, and state what type of galaxy the Milky Way is.

6. What kind of stars can explode as supernovae? Why is this?

7. Describe the sequence of elements created within stars by nuclear fusion. What is the heaviest element created during the red giant phase? Why is this?

8. Can our local star, the Sun, be a first-generation star that formed shortly after the big bang, or must it be a second-generation star that formed after one or more supernova explosions occurred? Discuss the reasoning for your choice.

9. Explain why some stars that are invisible to the unaided eye can be seen through a telescope.

10. According to the Hertzsprung-Russell diagram, what is the relationship between the luminosity and the temperature of a star?

CONNECTING TO LIFE/JOB SKILLS

Our knowledge of the universe is growing exponentially as new tools emerge to study the heavens. Careers in astronomy require extensive training in math and physics. Astronomers are also becoming increasingly computer literate because many remote sensing tools involve the use of image-processing software.

ANSWERS
SELF-TEST CONNECTION

Part A

1. absolute magnitude
2. Constellations
3. supernova
4. black hole
5. local group
6. hydrogen
7. light-year
8. main sequence
9. Nuclear fusion
10. parallax
11. Reflecting
12. subatomic particles
13. protostar
14. red giant
15. pendulum

Part B

1. **(b)**
2. **(d)**
3. **(e)**
4. **(b)**
5. **(e)**
6. **(c)**
7. **(a)**
8. **(b)**
9. **(d)**
10. **(e)**

Part C

1. True
2. False; local group
3. False; helium
4. False; Milky Way
5. True
6. False; galaxies
7. True
8. True
9. False; main-sequence stars
10. False; reflecting telescopes

CONNECTING TO CONCEPTS

1. The big bang theory states that the universe (time and space as we know them) began with an explosion during which matter formed and the universe began an expansion that is continuing today. The steady-state theory offers the view that the universe always was and always will be. It is no longer accepted as the best explanation of the history of the universe.

2. The big bang theory is supported by evidence of the ongoing expansion as observed by the redshift. It is also supported by the existence of background radiation, or leftover energy from this event, in all "corners" of the universe.

3. Items that may appear in your time line include the times of the big bang; the formation of the first stars, the first galaxies, and our solar system; and the appearance of life on Earth. Give some thought to developing a scale; perhaps 1 centimeter may equal 100 million years. Use whatever scale you choose consistently as you place the major events on your time line.

4. Remote sensing is the act of studying something without actually interacting with it directly. From our vantage point here on Earth, or even from space just a few hundred kilometers above, we cannot directly interact with any celestial objects.

We are therefore relegated to studying the universe using tools such as telescopes, radio telescopes, and cameras.

5. The four types of galaxies are spiral, barred spiral, elliptical, and irregular. The galaxies vary in shape. These shapes yield clues as to the age of each galaxy. The Milky Way is a spiral galaxy. Additional information on this topic can be found in the "Types of Galaxies" section in this chapter.

6. The largest or most massive stars will "go supernova." These stars possess the necessary mass to generate a gravitational field strong enough to support the heat and nuclear fusion process required for a supernova to occur.

7. The sequence is as follows: hydrogen to helium to carbon to oxygen to nitrogen to iron. Nuclear fusion ceases with the formation of iron. Not enough heat is generated to fuse iron, which has great stability.

8. The Sun must be a second-generation star. Prior supernova explosions were necessary to produce the heavier elements of which our solar system is composed.

9. Telescopes have greater light-gathering ability and power than the unaided eye.

10. The more luminous a star, the greater its surface temperature.

3

Our Solar System

WHAT YOU WILL LEARN

This chapter focuses on our solar system. In this chapter you will learn

- what the major components of our solar system are;
- how we group the planets;
- about Pluto's current status as a "dwarf planet";
- how the solar system formed;
- how our view of the solar system changed over time;
- about motions of the planets;
- about recent discoveries and where to locate the latest information on the Internet.

SECTIONS IN THIS CHAPTER

- Major Components of Our Solar System
- How Did the Solar System Form?
- An Early View of Our Solar System—The Ptolemaic System
- Contributions of the First Modern Astronomers
- Motions of the Planets
- Recent Discoveries/Special Topics
- History of Human Exploration of the Solar System
- Review Exercises for Chapter 3

Major Components of Our Solar System

THE SUN

The Sun, our nearest star, is effectively the heart of our solar system. The Sun and its proximity to Earth are of paramount importance to all life on this planet. Chapter 2, in part, addressed the question of how the Sun, like any star, produces heat and light. It was explained that the Sun produces energy by nuclear fusion and has been doing so as a relatively stable star for billions of years. The Sun is expected to remain in this state, its main-sequence stage, for several billion years to come.

The Sun's stability, combined with favorable conditions (to be discussed in Chapter 10) here on Earth, has allowed life on our planet to develop and flourish. It is worth noting again that Earth is the only place in the entire universe where we know with certainty that life exists. Were the relationship between Earth and the Sun even slightly different than it is, in all probability life, as we know it, would never have developed and could not exist here.

THE PLANETS

Earth is just one of the nine known objects traditionally identified as planets in our solar system. The ancients discovered some of these planets, and others were discovered relatively recently by using instruments not previously available. The planets within our solar system can be divided into two groups—the *inner* or *rocky core planets*, also known as the *terrestrial planets*, and the *outer* or *Jovian planets*, also called the *gas giants* (see Figure 3.1).

The rocky core planets include Mercury, Venus, Earth, and Mars. These planets are all relatively small, dense bodies, as compared to the Jovian planets. The rocky core planets possess an unusual abundance of heavy elements; the rest of the universe, in contrast, is composed primarily of hydrogen. The inner planets all have in common a rocky surface, or lithosphere. Each has an atmosphere, though the character and composition of these atmospheres vary significantly. A brief survey of each planet follows. Table 3.1 provides a tabular summary of interesting and relevant statistics for each planet.

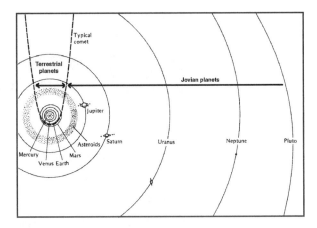

FIGURE 3.1 The solar system. Note the relative position of the inner planets versus the outer planets. Source: *Let's Review: Earth Science—The Physical Setting*, 2nd Ed., Edward J. Denecke, Jr., Barron's Educational Series, Inc., 2002.

TABLE 3.1
SOLAR SYSTEM DATA

Object	Mean Distance from Sun (millions of km)	Period of Revolution	Period of Rotation	Eccentricity of Orbit	Equatorial Diameter (km)	Mass (Earth = 1)	Density (g/cm^3)	Number of Moons
Sun	—	—	27 days	—	1,392,000	333,000.00	1.4	—
Mercury	57.9	88 days	59 days	0.206	4,880	0.553	5.4	0
Venus	108.2	224.7 days	243 days	0.007	12,104	0.815	5.2	0
Earth	149.6	365.26 days	23 hr 56 min 4 sec	0.017	12,756	1.00	5.5	1
Mars	227.9	687 days	24 hr 37 min 23 sec	0.093	6,787	0.1074	3.9	2
Jupiter	778.3	11.86 years	9 hr 50 min 30 sec	0.048	142,800	317.896	1.3	16
Saturn	1,427	29.46 years	10 hr 14 min	0.056	120,000	95.185	0.7	18
Uranus	2,869	84.0 years	17 hr 14 min	0.047	51,800	14.537	1.2	21
Neptune	4,496	164.8 years	16 hr	0.009	49,500	17.151	1.7	8
Pluto	5,900	247.7 years	6 days 9 hr	0.250	2,300	0.0025	2.0	1
Earth's Moon	149.6 (0.386 from Earth)	27.3 days	27 days 8 hr	0.055	3,476	0.0123	3.3	—

Source: The State Education Department, *Earth Science Reference Tables*—2001 Edition (Albany, New York: The University of the State of New York).

Mercury (see Figure 3.2), because of its small size and proximity to the Sun, has a razor-thin atmosphere. The Sun emits high-energy particles, often referred to as the *solar wind*. Near Mercury, the solar wind is strong enough to have removed all but a trace of the primordial atmosphere this planet may once have had. A visitor to Mercury, in theory, could walk on its cratered surface but would detect no real evidence of an atmosphere surrounding him- or herself.

NASA launched Messenger in August 2004. This ambitious project is designed to study Mercury. After launch, the probe flew by Earth in August 2005 and then by Venus in 2006 and 2007. Messenger's first flyby of Mercury occurred in January 2008. Messenger will ultimately be placed in orbit around Mercury in 2011. Through the intensive study of Mercury, we hope to learn more about the formation of our solar system and perhaps even life in the solar system. Messenger is

FIGURE 3.2 Photograph of Mercury, closest planet to the Sun. Craters are Mercury's most striking surface feature, along with no distinguishable evidence of an atmosphere. Source: NASA.

already contributing to our body of knowledge about Mercury! In a recent flyby, Messenger detected "magnetic field tornadoes" stirred up by the solar wind, capable of reaching Mercury's surface and possibly offering an explanation for why Mercury still has any atmosphere remaining. Given Mercury's proximity to the Sun and limited gravity (only 38 percent of that experienced here on Earth), one would expect Mercury to have lost its atmosphere eons ago. Scientists are beginning to understand how the solar wind may actually be acting to replenish Mercury's atmosphere, where it has the opposite effect on the other three terrestrial planets (Venus, Earth, and Mars).

NOTES:
You can read more about Messenger here: *www.nasa.gov/mission_pages/ messenger/main/index.html.*

In 2013, NASA will launch MAVEN, the Mars Atmosphere and Volatile Evolution Mission, a project that may explain the role of the solar wind and its impact upon the Martian atmosphere.

Venus (see Figure 3.3), in contrast, has a very thick atmosphere composed largely of carbon dioxide and sulfuric acid. This very dense and corrosive atmosphere has the ability to trap large amounts of heat, making Venus the planet with the hottest surface conditions despite the fact that it is nearly twice as far from the Sun as Mercury. Neither Mercury nor Venus is likely to be visited by humans at any time in the near future; conditions are too forbidding on the surfaces of these planets.

The only significant similarity among Earth, Venus, and Mercury is the terrestrial, or rocky, surface conditions. Radar scanning of Venus's surface, as well as direct observation by a Russian craft that landed on Venus and took pictures, reveals a barren, desolate, solid surface with mountains and flat plains (Figure 3.3). Much of the surface is covered by relatively young igneous rock that appears to have been produced during a major volcanic event on the order of 300 million years ago. Mercury's surface is also rocky in nature and is heavily cratered.

In contrast, Earth (see Figure 3.4), the third planet from the Sun, has an atmosphere composed primarily of nitrogen and oxygen. Earth is the only planet known to have a hydrosphere and, as noted earlier, a biosphere. Liquid water may be quite rare throughout the universe, as H_2O has a narrow temperature range where it exists as a liquid. At temperatures below 0°C, H_2O exists as ice, and at temperatures above 100°C it exists as water vapor, a clear colorless gas.

FIGURE 3.3 Photograph of Venus taken with filters to highlight the thick banding of clouds in its atmosphere. Source: Courtesy of NASA/JPL/Caltech.

FIGURE 3.4 Earth, the crown jewel of the solar system, as seen from space. Source: NASA.

Mars, the outermost of the rocky core planets, has a thin carbon dioxide atmosphere covering its iron-oxide-rich surface. It is this iron oxide surface that gives Mars its distinctive reddish appearance to observers at night. Mars has a varied terrain. Olympus

Mons is a major feature on the Martian surface. A volcano with a height of nearly 27 kilometers, Olympus Mons eclipses any similar feature on Earth.

Mars is 50 percent farther from the Sun than is Earth and therefore is much colder. Of all the other planets in the solar system, Mars is the most likely to be visited by humans in the foreseeable future.

In fact, Mars may once have harbored life. Recent studies of meteorites here on Earth suggest that they came from Mars and that the traces of ancient fossilized bacteria found within them may have originated with the meteorites on Mars.

> **REMEMBER**
> Want evidence from NASA? Visit http://nssdc.gsfc.nasa.gov/planetary/marslife.html.

More than a century ago telescopes became powerful enough to reveal surface details on Mars. Astronomers were energized by their discoveries of what appeared to be canals and streambeds, features that could have been formed only by running water. Various unmanned satellites have been designed to journey to Mars to study the planet and to search for clues as to where this water went. The first was Mariner 4, which flew by Mars and took pictures in the summer of 1965. In 1976, Viking 1 became the first craft to land on another planet when it touched down on July 20.

The first satellite visits to Mars only added to the mystery, as evidence of water was widespread, but water itself was nowhere to be found. Only in 2002 was it discovered that large amounts of water in the form of ice are located just beneath the rocky, dusty surface. These findings are treated in greater detail in a later section of this chapter headed "Large Quantities of Ice Found on Mars."

Mars is the only other rocky core planet found to have any natural satellites. Mars's two moons, Phobos and Deimos, are quite small and possibly originated as asteroids that were later trapped by the planet's gravitational field. This hypothesis has merit, as Mars is located immediately inside the asteroid belt, which is discussed in depth on page 64.

Beyond the asteroid belt, the Jovian planets, or gas giants, rule the solar system. The Jovian planets—Jupiter, Saturn, Uranus, and Neptune—offer an amazing contrast to the rocky core planets. The Jovian planets are huge, much larger than the rocky core planets, and are composed primarily of gases. Consider the fact that Jupiter, the largest body in our solar system other than the Sun, could, if hollowed out, contain within it more than 100,000 Earth-sized planets. Similarities among the four gas giants include thick atmospheres composed primarily of hydrogen and helium, with minor concentrations of methane, ammonia, and water vapor and small liquid or solid cores.

Lighter elements prevail on these planets, in comparison to those found on the rocky core planets. As a result, planetary densities are close to the density of water, or only about one-fifth of the densities on the inner planets. Each Jovian planet has a ring system surrounding it (Fig. 3.1); however, Saturn stands out with a distinct set of rings composed primarily of rock and ice (Fig. 3.6). Saturn's ring system is the only system visible from Earth. The other Jovian planets' ring systems were discovered only recently as unmanned craft completed missions to these planets. From the late 1970s until the early 1990s, unmanned probes dubbed Voyagers 1 and 2, designed to do "flybys" of Saturn and the other gas giants, took spectacular photographs revealing the previously unknown extensive ring system surrounding Saturn, as well as the ring system of each of the other Jovian Planets.

Pluto, the ninth planet, is located about 40 times farther from the Sun than is Earth. Pluto does not fit well into either of the above groupings of planets; recent debates among astronomers have even challenged the classification of Pluto as a planet. Pluto is small, smaller even than Mercury, the smallest of the rocky core planets, and was not discovered until 1930. Since then, scientists have identified Pluto's orbit as

FIGURE 3.5 Jupiter's Ring System—Nearly invisible, thin rings, likely composed of ice crystals and dust particles. (NASA)

being highly eccentric or elliptical, so eccentric, in fact, that at times Pluto is actually closer to the Sun than is Neptune. Despite its small size, Pluto has a moon,

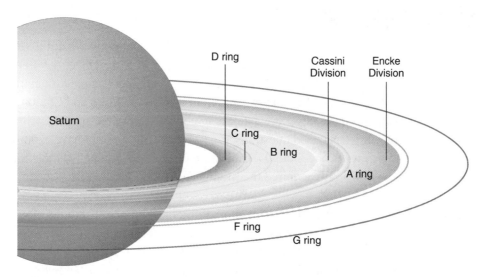

FIGURE 3.6 Saturn's Ring System—Saturn's robust ring system is the only ring system visible from Earth.

Charon, which is so large in proportion to Pluto that the two form almost a dual planet system.

Recent research suggests that Pluto may be the largest of a group of perhaps thousands of cold, icy worlds that orbit at the outer reaches of our solar system. This region is known as the Kuiper belt. In 2003, another Kuiper Belt object now known as Eris was discovered. Eris is 27 percent more massive than Pluto, and its discovery has prompted a new classification of solar system objects known as dwarf planets, among which Pluto is now one—and as noted is not the largest of these objects. Since the 2006 reclassification, additional objects have been named. These objects may contain a fossilized record of the birth of our solar system!

NOTES:
Interested readers may find additional details about these discoveries here: *www.gps.caltech.edu/~mbrown/dwarfplanets.html.*

The afternoon of January 19, 2006, marked the launch of New Horizons, an ambitious mission designed to explore Pluto and the Kuiper Belt. Although New Horizons is not due to reach Pluto until 2015, the mission is already making headlines. During a flyby of Jupiter and one of its moons, Io, in 2007, sophisticated instruments confirmed Io as the most geologically active body in the entire solar system, mapping some 20 changes from the Galileo mission just a few years earlier. New Horizons should reach Pluto before the dwarf planet's atmosphere freezes in the extreme cold of the outer solar system.

NATURAL SATELLITES

Each planet, with the exceptions of Mercury and Venus, has one or more natural satellites, or moons, revolving around it. In most cases, if not all, the planet's gravitational field has captured each moon. Virtually all of the satellites are much smaller than the planets about which they orbit. The one exception, in addition to Pluto and Charon, is our own Moon, which is quite large as a satellite in comparison to the size of Earth. This unique relationship has raised questions as to whether Earth might have captured the Moon within its gravitational field. Debates have produced a number of most interesting theories as to how Earth acquired the Moon. These theories are discussed in Chapter 4.

The Jovian planets, in particular, Jupiter and Saturn, have numerous moons of varying sizes. Some of these moons are nearly as large as Earth and have quite interesting environments. Europa, Io, and Titan, while not the largest, are perhaps the most interesting satellites. Europa and Io revolve about Jupiter (Fig. 3.10); Titan revolves about Saturn (Fig. 3.9).

FIGURE 3.7 Natural Satellite.
Courtesy: NASA.

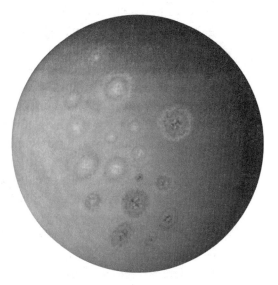

FIGURE 3.8 Bright spots on Io indicate areas where volcanic vents are spewing hot lava. Evidence suggests the lava may actually be hundreds of degrees Celsius hotter than lava eruptions observed on Earth.

FIGURE 3.9 Artist's rendition of Titan's surface with a lake of liquid hydrocarbons and clouds raining hydrocarbons into it. Courtesy: JPL.

Images captured by Voyagers 1 and 2 in 1979 reveal Europa to be a fascinating frozen world with a surface of ice covering an ocean beneath. Considering the relatively recent discovery of life deep within Earth's oceans, in a zone with near-freezing temperatures and no sunlight, scientists are hopeful that someday life may be discovered under Europa's thick layer of surface ice.

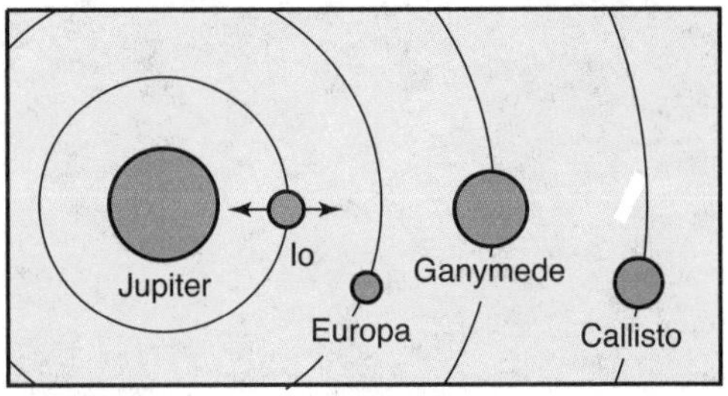

Io is perhaps the most tectonically active body in the solar system. Voyager 2 actually photographed an active volcanic eruption during a flyby of this moon. To envision just how unique this event was, consider the likelihood of observing a volcanic eruption in action on any given day here on Earth.

FIGURE 3.10 Jupiter and Ganymede (with help from Callisto and Europa) act to pull Io apart by creating tremendous gravitational stress on this Jovian satellite.

Io's extreme tectonic activity can be explained by the fact that it is being pulled apart by gravitational stress. Io's orbit is located between Jupiter and three of Jupiter's outer moons; the four moons are known collectively as the Galilean satellites—the first Jovian moons discovered by Galileo. Io's unique position (see Figure 3.10) between Jupiter and the other Jovian moons creates considerable gravitational stress within Io, thus producing the heat energy responsible for this satellite's extreme volcanic activity.

Titan, the largest satellite orbiting Saturn and the second largest satellite in our solar system, is approximately the same size as Mercury. Titan is the only satellite in our solar system to have a substantial atmosphere, actually 150 percent thicker than Earth's. Its composition is mostly nitrogen, with some methane making the surface environment quite different from that on Earth.

ASTEROID BELT

The asteroid belt is an interesting feature of our solar system that separates the inner from the outer planets. It is composed of tens of thousands of chunks of rock that range in size from about 1 kilometer to hundreds of kilometers in diameter. The largest asteroid, Ceres, has a diameter of about 1000 kilometers. While most of the asteroids remain in orbit between Mars and Jupiter, some are in highly eccentric orbits that take them quite close to Earth and our Moon. Recent cratering on the Moon is probably a result of collisions with these asteroids.

The asteroid belt is thought to be material that never accreted to form a planet, as the other inner planets were thought to have formed, or a planet may have formed and then been pulled apart as a result of gravitational stress when it was caught between Jupiter's and the Sun's gravitational fields. Other theories suggest that several small planets coexisted between Mars and Jupiter and that collisions between these planets produced the asteroid belt.

Launched in September 2007, the Dawn Mission is charged with characterizing the conditions and events that marked the solar system's earliest days by exploring in detail two of the largest bodies in the asteroid belt known as Ceres and Vesta. It is hoped that over time, the Dawn Mission will provide answers to questions about this mysterious region within our solar system.

METEOROIDS

Meteoroids are among the smallest members of our solar system. Most are the size of sand grains, though some are much larger. Each year Earth experiences several meteor showers, which produce a show of "shooting stars," as they are often called. Meteor showers often occur with swarms of short-term comets, leading to the theory that most meteors consist of materials lost by comets. Although most meteoroids burn up as they fall through our atmosphere, some survive and are known as meteorites when found here on Earth. The largest meteorites can produce craters on Earth's surface; the most famous is the Meteor Crater in Arizona. This crater has a diameter of about 1.2 kilometers and is 170 meters deep. Probably the meteoroid that created this crater hit Earth within the last 20,000 years, as our planet's dynamic surface would erase the effects of such an impact after a greater period of time.

The meteorite that created the Meteor Crater was small, however, compared to the meteoroid that is thought to have caused the extinction of the dinosaurs and of 95 percent of all other life-forms about 65 million years ago. Giant meteoroids such as this one, large enough to create craters hundreds of kilometers across, impact Earth rather infrequently, but cause great change, such as mass extinctions of life-forms, whenever they do. Fortunately, millions of years pass between these events on Earth. (Fig. 3.11)

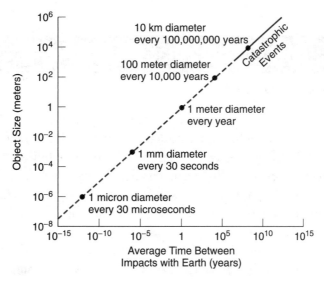

FIGURE 3.11 Object Size vs. Avg. Time Between Impact Graph—Large objects can impact Earth and cause catastrophic events, but fortunately, millions of years pass between such incidents.

Source: www.tulane.edu/~sanelson/geol204/impacts.htm.

COMETS

Comets are extremely interesting and somewhat unpredictable objects. Often likened to
dirty snowballs, they are composed of a
variety of frozen gases, including water
vapor, ammonia, methane, and carbon
dioxide, along with small pieces of rocky
and metallic materials. These bodies are
locked into highly elliptical orbits about the
Sun, taking them at times far outside Pluto's
orbit and at other times quite close to the
Sun. When a comet approaches the Sun, the
frozen gases begin to vaporize, forming a
glowing head called a *coma* and often a tail
that can stretch for millions of kilometers.
Halley's comet (see Figure 3.12) is probably
one of the most spectacular comets, appearing on a regular 75-year schedule. Despite a
comet's small size, its glowing coma and
long tail can put on quite a celestial show for
a period of days or weeks as the comet
passes by Earth on its way around the Sun.

FIGURE 3.12 Halley's Comet provides a spectacular sky show as it passes by Earth every 75 years.

The next appearance of Halley's Comet will occur in 2061.

How Did the Solar System Form?

NEBULAR HYPOTHESIS

The story of our solar system's formation is a fascinating one. You are about to read a
relatively detailed account of likely events that transpired well before humans ever
walked this planet. Since these events occurred on the order of billions of years ago,
you might ask on what evidence this account is based. The search for evidence
involves studying the composition of the solar system today, the ways in which elements found in our present solar system probably formed, and related processes that
can be observed elsewhere in the universe today.

Let us begin to address the question as to how our solar system formed by identifying some known physical facts about this system. We know, for example, that all nine
planets are located in one relatively flat plane in space, orbit the Sun in the same direction, and can be grouped into two very different types based upon composition. We
know that the Sun is composed primarily of hydrogen and helium, elements that are
not highly abundant on the inner planets but are much more abundant on the Jovian
planets. Finally, our best estimates suggest that everything in the solar system formed

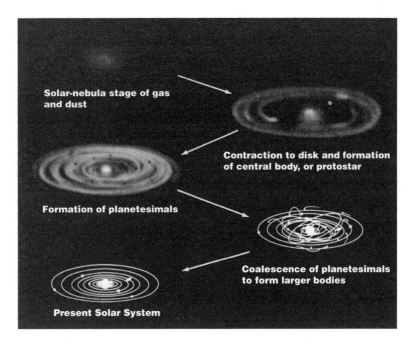

FIGURE 3.13 The formation of the solar system from a solar nebula to our present solar system. Source: *Exploring the Cosmos,* Louis Berman and J.C. Evans, Little, Brown, 1986.

around the same time. A satisfactory theory should ideally account for all of these observations. The commonly accepted theory that attempts to explain how our solar system came to be is known as the *nebular hypothesis.*

The nebular hypothesis (see Figure 3.13) proposes that our solar system began to take shape when a *nebula*, or huge rotating cloud of dust and gases, began to contract. This contraction was probably initiated by a shock wave created by a "nearby" supernova explosion. In fact, the nebula itself owed its existence to matter produced by earlier supernovae. This sequence of events is considered quite likely as heavier elements, those known to form only during a supernova, are found in relative abundance throughout our solar system. The first supernova thus, in essence, created the matter that eventually coalesced to form Earth and everything else in our solar system. The second supernova created the shock wave that initiated the contraction of the nebula. This contraction occurred on the order of 5 billion years ago.

Gravitational forces would have impelled most of the matter in this nebula toward the center. Next, gravitational compression caused temperatures to rise near the center to a point where nuclear fusion began to occur. When fusion began, our closest star, the Sun, was effectively "born."

Away from the center, some dust and gases remained in orbit around a now flattening disk. The dust particles began to coalesce by engaging in constructive collisions that produced larger and larger planetesimals, that is, small, solid celestial bodies. As the planetesimals grew (see Figure 3.14), their gravitational attraction increased and they became more effective at capturing additional material that was still orbiting the "protosun," or early Sun.

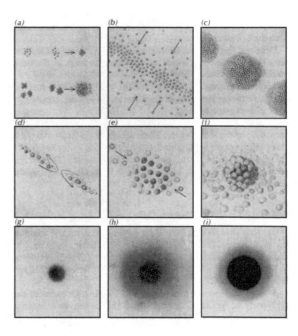

FIGURE 3.14 Planets form by the process of accretion. Planets form as a result of "constructive collisions," in which small grains stick together to form larger particles. Small grains collide in the primordial solar system (a). The growing particles fall toward the plane of the original cloud (b), forming a loose disk of material. The disk breaks up into asteroid-sized bodies (c), which cluster together (d), collide (e), and coalesce (f) into planet-sized bodies (g). The planet-sized bodies have enough gravity to collect gas from the nebula (h). The end result is the formation of a primordial planet (i). Source: *The Origin and Evolution of the Solar System,* A. G. W. Cameron, Scientific American, Inc., 1975. All rights reserved.

THE INNER PLANETS VERSUS THE OUTER PLANETS

Conditions near the Sun were quite hot, so volatiles, or lighter elements, were unable to condense, whereas farther away from the Sun, condensation of volatiles was possible. This fact explains the difference between the composition of the gas giants—Jupiter, Saturn, Uranus, and Neptune, which possess high percentages of light elements, including hydrogen and helium—and that of the rocky core planets, which have an abundance of heavier elements, including iron and nickel. Iron and nickel will condense at much higher temperatures than hydrogen and helium, and thus they solidified on the planets that formed closer to the Sun.

The presence of lighter materials on Earth, including the gases in our atmosphere and the water on our planet, is attributed to cometary collisions that occurred sometime after this initial planetary accretion process. Any early atmosphere of Earth's was swept away by the solar wind. This view is supported by the lack of atmosphere on Mercury; the solar wind prevented an atmosphere from ever accumulating near that planet's surface.

The rotating and flattening nebula view proposed by the nebular hypothesis helps to explain the observed orbital motions of the planets and the flat-plane appearance of the solar system. As this discussion indicates, the nebular hypothesis also accounts for the compositional differences observed between the inner and outer planets.

An Early View of Our Solar System— The Ptolemaic System

Humans have long wondered how the solar system is organized and how it formed. One of the earliest perspectives on the organization of the solar system that gained

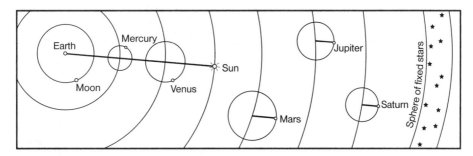

FIGURE 3.15 Ptolemy's geocentric model of the universe. This model was accepted for well over 1,000 years. Source: *Horizons: Exploring the Universe,* Michael A. Seeds, Wadsworth, 1987.

wide acceptance dates back to the second century. Ptolemy, a Greek astronomer, presented a *geocentric*, or Earth centered, view so clearly that it lasted well into the seventeenth century. Ptolemy was probably not the first to espouse this view, but his thirteen-volume treatise, *Almagest* or *The Great Work*, has been preserved over the centuries. Ptolemy's model was accepted largely because it accounted for the observable motions of the planets.

The key features of Ptolemy's geocentric model (see Figure 3.15) included a motionless Earth at the center of the known universe, surrounded by the Sun and the known planets moving about Earth in perfectly circular orbits. The Greeks considered the circle to be a pure and perfect shape; thus all orbits around Earth were deemed to be perfectly circular.

Another key feature of Ptolemy's model was the use of *epicycles*, or small, circular suborbits, to explain the apparent retrograde motion of the planets against the background of stars. When viewed nightly, the planets appear to be moving eastward against the background of much more distant stars; however, as a result of a combination of each planet's motion and Earth's motion around the Sun, a particular planet being observed appears to "retrograde" westward for a period of time. To visualize this effect, consider Earth and Mars. Because Earth is closer to the Sun, it travels faster in its orbit than does Mars. Therefore, at some points during an Earth year, our planet will "overtake" Mars as we orbit the Sun. When that happens, Mars will appear to retrograde, that is, move westward, in our sky. As Figure 3.16 indicates, Ptolemy developed a rather elaborate system to account for this westward drift.

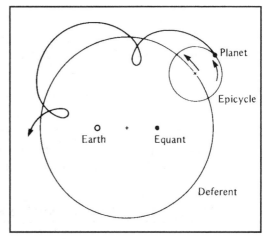

FIGURE 3.16 Ptolemy's universe. Ptolemy added epicycles to Aristotle's model to explain the observed retrograde motion of the planets and the apparent changes in their diameter. Source: *Horizons: Exploring the Universe,* Michael A. Seeds, Wadsworth, 1987.

Contributions of the First Modern Astronomers

NICOLAUS COPERNICUS AND THE HELIOCENTRIC THEORY

Ptolemy's geocentric model was not seriously challenged until the sixteenth century. Then a Polish mathematician named Nicolaus Copernicus (see Figure 3.17) described a *heliocentric model* (see Figure 3.18) that placed the Sun at the center of our solar system, with Earth orbiting the Sun while rotating on its axis. Proof of Earth's rotation would not be secured for over 300 years, yet Copernicus believed this to be the key to explaining the observed motions in the heavens.

FIGURE 3.17

Nicolaus Copernicus. Source: *Astronomy: The Cosmic Journey,* William K. Hartman, Wadsworth, 1987.

It is interesting that Copernicus first discreetly distributed his writings in 1514, but did not publish his work titled *On the Revolutions of the Heavenly Spheres* until he was virtually on his deathbed in 1543. It is likely that Copernicus dedicated his writings to the pope and waited to publish until the end of his life for fear of how the Catholic Church would respond to his new and radical theory. Furthermore, he was probably concerned about how the academic world would view his heliocentric model as it conflicted with the physics of Aristotle—the mainstream view of physics at that time. According to Aristotle, an object when released falls to Earth as a result of its being drawn to the center of the Universe, which, according to Ptolemy's model, was Earth. The problem here, of course, is that if the Earth is no longer the center of the universe, as described by Copernicus, why do objects, once released, fall to Earth?

Concerns aside, the enunciation of the heliocentric model marked the beginning of a scientific revolution, and caused astronomers to view the universe as much larger than previously thought. Copernicus's heliocentric model was far simpler than the geocentric model because fewer complex theories were required to explain the observed motions in the heavens. Copernicus's model, however, retained perfect circles for planetary movements and therefore still required epicycles, though smaller than those generated by Ptolemy. Refer to Figure 3.19 for a depiction of Copernicus's explanation of retrograde motion and his use of epicycles.

TYCHO BRAHE AND STELLAR PARALLAX

Tycho Brahe (see Figure 3.20), was a Danish astronomer whose detailed observations during the late sixteenth century led to the great works of Kepler, who was actually hired by Brahe in 1600 during the final year of his life. Brahe's contributions include cataloging over 1000 stars, far more than any astronomer before him. Also, he designed and built instruments to measure celestial motions throughout the year. He was not a supporter of the Copernican model of the universe; however, his model eventually incorporated some of Copernicus's ideas.

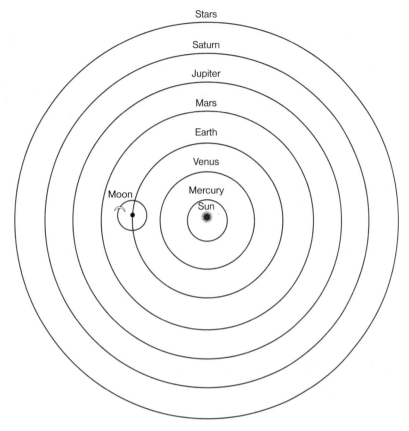

FIGURE 3.18 The Copernican heliocentric universe. Copernicus proposed a Sun-centered model in which all planets and stars moved in perfect circles around the Sun. Source: *Discovering Astronomy,* Robert D. Chapman. W. H. Freeman, 1978. Used with permission.

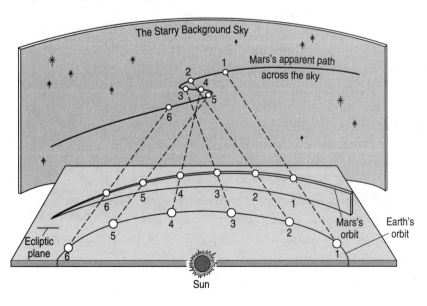

FIGURE 3.19 Copernicus's simple, forthright explanation of retrograde motion. Both Earth and Mars move in a continuous path, but the inner planet (Earth) covers more of its orbit in the same period, changing its point of view toward the outer planet (Mars). Source: *Astronomy: The Cosmic Journey,* William K. Hartman, Wadsworth, 1987. Used with permission.

Brahe used his instruments in an attempt to identify *parallax*, that is, the angular difference in direction of a celestial body as measured from two points on Earth's orbit. He reasoned that, if Earth were indeed orbiting the Sun, then, as Earth moved through the heavens, at times it would be closer to certain constellations and, at these times, those constellations would appear larger. His instrumentation was not sensitive enough to detect these changes, however, and therefore he rejected the heliocentric model, preferring not to oppose Aristotlean physics, and formulated his own model of the solar system. His model included a stationary Earth at the center of the universe, with the Sun and Moon orbiting Earth, but all the other planets orbiting the Sun.

His observations were sufficiently precise to enable Kepler, in the years following Brahe's death, to determine correctly that planetary orbits were in fact elliptical and not the perfect circles previously accepted.

FIGURE 3.20 Tycho Brahe (1546–1601) Source: *Astronomy: The Cosmic Journey*, William K. Hartman, Wadsworth, 1987. Used with permission.

JOHANNES KEPLER AND THE LAWS OF PLANETARY MOTION

Johannes Kepler (see Figure 3.21), a mathematician who grew up in what today is Germany, was a devoutly religious man who, paradoxically, was excommunicated in 1612 for his views concerning the universe and Earth's place in it. Kepler's first written work, published at the age of 25 in 1596, eventually earned him his apprenticeship with Brahe in 1600. In his *Mysterium Cosmographicum*, or *Mysteries of the Cosmos*, Kepler argued that five regular solids all could fit within the orbits of the planets. This was an important step toward understanding the geometry of planetary motions. Kepler also used his book to raise nine important questions or points designed to challenge current thinking. For instance, Kepler noted that the Copernican theory explains why Venus and Mercury are never seen very far from the Sun (they lie between Earth and the Sun), whereas the geocentric theory has no explanation for this fact.

After Brahe's death in 1601, Kepler was appointed to the highly prestigious position of Imperial Mathematician. He continued his research, and in 1609 developed his first law of planetary motion. This law states that planets move in elliptical orbits about the Sun (see Figure 3.22), which is at

FIGURE 3.21 Johannes Kepler (1571–1630) Source: *Astronomy: The Cosmic Journey*, William K. Hartman, Wadsworth, 1987. Used with permission.

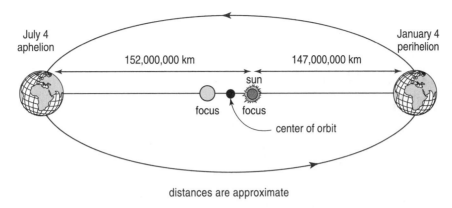

distances are approximate

FIGURE 3.22 View of Earth's elliptical orbit with the Sun at one focus. Earth is closest to the Sun at perihelion and farthest away at aphelion. Distances are approximate. Source: *Let's Review: Earth Science—The Physical Setting,* 2nd Ed., Edward J. Denecke, Jr., Barron's Educational Series, Inc., 2002.

one of the foci of the orbits. Kepler's conclusions were based upon years of observation of the planet Mars and included data collected by Brahe, which he knew to be extremely accurate. Current models of his day, those that relied upon circular motion, could not explain the observed positions of Mars at certain times throughout the year.

In 1619, Kepler added two additional laws of planetary motion. The second law describes the velocities of planets throughout their orbits and states that planets move fastest when they are closer to the Sun and slower when they are farther from it (see Figure 3.23). The third law states that orbital velocity decreases as orbital radius increases. Thus the planets closer to the Sun are moving faster than the planets at great distance from it (see Figure 3.24). The one major issue that Kepler did not address was the nature of the invisible force that keeps all the planets in orbit about

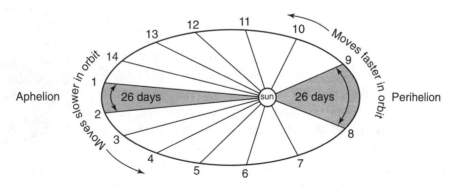

FIGURE 3.23 Kepler's law of equal areas. A planet sweeps out equal areas of its elliptical orbit in equal periods of time. Since the planet travels less distance in the 26 days it takes to move from position 1 to 2 on the ellipse above, it is traveling slower than when it moves from position 8 to 9. Source: *Let's Review: Earth Science—The Physical Setting,* 2nd Ed., Edward J. Denecke, Jr., Barron's Educational Series, Inc., 2002.

the Sun. This insight would wait for Newton, later in the seventeenth century.

GALILEO GALILEI AND THE TELESCOPE

Galileo Galilei (see Figure 3.25) was certainly a giant of his time, and many volumes have been written about him. The intent here is to briefly summarize his contributions to the scientific world and our understanding of the universe. Galileo was born in Italy in 1564, studied mathematics, and became a lecturer in 1588 at the age of 24. Just three years later, he became a full professor, despite his demonstrated talent for controversy. An early illustration of this talent can be seen in his research on pendulums, where he showed that a pendulum's motion is not consistent with the physics of Aristotle. Galileo is considered a pioneer of modern experimental scientific method.

It was Galileo's use of the recently invented telescope for scientific study that enabled him to earn his place in history. In 1609, only one year after Hans Lippershey's invention of the telescope, Galileo built several of these instruments that improved upon Lippershey's design. During the winter of 1609–1610, he made his now famous observations and discoveries. He identified mountains and what he thought were seas on the Moon, he described the Milky Way as being composed of millions of individual stars, he discovered the rotational period of the Sun (28 days) by observing sun spots, and—the most controversial discovery of all—he revealed that Jupiter is itself a center of revolution as it has four small moons (the number of Jovian moons identified by Galileo; refer to Figure 3.10) orbiting it. Even today, these four moons are often referred to as the Galilean satellites of Jupiter.

Despite initial acceptance of Galileo's views by the Renaissance church, Galileo was tried and convicted of heresy before the Roman Inquisition in 1633. As a result he was forced to recant his support of Copernicus and the heliocentric theory. Also, he was condemned to house arrest and prohibited from publishing. His later works were published after being smuggled out of Italy.

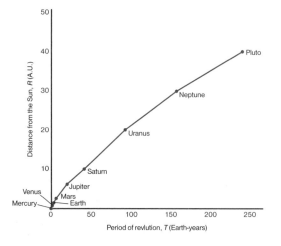

Planet	T (period of revolution in Earth-years)	R (distance from the Sun in AU)
Mercury	0.24	0.39
Venus	0.62	0.72
Earth	1.0	1.0
Mars	1.88	1.52
Jupiter	11.86	5.23
Saturn	29.46	9.54
Uranus	84.0	19.18
Neptune	164.8	30.05
Pluto	247.7	39.44

FIGURE 3.24 Relationship between period of revolution and distance from the Sun. Source: *Let's Review: Earth Science—The Physical Setting,* 2nd Ed., Edward J. Denecke, Jr., Barron's Educational Series, Inc., 2002.

FIGURE 3.25 Galileo Galilei (1564–1642). Source: *Astronomy: The Cosmic Journey,* William K. Hartman, Wadsworth, 1987. Used with permission.

SIR ISAAC NEWTON AND GRAVITY

Isaac Newton (see Figure 3.26), of gravity fame, deserves mention here, as he realized that all celestial bodies have gravitational attraction for one another. The attraction between two bodies is proportional to the product of their masses and inversely proportional to the square of the distance between them. In other words, the greater the total mass of the two objects and the closer their proximity, the greater the gravitational attraction between them. The Sun is the most massive object in our solar system and as such has the greatest gravitational attraction, effectively holding all the planets and minor members of the solar system in orbit about it.

Newton's laws of motion, first published in 1686 in his book *Principia*, form the basis for modern physics and engineering. These laws were a radical departure for the time. Newton identified an invisible force called *gravity* and explained that an object falls to Earth as a result of the pull of gravity. He understood that all objects have gravity and that the larger the object, such as the Sun, the greater its gravitational attraction (see Figure 3.27). Before Newton, the prevailing view held that all objects were attracted to Earth. This view explained not only why objects fell toward Earth when dropped, but also why Earth was (as then believed) the center of the universe. Since all objects, including other planets and the stars, were drawn toward Earth, it had to be the center of the universe.

FIGURE 3.26 Issac Newton (1642–1727). Source: *Astronomy: The Cosmic Journey,* William K. Hartman, Wadsworth, 1987. Used with permission.

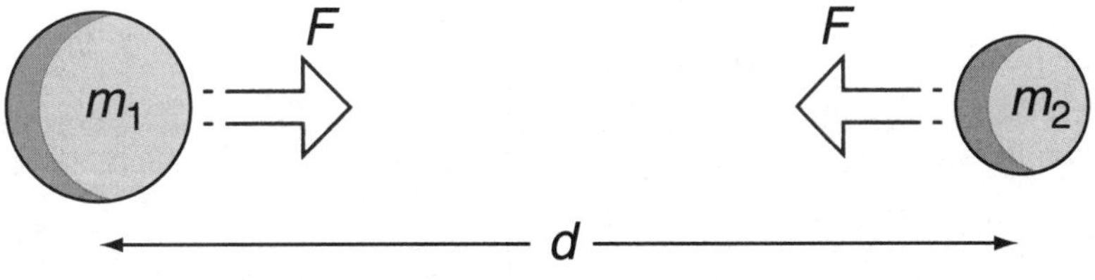

$$F = G \frac{m_1 m_2}{d^2}$$

FIGURE 3.27 Newton's law of gravity. In this equation, F is the force of gravity acting between two masses, G is the gravitational constant, m_1 and m_2 are the masses, and d is the distance between them. Source: *Astronomy: A Self-Teaching Guide,* 4th Ed., Dinah L. Moché, John Wiley, 1998. Used with permission.

Motions of the Planets

ROTATION

The planets engage in several types of motions, one of which is rotation. *Rotation* is the turning or spinning of a body on its axis (see Figure 3.28). All planets and their satellites are known to rotate on their axes. Earth rotates at a rate such that it completes one rotation approximately every 24 hours. This is the basis for our definition of a day. Although Copernicus proposed the notion that Earth rotates, he did so only to simplify the accepted model of the universe at that time; he offered no proof of Earth's rotation. It was not until the mid-nineteenth century, in 1851, that Jean Foucault used a free-swinging pendulum to prove that Earth does, in fact, rotate on its axis.

To visualize the Foucault pendulum (see Figure 3.29), imagine a free-swinging pendulum over the North Pole. Bear in mind that, once in motion, the pendulum will continuously swing in the same plane unless acted upon by an outside force. Imagine further that a sharp pointer capable of marking the frozen surface is attached to the base of the pendulum. As Earth turns beneath the pendulum, the marks on the snow-covered surface indicate that the pendulum is slowly and continuously changing position. After 24 hours, a complete circle has been marked in the ice. The Earth has completed one full rotation under the pendulum.

Foucault himself, of course, did not conduct this experiment at the North Pole; his work was done at the Pantheon in Paris. Similar pendulums can be found in major museums in most American cities. Examples of sites where pendulums can be seen include the Franklin Museum in

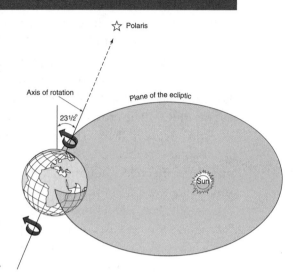

FIGURE 3.28 Rotation. Earth rotates from west to east once every 24 hours around an axis that runs through the Poles. Source: *Let's Review: Earth Science—The Physical Setting,* 2nd Ed., Edward J. Denecke, Jr., Barron's Educational Series, Inc., 2002.

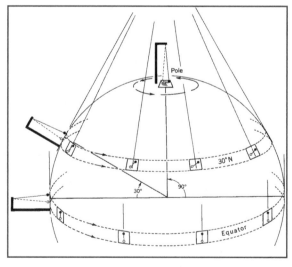

FIGURE 3.29 The Foucault pendulum. Source: *Let's Review: Earth Science—The Physical Setting,* 2nd Ed., Edward J. Denecke, Jr., Barron's Educational Series, Inc., 2002.

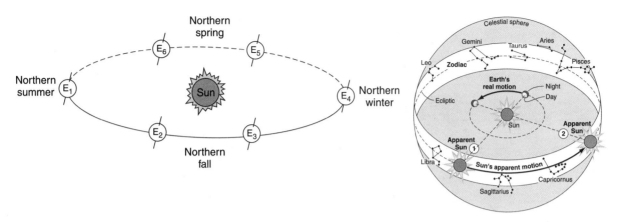

FIGURE 3.30 Revolution. (a) The constant orientation of Earth's axis as the planet orbits the Sun. Source: *Astronomy Explained*, Gerald North, Springer-Verlag, 1997. Used with permission. (b) Earth revolves around the Sun once every 365.25 days. During each season, different parts of the universe are visible at night. Source: *Astronomy: A Self-Teaching Guide,* 4th Ed., Dinah L. Moché, 1993, John Wiley, 1998. Used with permission.

Philadelphia, the Smithsonian in Washington, D.C., and the Museum of Science and Industry in Chicago.

REVOLUTION

Revolution is defined as the motion of a body, such as a planet or a satellite, along a path around some point in space. Clear examples include Earth and all of the other planets revolving around the Sun.

Evidence for revolution (see Figure 3.30) includes the fact that the positions of the constellations change slightly from night to night. These changes are most noticeable from season to season as some constellations, Orion for instance, are quite prominent in the sky during one season but not in another. Orion is a clearly identifiable feature in the Northern Hemisphere's winter sky, but is not easily seen in the summer sky. Additional evidence can be found by studying parallax, which is discussed at some length in Chapter 2.

Each satellite revolves around its own planet. Thus, Earth's Moon essentially revolves about Earth. This statement requires clarification, however, as Earth and the Moon both revolve about a center of mass that is located just beyond Earth's surface. This situation exists because the Moon has a significant mass when compared to the mass of Earth. In contrast, the moons of Jupiter have so little mass in comparison to the mass of Jupiter that they can truly be said to revolve about Jupiter.

PRECESSION

Precession is a very gradual movement exhibited by Earth and other planets (see Figure 3.31). Precession is a gradual change in the orientation of Earth's axis—essentially, a "wobble" in the axis. Earth's axis is tilted 23.5° from the vertical in space, causing Earth,

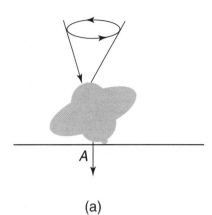

(a)

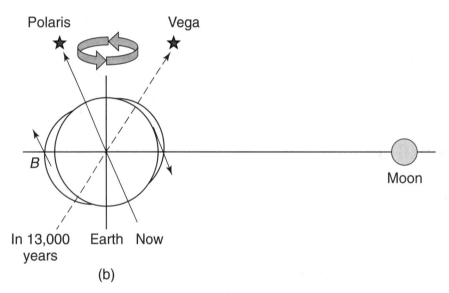

(b)

FIGURE 3.31 Precession. Precession causes Earth's axis to wobble like a top. (a) The force of gravity (*A*) on the spinning top tries to tip it over. The result is precession. If the top were not spinning, it would simply topple over. (b) The net force of the Moon (*B*) on Earth's tidal bulge tries to pull the planet's equatorial plane into the new ecliptic plane. Again, the result is precession. Like the top, if Earth were not rotating, its equator would correspond to the ecliptic. Source: *Discovering Astronomy,* Robert D. Chapman, W.H. Freeman, 1978.

by chance, to point toward Polaris, the North Star. As Earth precesses, it slowly turns away from Polaris and points toward Vega in space. In about 12,000 years Vega will essentially be the North Star! Earth will complete its current precession by A.D. 28,000, at which time Polaris will again be the North Star. The rate of precession is too slow in terms of the human life span to consider any star other than Polaris as the North Star.

This gradual precession may play a role in climate change. Seasons are caused by the planet's tilt, which causes the Sun's rays to be more (or less) direct at various times throughout the year. Since the winter season in the Northern Hemisphere currently coincides with perihelion, or Earth's closest approach to the Sun, the winter season is not quite as cold as it could be. As Earth precesses from the current position, eventually winter will coincide with aphelion, or Earth's most distant position from the Sun. At that time, winter temperatures will be colder than they are now. The same argument can be made for the summer season, as aphelion currently occurs in July.

Discussion of this issue is limited to the Northern Hemisphere, where most of Earth's landmass is found. The abundance of surface water in the Southern Hemisphere limits any variations in climate from season to season and would also limit the effects caused by Earth's precession. A more extensive discussion of seasons appears in Chapter 13.

Recent Discoveries/Special Topics

OTHER SOLAR SYSTEMS

The universe is filled with phenomena just waiting to be discovered. We humans are relegated to remote observation of our galaxy and the universe, as distances are too great to even think about sending manned missions or unmanned probes anywhere outside our own solar system. However, by using increasingly sophisticated telescopes, including the Hubble Space Telescope (HST), which is stationed outside Earth's atmosphere, we can now study distant objects and learn more than we ever could before. Earth's atmosphere has the effect of blurring our view of space, but the HST has enabled us to move beyond that barrier.

One of the most exciting recent achievements is the discovery of planets orbiting other stars. Astronomers, including Carl Sagan of *Cosmos* fame, have long suggested that ours cannot be the only solar system in the known universe; in fact, there must be billions of solar systems and billions of other planets. We now have proof!

Careful analysis of slight wobbling in distant stars reveals the presence of planets orbiting these stars. Most of these planets are not visible, but astronomers can infer their sizes, distances from their "parent" stars, and, in a few cases, even their chemical compositions. To date, more than 353 extra-solar planets have been identified. Most of these are large, Jupiter-like bodies orbiting their stars, but recently a few smaller planets have also been detected. The race is on to find an Earth-like planet that may even have life!

Late in 2001, astronomers using the HST detected sodium in the atmosphere of an unusually large extra-solar planet orbiting quite close to its parent star, HD 209458. As the planet "transits" or passes between that star and Earth every 3.5 days, astronomers were able to study changes in the star's "spectra" or energy profile, thereby enabling them to identify the chemical makeup of the planet. Kepler, NASA's newest tool, launched in March 2009, is designed to study 100,000 stars for signs of planets. Although gas giants, ice giants, and hot super-Earths have been identified, the great challenge is to find an Earth-like planet in the "habitable" zone where conditions similar to our own planet may exist and hence the possibility of life.

The HST has been instrumental in improving our understanding of the universe and the phenomena contained within it. In the summer of 2002, scientists at Rutgers University, using the HST and the National Science Foundation's Very Large Array (VLA) radio telescope, found compelling evidence that supermassive black holes, found in the hearts of large galaxies, collide when galaxies merge. This finding confirms that two black holes coalesce rather than remain as a binary system.

VENUS AND ITS RUNAWAY GREENHOUSE EFFECT

Within our own solar system, the planet Venus has an interesting story to tell. This story has relevance to us here on Earth because Venus is our nearest planetary neighbor in space and has many physical similarities to Earth. Venus is several million miles closer to the Sun than Earth is, but this fact alone does not explain its fiery hot surface temperatures of nearly 450°C (900°F).

Venus owes its hellish surface environment to an atmosphere rich in carbon dioxide, a gas known to trap solar energy. The same gas is found here on Earth, but is present in much lesser quantities. In fact, were it not for a small amount of this greenhouse gas, Earth would be too cold to support life. In regard to carbon dioxide, it appears that a little is vital, but too much can be deadly. The question here on Earth is, of course, as we burn fossil fuels and dump other greenhouse gases into our atmosphere, will Earth eventually suffer a fate similar to that of Venus? No one knows what caused Venus to have a runaway greenhouse effect, and no one can state with certainty how our industrial activities, currently impacting the atmospheric chemistry of our planet, will affect the global climate and subsequently the habitability of Earth.

Most recently, Venus was explored by the Magellan craft. Launched in May 1989, Magellan arrived some 15 months later and began radar mapping the surface of the planet. By 1994, 98 percent of the surface had been mapped. The mapping project revealed a lava-covered surface because over 85 percent of Venus's surface has been impacted by volcanic activity. Plate tectonics, a very active process on Earth, seems to be absent on Venus.

LARGE QUANTITIES OF ICE FOUND ON MARS

Mars has long been an ongoing candidate for the discovery of evidence of ancient remnants of simple life-forms. A recent discovery by the Mars Odyssey spacecraft has refueled hope among some researchers. Using a gamma ray spectrometer, the spacecraft detected large quantities of water ice, enough to fill Lake Michigan twice. Scientists speculate that much more remains to be found.

The gamma ray spectrometer is designed to detect hydrogen based upon the intensity of the gamma rays emitted and the intensity of the neutrons affected by the hydrogen. In soil up to 1 meter beneath the surface, large amounts of ice were identified. The average depth of the ice is 60 centimeters beneath the surface at approximately 60° South of the Martian Equator, and 30 centimeters beneath the surface at latitudes closer to the Martian South Pole.

Equipped with panoramic cameras and an assortment of soil-sampling tools, two robotic geologists called Spirit and Opportunity landed on Mars in early 2004. This long-term mission, continuing through the writing of this new edition, has four scientific goals: to determine whether life ever arose on Mars, to characterize the climate of Mars, to study the geology of Mars, and to prepare for future human exploration. After an extensive array of photographs and analyses of samples, evidence found in craters indicates that Mars's past includes the presence of liquid water, high winds, and sand storms that acted to shape the planet's surface.

NOTES:
NASA maintains a Mars Rover Exploration Web site at http://marsrovers.nasa.gov/home/index.html.

COMET SHOEMAKER-LEVY AND ITS IMPACT UPON JUPITER

Comets, as noted earlier, are minor members of the solar system that travel in highly elliptical orbits about the Sun. Comet Shoemaker-Levy made headlines in July 1994 when it slammed into Jupiter with a force equal to 6 trillion tons of TNT. A similar impact here on Earth would have been devastating. The collision was clearly the most dramatic event in our solar system ever to be observed in real time by humans. Comet Shoemaker-Levy had been discovered a year earlier at California's Mount Palomar observatory.

History of Human Exploration of the Solar System

U.S. CONTRIBUTIONS

The United States has clearly been a leader in space exploration after some early accomplishments and "firsts" by the former Soviet Union. The Soviets put the first satellite and then shortly afterward the first man in space beginning in the late 1950s. In 1961, however, President John F. Kennedy challenged the nation to land a man on the Moon by the end of that decade. With President Kennedy's initiative, the space race had truly begun. The Soviets soon realized that in the quest for the Moon they were outdone by our resolve and technology and turned their attention toward the unmanned study of Venus and other less ambitious projects.

The 1960s saw a series of U.S. manned missions, all designed to culminate with landing a man on the Moon. A series of unmanned satellites designed to aid in the study of our own planet was also launched. For example, the first weather satellites were launched during this era. People saw the first pictures of the clouds and weather systems that cover our planet, images that we now see regularly on the evening television news or via the Internet.

The manned missions, however, were the truly exciting part of the U.S. space program throughout the 1960s. The first, Project Mercury, was charged with sending a man into space and returning him safely after achieving Earth orbit. John Glenn was the first American to orbit Earth in space; this occurred on February 20, 1962. Glenn's craft, the *Friendship 7*, made three complete orbits around Earth before returning home safely.

Project Mercury was followed by Project Gemini, appropriately named as two men were carried into space on each mission. As with Mercury, the goal was to place the pair into a successful Earth orbit and then return them safely home. The most ambitious of all the programs, Project Apollo, suffered a severe setback in 1967 when a disastrous explosion occurred during a test launch and three men lost their lives. After several months of reengineering the command capsule (the astronauts' "home" during their travel in space) and the atmosphere within it, the Apollo program began in earnest in 1968 with the successful launch of Apollo 7.

Rapid progress ensued after Apollo 7 successfully completed its mission—to orbit Earth and return its crew of three men home safely. Apollo 8 was the first manned mission to orbit the Moon; on Christmas Eve, 1968, residents of Earth were treated to pictures of our planet from a distance of 250,000 miles. It has been said that the Apollo 8 mission and those pictures provided the finest possible conclusion to the turbulent year 1968. The successful missions of Apollos 9 and 10 led to the momentous event of Apollo 11, when Neil Armstrong and Buzz Aldrin landed on the Moon. On July 20, 1969, the first photos of humans walking on a surface other than Earth's were beamed home, again for all humankind to see and marvel at.

There were several more Apollo missions after Apollo 11. In 1972, however, the entire program came to an end with Apollo 17 when Congress cut NASA's funding and the public began to view trips to the Moon as routine expeditions. Three more Apollo missions that had been planned were abandoned when the funding was cut. Saturn 5 rockets had already been built for these missions and are on display at Cape Kennedy in Cape Canaveral, Florida; the Johnson Space Center in Houston, Texas; and the U.S. Space and Rocket Center in Huntsville, Alabama. The rocket in Huntsville is actually wired and "flight ready."

That same year, 1972, saw the birth of the space shuttle program. Although flights did not begin until the 1980s, the vision of a reusable craft was born just as the Apollo program was ending. Part of the motivation behind the space shuttle program was cost saving, a perspective that NASA hoped would promote congressional support for funding the program. Among its accomplishments, the space shuttle is responsible for the installation of the Hubble Space Telescope. This mission and many others have collectively added to human knowledge and appreciation of our universe.

In our desire to maintain a human presence in space, however, we are periodically reminded of the kinds of risks and dangers our astronauts face. The space shuttle program has not been immune from tragedy. On January 28, 1986, the shuttle *Challenger* exploded shortly after takeoff. The cause was determined to be an O-ring failure combined with cold weather. On February 1, 2003, the shuttle *Columbia* broke up on reentry to Earth.

Columbia's breakup resulted from damage sustained during launch when a foam insulation tile about the size of a laptop case broke off the shuttle's external propellant tank and stuck to the left wing, damaging the shuttle's thermal protection system. Heat generated during the reentry caused the failure. Since then, new procedures have been implemented to minimize the risk of a reoccurrence.

Columbia's mission was a rather routine one in which a number of scientific experiments were conducted during the course of the astronauts' two weeks in space. Perhaps the only aspect of its last flight that caused mass media attention before February 1, 2003, was the presence of the first Israeli astronaut in space, Ilan Ramon. Although the shuttle program has become a model of international cooperation and collaboration, if it were not for Ramon's presence on the shuttle, most Americans would not have even been aware of the fact that it *was* in space. The shuttle program has become so seemingly routine and successful that we collectively lose sight of the risks involved in space travel.

More information can be obtained at the following Web sites: *www.shuttlepresskit.com/ISS_OVR/* and *http://spaceflight.nasa.gov/station/*.

SOVIET/RUSSIAN CONTRIBUTIONS

The Soviets hold title to launching the first satellite into space, *Sputnik I*, on October 4, 1957. *Sputnik* remained in orbit for nearly 3 months. The launch of *Sputnik* during the Cold War era gave residents of the United States cause for concern about our technological lead over the Soviets. *Sputnik* sent a simple beeping signal back to Earth as it orbited the planet. Some U.S. citizens were suspicious about the nature of the beeping and the kind of information that *Sputnik* was gathering.

On March 17, 1958, the United States followed the Soviet lead by launching *Vanguard I*. The Soviets again shocked the American public, however, by putting the first human in space on April 12, 1961. *Vostok I* carried Yuri Gagarin into space for a total voyage of 1 hour and 48 minutes. He returned to Earth after completing one orbit.

Shortly thereafter, on May 5, 1961, America sent its first astronaut, Alan B. Shephard, into space aboard *Freedom 7*. American progress was rapid thereafter, and John Glenn became the first American to orbit Earth some 10 months later.

The Soviets are credited with several additional "firsts" in space during the rest of the 1960s and the 1970s, but they also suffered some major setbacks, including the loss of cosmonauts on more than one ill-fated mission. After the race to the Moon was essentially lost to the Americans, the Soviets focused their efforts on landing an unmanned probe on Venus, along with building the first space station.

In 1971, the Soviets launched *Salyut I*, the first of their space stations. They followed it with several others in the coming years, culminating with the launch of *Mir* in 1986. When *Mir* was initially placed in service, it represented another Soviet initiative in space. Designed to last 5 years, *Mir* actually remained in service until March 2001. Highlights of *Mir*'s existence include housing Valeri Polyakov for 14 months between January 1994 and March 1995. This record remains as the longest time spent by any human in space. Also in 1995 shuttle astronauts first docked with *Mir* in preparation for future joint U.S.-Russian missions aboard the international space station. Additional information about *Mir* may be obtained at the following Web site: *www.russianspaceweb.com/mir.html*.

FUTURE ENDEAVORS

As of this writing in June 2009, NASA has selected two scientific investigations: one that will focus on the examination of Mars's interior and the other on Mercury's razor-thin atmosphere. Each project will advance our understanding of the terrestrial planets and will be conducted in alliance with the European Space Agency.

Farther down the road, NASA is planning a return to the Moon, with the goal of building a sustainable long-term human presence. The next fleet of vehicles that will replace the aging space shuttle fleet may be able to carry humans to the Moon and possibly to Mars or even beyond.

NASA has been actively engaged in a series of missions, known collectively as "Mission to Planet Earth." These missions, carried out mostly by unmanned satellites orbiting Earth, are designed to monitor various systems of the planet and to search for interactions between systems. Several missions are discussed in Chapter 11. These missions are engaged in the study of the oceans, ocean-atmosphere interactions, biotic activity within the oceans and its impact of life on land, and other phenomena. A particularly useful Web site designed for the study of our planet is located at *www.earth.nasa.gov/*.

When someone asks, "Are the results worth the cost?" we have to remember that it is in our nature as a species to explore the unknown. The space program has also brought us many tangible benefits, including medical monitoring, earth observation, and communication and navigation tools. Velcro is also a practical by-product of our space program.

Related Internet Resources for Great Images and Information

www.nineplanets.org

"The Nine Planets: A Multimedia Tour of the Solar System" is one of the oldest and perhaps best Web sites on this topic on the entire Internet. The site is loaded with images, information, and statistics about the planets and everything else in our solar system.

http://solarsystem.nasa.gov/planets/index.cfm

Created by NASA, "Solar System Exploration" is essentially a photo journal that provides spectacular photography of the planets in our solar system.

http://space.jpl.nasa.gov/

The Jet Propulsion Laboratory is a NASA facility involved in cutting-edge space science research. The "Solar System Simulator" is described by the makers of this Web site as a "spyglass on the cosmos." Students can use the tools on this site to customize their own images of any planet or satellite.

www.solarviews.com/eng/homepage.htm

"Views of the Solar System" is a highly visual site with many images and a wealth of information about all major components of our solar system.

REVIEW EXERCISES FOR CHAPTER 3

WORD-STUDY CONNECTION

ammonia	Jovian planets	precesses
asteroid belt	Jupiter	precession
atmosphere	Kuiper belt	protosun
carbon dioxide	lithosphere	retrograde
Celsius	Mars	revolution
Charon	Mercury	rocky core planets
coma	meteorite	satellite
comet	meteoroid	Saturn
corrosive	methane	solar wind
crater	Moon	spectra
Earth	nebula	Sun
epicycles	nebular hypothesis	supernova
Europa	Neptune	tail
geocentric	nitrogen	telescope
gravity	North Star	Titan
greenhouse effect	outer planets	Uranus
heliocentric	oxygen	Venus
helium	parallax	volatile
hydrogen	pendulum	*Voyager 1*
inner planets	planetesimals	*Voyager 2*
International Astronomical Union	Pluto	water
Io	Polaris	water vapor

SELF-TEST CONNECTION

PART A. Completion. Write in the word or words that correctly complete the statement.

1. Polaris is otherwise known as the _____.

2. Earth revolves about the Sun, but _____ on its axis.

3. Ptolemy used _____ to explain the planets' retrograde motion that he observed.

4. Pluto and _____ are often referred to as a "dual planet" system.

5. When a meteoroid collides with our planet's surface, a _____ impact is said to have occurred.

6. According to the nebular hypothesis, _____ collided constructively to ultimately form the planets.

7. The _____ may be the main reason that Mercury has no atmosphere.

8. Venus is best known for its runaway _____.

9. As Earth _____, Polaris will cease to function as the North Star.

10. The _____ is located between Mars and Jupiter in our solar system.

11. The Jovian planets have thick atmospheres, believed to be composed largely of ammonia and _____.

12. The only major system that all of the rocky core planets share in common is a _____.

13. The _____ is the agency that has decided to retain Pluto as a planet despite its significant differences from the other planets.

14. Copernicus is credited with developing the _____ theory, which places the Sun at the center of the solar system.

15. Pluto may be a large member of a group of icy worlds located in the _____.

PART B. Multiple Choice. *Circle the letter of the item that correctly completes the statement.*

1. The Sun produces energy by
   (a) nuclear fission
   (b) nuclear fusion
   (c) chemical combustion
   (d) both a and b
   (e) both a and c

2. _____ is the gas primarily responsible for trapping heat near the surface of Venus.
   (a) Hydrogen
   (b) Oxygen
   (c) Carbon dioxide
   (d) Nitrogen
   (e) Methane

3. Earth is the only planet to have
   (a) an atmosphere
   (b) a lithosphere
   (c) a hydrosphere
   (d) a cryosphere
   (e) all of the above

4. The planet most likely to be visited next by humans is
   (a) Mercury
   (b) Venus
   (c) Moon
   (d) Mars
   (e) Jupiter

5. A planet is likely to have several moons if it
   (a) is close to the Sun
   (b) is far from the Sun
   (c) has great mass
   (d) has little mass
   (e) is lucky

6. One of the most interesting satellites discovered thus far is Europa. Its most distinctive feature is
   (a) a thick atmosphere
   (b) an icy surface with an ocean, possibly composed of water beneath the ice
   (c) extensive volcanic activity
   (d) small forms of life
   (e) an extensive ring system

7. The inner planets are composed of _____ elements than the outer planets. This may have occurred as these elements condense at _____ temperatures.
    (a) heavier, higher
    (b) heavier, lower
    (c) lighter, higher
    (d) lighter, lower
    (e) modern, all

8. Ptolemy developed the _____ theory, which placed Earth at the center of the universe and had other bodies orbiting Earth in _____ orbits.
    (a) heliocentric, circular
    (b) heliocentric, elliptical
    (c) geocentric, circular
    (d) geocentric, elliptical
    (e) none of the above

9. Kepler relied upon data from _____ to establish that the planetary orbits were elliptical.
    (a) Copernicus
    (b) Brahe
    (c) Newton
    (d) Ptolemy
    (e) Galileo

10. _____ orbit the Sun in highly elliptical orbits and exhibit a tail or coma when they are near the Sun.
    (a) Comets
    (b) Planets
    (c) Meteoroids
    (d) Asteroids
    (e) Meteorites

PART C. Modified True/False. *If a statement is true, write "true" for your answer. If a statement is incorrect, change the <u>underlined</u> expression to one that will make the statement true.*

1. <u>Venus</u> has a thick atmosphere composed largely of carbon dioxide and sulfuric acid.

2. Venus and <u>Mars</u> are unlikely to be visited by humans in the future because of their extremely high surface temperatures.

3. Earth is the only planet known to have <u>liquid</u> water on its surface.

4. The Jovian planets are generally <u>more</u> dense than the rocky core planets.

5. <u>Neptune</u> is one of the Jovian planets.

6. Pluto has a highly elliptical orbit; when Pluto is not the outermost planet, <u>Uranus</u> is.

7. Shooting stars are produced by <u>comets</u> entering Earth's atmosphere.

8. Nickel and iron cores have been found in each of the <u>Jovian</u> planets.

9. In Ptolemy's geocentric model, epicycles were necessary to explain <u>retrograde</u> motion by the planets.

10. The <u>heliocentric model</u> places the Sun at the center of the solar system and has Earth, along with the other planets, orbiting the Sun.

CONNECTING TO CONCEPTS

1. Compare the heliocentric and geocentric models of the solar system. When was each developed? How was the heliocentric model first received?

2. Cite evidence that supports the theory that all the planets formed around the same time.

3. Why was Brahe unwilling to accept the heliocentric model, even though Copernicus had proposed it a few decades earlier?

4. Io, a satellite of Jupiter, has been identified as the most tectonically active body in the solar system. What forces are causing this behavior on Io?

5. Why is the Hubble space telescope such a valuable tool to astronomers?

6. Explain what keeps planets in their orbits around the Sun.

7. What appears to be retrograde (backward) and forward movements of the planets is not their motions, but the motion of Earth. Earth's motion alone is able to explain many different celestial phenomena.

   a. State one assumption made by Copernicus that does not agree with our current heliocentric model of the universe.
   b. State two commonsense objections that people at the time had to the idea that Earth rotates and revolves.
   c. Which part of the geocentric model of the universe did Copernicus keep as part of his new heliocentric model?

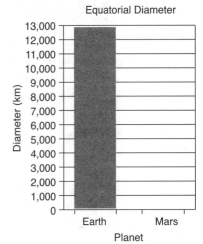

8. The bar graph to the right shows the equatorial diameter of Earth. Construct the bar that represents the equatorial diameter of Mars.

CONNECTING TO LIFE/JOB SKILLS

We are still in the early stages of exploring our solar system. Many great achievements are yet to come. Qualified astronomers, engineers, and physicists will be needed to work on projects such as a mission to explore Pluto, a mission to visit and possibly land on Europa, and a manned mission to Mars. All of these missions will require elaborate support teams with expertise in many areas.

ANSWERS
SELF-TEST CONNECTION

Part A

1. North Star	6. planetesimals	11. methane
2. rotates	7. solar wind	12. lithosphere
3. epicycles	8. greenhouse effect	13. International Astronomical Union
4. Charon	9. precesses	14. heliocentric
5. meteorite	10. asteroid belt	15. Kuiper belt

Part B

1. **(b)**	3. **(c)**	5. **(c)**	7. **(a)**	9. **(b)**
2. **(c)**	4. **(d)**	6. **(b)**	8. **(c)**	10. **(a)**

Part C

1. True	5. True	8. False; inner or rocky core
2. False; Mercury	6. False; Neptune	9. True
3. True	7. False; meteoroids	10. True
4. False; less		

CONNECTING TO CONCEPTS

1. The geocentric model placed Earth at the center of our solar system (and our universe). Elaborate models were created in attempts to explain the apparent motions of celestial objects, including stars and planets. The geocentric model is at least 2000 years old. The heliocentric model was first proposed in the sixteenth century and was not well received. This model challenged medieval churches' teachings, as it placed the Sun at the center of the solar system, with Earth just another planet orbiting the Sun.

2. Evidence supporting the nebular hypothesis includes the facts that all planets and the Sun are the same age and that the planets all orbit the Sun in the same direction

and in the same plane. These are consistent with the idea that the solar system formed out of a spinning disk, the center of which became the Sun and outside of which the planets formed.

3. Brahe used his instruments in an attempt to identify parallax. He reasoned that, if Earth were indeed orbiting the Sun, then, as it moved through the heavens, at times Earth would be closer to certain constellations and, at these times, the constellations would appear larger. His instrumentation was not sensitive enough to detect these changes, however, and therefore he rejected the heliocentric model.

4. Io is being pulled apart gravitationally as a result of its position between Jupiter and the three large outer Jovian satellites. The gravitational stress is creating internal heat that produces magma and results in frequent volcanic activity.

5. The Hubble space telescope is located in space about 500 kilometers above Earth's surface and most of the atmosphere. Our planet's atmosphere obscures our view of the rest of the universe. With the view provided by this telescope, scientists and society in general have better views of the universe than were ever available before.

6. Each planet is in a stable orbit around the Sun. The forces exerted by a planet's tendency to move outward are balanced by the forces pulling the planet toward the Sun.

7. a. Copernicus assumed that the orbits of all planets were perfect circles.
    b. Objections included (1) the fact that people couldn't feel the rotation of the planet and (2) the church's teaching that Earth was the center of the known universe.
    c. Epicycles were kept. They were mathematically necessary because of Copernicus's desire to retain perfect circular orbits.

8. The bar should reach to a position just below the 7000-kilometer mark.

The Earth–Moon System

4

WHAT YOU WILL LEARN

This chapter focuses on the Earth–Moon system. In this chapter you will learn

- what major systems are found on the Moon and surface features on the Moon;
- the unique relationship between Earth and the Moon;
- what effects the Moon has upon Earth;
- lunar phases and why we see what we see here from Earth;
- the difference between a sidereal month and a synodic month;
- how eclipses occur.

SECTIONS IN THIS CHAPTER

- Earth and the Moon—A Comparison
- Importance of the Moon
- Lunar and Solar Eclipses
- Internet Resources to Lunar Phase and Tide Data
- Review Exercises for Chapter 4

Earth and the Moon—A Comparison

MAJOR SYSTEMS AND PROCESSES

Earth is the only celestial body in the entire solar system to have all of the five major systems discussed in Chapter 1: an atmosphere, a hydrosphere, a geosphere, a cryosphere, and a biosphere. The Moon has only two of these major systems, a geosphere and a cryosphere. Recently, a relatively minor amount of frozen water has been discovered deep within some lunar craters and beneath the regolith-covered surface of the Moon (see Figure 4.1), so technically Earth's satellite has a cryosphere.

On Earth, the interaction between the five major systems makes our planet what it is—a dynamic place teeming with life in virtually every "nook and cranny." The Moon, in comparison, is much less dynamic. In fact, the static state of the lunar surface makes it a great natural laboratory in which to study the history of our solar system. Studies of the Moon have yielded clues as to the formation of both celestial bodies, Earth and the Moon. The lunar surface is much older than Earth's surface. Erosional processes on Earth regularly reshape its surface and erase evidence of earlier events, whereas the Moon has remained largely unchanged for billions of years.

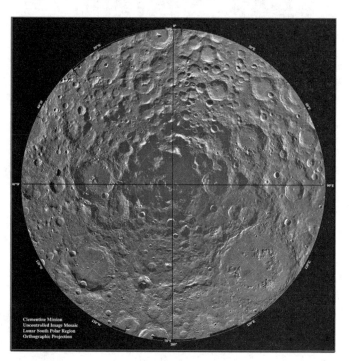

FIGURE 4.1 In March 1998, Lunar Prospector discovered potentially as much as 6 trillion kilograms (6.6 billion tons) of ice, buried beneath about 40 centimeters (18 in.) of regolith in areas near the Moon's North and South poles. Source: NSSDC/NASA.

SURFACE FEATURES

Earth's surface is in a constant state of change. Surface materials are continually broken down by weathering and erosional agents into sediments and soils. Scars left by meteorite impacts here on Earth are worn away relatively quickly as new features replace them. Moreover, Earth is characterized by widely varying surface features from one locale to another. Arctic regions have an abundance of surface ice, and tropical jungles are common in equatorial regions.

FIGURE 4.2 This image depicts lunar maria (smooth, less cratered, dark regions) and highlands, which are lighter, heavily cratered regions. The relative absence of craters in the lunar maria suggests that the maria are younger than the highlands. More recent analysis suggests that the maria are lower regions that were flooded by lava flows.

The lunar surface, in stark contrast, is scarred by ancient meteorite impacts. These scars have remained essentially unchanged for eons. Weathering agents such as wind, water, ice, and even various forms of life are not found on the Moon; therefore, surface scars and other features described below remain. Some lunar features are estimated to be as old as the Moon itself.

The Moon's major features, perhaps noted first by Galileo through his telescope, include dark lowlands and bright, densely cratered highlands (see Figure 4.2). The lowlands are also, although somewhat less, cratered and are referred to as "maria," or "seas." They may have formed when asteroids penetrated the lunar surface sufficiently to cause basaltic, that is, dark-colored, lava, rich in iron, to form and flow along the surface. Alternately, the maria may have formed on a young lunar surface as a result of surface melting. The highlands, which make up the majority of the lunar surface, are heavily cratered mountainous regions. Some of the highest lunar peaks approach the height of Mount Everest.

The entire lunar surface is covered by *regolith*; this is a layer of unconsolidated debris that has formed as a result of a few billion years of bombardment. The regolith measures about 3 meters (10 ft) thick at the sites in the maria where the Apollo astronauts landed (see Figure 4.3).

RELATIVE SIZE–A UNIQUE RELATIONSHIP

The Moon is the second largest satellite in the solar system in relation to the planet it orbits (see Figure 4.4). The circumference of the Moon is over one-fourth that of Earth's. In all other planet-satellite relationships but one, the satellite's circumference is a small fraction of the circumference of the planet it orbits. This fact has led to a number of interesting debates about how the Moon became Earth's only natural satellite.

While the Moon's actual origin is unknown, there are several theories worth exploring. The theory most widely accepted today could be termed the *catastrophic collision hypothesis*. This theory suggests that a large celestial body may have collided with primordial (early) Earth, causing a significant portion of the planet's mantle and crust to liquefy and a large chunk to break off into space. Earth's gravitational field then captured much of this material, which, as it cooled and solidified, formed our planet's only natural satellite (see Figure 4.5).

FIGURE 4.3 A dusty lunar rover carrying the Apollo 17 astronauts; footprints are visible in the soft lunar regolith. Courtesy: NASA.

Earth–Moon Relative Size: Comparison of Diameters

Earth
12,700 km

Moon
3,450 km

FIGURE 4.4 The relative sizes of Earth and the Moon.

Apollo astronauts gathered important evidence supporting the catastrophic collision hypothesis. They found moon rocks to have an average density 3.3 times that of water. This density is quite comparable to an average of the densities of Earth's mantle and crust, two layers that would have been liquefied in such a collision as the theory describes. Also, the lunar rocks analyzed were consistent with the composition of Earth's mantle as they were relatively poor in iron, which is found only in our planet's core.

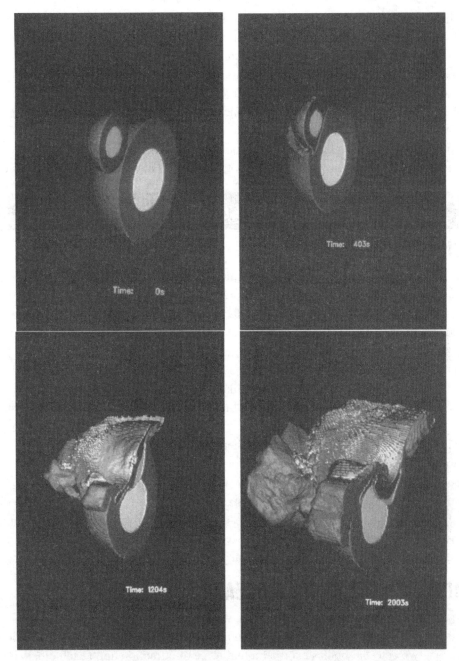

FIGURE 4.5 Computer simulation of the formation of the Moon by a giant impact. Source: M. E. Kipp, Sandia National Labs. Used with permission.

Another theory that attempts to explain how Earth "acquired" the Moon suggests that the Moon was traveling through space and was captured by Earth's gravitational field. Considering the relative size of each celestial body, however, this is considered unlikely, as Earth does not have a strong enough gravitational field to capture an object as large as the Moon as it is flying by in space.

Another theory regarding the Moon's formation postulates simultaneous formation, with the Moon forming in orbit about Earth. Yet another suggests that early Earth was spinning so fast that a chunk of it broke off and began to orbit the rest. Neither of these is as well accepted, however, as the catastrophic collision hypothesis. One reason for their rejection is that Earth's interior is quite different from the Moon's. Earth has a dense, molten core composed largely of iron and nickel. The Moon's interior structure, however, is quite similar to Earth's crust and mantle. The catastrophic collision hypothesis explains adequately why the Moon is made only of crustal and mantle material. A collision of this nature would not have tapped any material from Earth's core.

Importance of the Moon

SCIENTIFIC STUDY

The Moon serves as an important astronomical laboratory. Its lack of weathering and erosional activity has left its surface well preserved. The surface features and materials contain clues to the origins of our planet and of the greater solar system. Also, the marked similarity between the compositions of the Moon and of our planet's crust and mantle may help scientists to understand our own planet better.

The Moon's ancient and cratered surface helps us to understand the nature of the intense bombardment by meteoroids that likely impacted all planets and satellites early in the solar system's development. Much of the Moon's surface reveals events that occurred over 3 billion years ago!

The Moon completely lacks an atmosphere; thus should any nation choose to establish a lunar base station and observatory, the lunar surface would provide an opportunity to observe the heavens clearly without the distorting effects of an atmosphere. Gravity on the Moon is one-sixth of Earth's as a result of our satellite's lesser mass. This feature could be another advantage of establishing such a station in the future. With less gravity to overcome, vehicles could launch from the Moon for more distant destinations with less fuel than is required to launch from Earth.

IMPACT ON THE EARTH'S OCEANS

One of the most interesting interactions between Earth and its nearest celestial neighbor involves the Moon's impact upon our oceans. Newton's laws of gravitation state that two celestial bodies exert a mutual force of attraction upon each other because of gravity. This force of attraction is proportional to the relative masses of the bodies and inversely proportional to the square of their distance.

Earth's gravity is responsible for keeping the Moon in orbit about Earth. The Moon, however, has its own gravitational pull on Earth. This force is weaker than Earth's because a celestial body's force of attraction is dependent in part upon the mass of the body, and the Moon is smaller than Earth.

The Moon's gravity is strong enough, however, to have a pronounced effect, known as *tidal motion*, on Earth's oceans. Tidal motion can be described as the daily changes in the elevation of the ocean surface. A rhythmic rise and fall along all coastlines has been observed since humans began to notice such things. Tides should not be confused with waves, which crash on the beach. Tides are highly predictable, and those who are interested can consult tables that list the times of high and low tides for coastal waters and nearby rivers for months in advance.

By studying tidal patterns, one can see that there are actually two tidal bulges on the planet at any one time. One bulge is a direct result of the oceans' being pulled toward the Moon; however, a tidal bulge on the opposite side of Earth (the side facing away from the Moon) also exists as a result of an inertial lag. In between these two bulges, low tides are experienced.

Waves, in contrast, are created by the atmosphere's interaction with the ocean surface. As winds vary from day to day, waves will change over time and are not highly predictable. Although tides can be enhanced by wind action, their basic patterns are quite predictable. Figure 4.6 illustrates spring and neap tides as they occur throughout each month as the Moon orbits Earth. Tides are discussed in detail in Chapter 12.

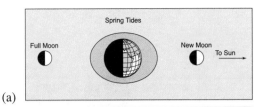
(a)
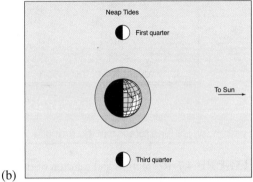
(b)

FIGURE 4.6 Spring and neap tides. (a) When the Sun and the Moon pull in the same direction, their tidal forces combine and tidal bulges on Earth are larger. Spring tides occur at new Moon or full Moon. (b) When the Sun and the Moon pull at right angles, their tidal forces do not combine and tidal bulges are much smaller. Neap tides occur at the first- and third-quarter lunar phases. Source: *Let's Review: Earth Science—The Physical Setting*, 2nd Ed., Edward J. Denecke, Jr., Barron's Educational Series, Inc., 2002.

LUNAR MOTIONS

The Moon orbits Earth at such a rate that it completes one orbit approximately every month. The Moon's orbit can best be described as somewhat elliptical with an average distance from Earth of 384,000 kilometers (250,000 mi). More detail about lunar revolution can be found in the section headed "Sidereal Versus Synodic Month" later in this chapter.

The Moon rotates on its axis as it revolves about our planet. Since the Moon is gravitationally locked to Earth, its period of rotation equals its period of revolution (see Figure 4.7). In other words, in the time required for the Moon to complete one revolution, one rotation is also completed. Because of this relationship in time, we always see the same side of the Moon from our Earth-bound perspective. The only humans who have directly observed the "other side" of the Moon (see Figure 4.8) are the Apollo astronauts who landed on the Moon or orbited it. This elite group includes most of the Apollo missions from Apollo 8 through Apollo 17, which occurred between the years 1968 and 1972.

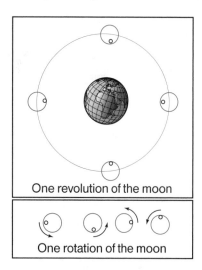

FIGURE 4.7 Revolution and rotation of the Moon. The Moon completes one rotation in the same time it makes one revolution. Therefore, the same side of the Moon always faces Earth. Source: *Let's Review: Earth Science—The Physical Setting*, 2nd Ed., Edward J. Denecke, Jr., Barron's Educational Series, Inc., 2002.

FIGURE 4.8 Photo taken from Apollo 8 spacecraft. View is beyond the eastern limb of the Moon as viewed from Earth. Source: NASA.

PHASES OF THE MOON

The most apparent change in the Moon's appearance each month involves the phases through which the Moon passes. On any given evening, the appearance of the Moon, or its phase, is dependent upon the relative positions of Earth, the Moon, and the Sun. As with any spherical celestial body, one-half of the planet or satellite will be illuminated by the Sun (or any other star that it is orbiting) at any given moment. The key to what we see here on Earth is how much of the illuminated portion is visible from our perspective. During a *full Moon*, when we see the entire illuminated portion of the Moon, the Sun, Earth, and the Moon are all in direct alignment, with the Moon "behind" Earth (see Figure 4.9). Then, for an entire night, from sunset to sunrise, we can see the entire portion of the Moon that is reflecting some of the sunlight reaching its surface.

As the Moon continues to revolve around Earth, our satellite will pass through several phases;

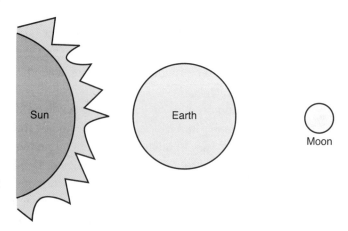

FIGURE 4.9 Relative positions of Earth, Moon, and Sun during the full Moon phase, as viewed from Earth.

the full Moon, the *waning (or "old") gibbous*, the *third quarter*, and the *waning crescent* occur, in that order, before the *new Moon* phase is reached. The Moon is not visible from Earth during the new Moon phase, as the illuminated portion of the Moon is facing the Sun and the dark side is facing Earth. Figure 4.10 illustrates how this situation is produced.

After the new Moon phase, the Moon proceeds through a similar series of waxing (or "new") phases before the next full Moon approximately 2 weeks later. These stages include the *waxing crescent, first quarter,* and *waxing gibbous* (see Figure 4.11).

As the Moon proceeds through its various phases, the times of moonrise and moonset will vary. The easiest alignment to visualize occurs during the full Moon phase, when the Moon is located directly "behind" Earth, with the Sun "in front of" our planet. This alignment results in moonrise at sunset and moonset at sunrise. Consider this: If you were able to see the moon during its new Moon phase, the Moon would rise with the Sun at sunrise and would set with the Sun at sunset. Study Figure 4.12 to prove this to yourself. Since the Moon is revolving about Earth, the times the moon rises and sets are controlled by the relative positions of the two celestial bodies. During the quarter phases, for example, the Moon rises during the daylight hours and sets at night (last quarter), or rises late at night and sets in the morning (first quarter). Tides are also affected by lunar motion. A more detailed discussion of their timing is given in Chapter 12.

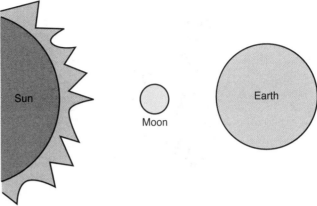

FIGURE 4.10 Relative positions of Earth, Moon, and Sun during the new Moon phase, as viewed from Earth.

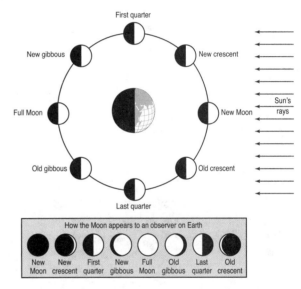

FIGURE 4.11 The phases of the Moon as observed during approximately a one-month period of time.

SIDEREAL VERSUS SYNODIC MONTH

How much time does the Moon take to complete one revolution about Earth? The best answer is "It depends." Although that reply may seem confusing, Figure 4.13 shows

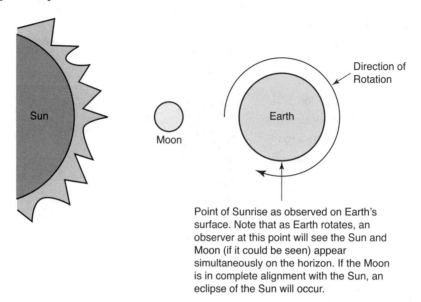

FIGURE 4.12 Earth, Moon, and Sun in alignment during the new Moon phase. Sunrise would occur with moonrise, if it could be seen during this phase.

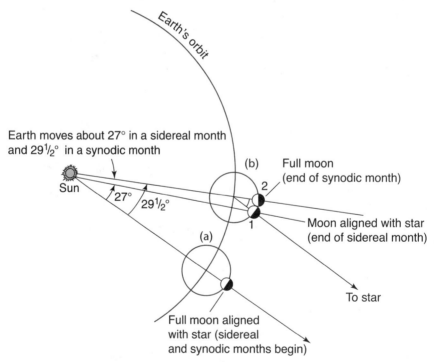

FIGURE 4.13 Difference in the duration of sidereal and synodic months. (a) The full Moon is observed to be near some bright star. (b) Roughly 27⅓ days later, at position 1, the Moon is again near the same bright star, completing a *sidereal month*. Because Earth and the Moon have moved together around the Sun during that time, the Moon must revolve for 2⅙ more days before the Moon, Earth, and the Sun are again in alignment so that the Moon is full again (position 2), completing a cycle of phases, or a synodic month. Thus, the time between full Moons, or the synodic month, is 27⅓ + 2⅙ or 29½ days. Source: *Let's Review: Earth Science—The Physical Setting*, 2nd Ed., Edward J. Denecke, Jr., Barron's Educational Series, Inc., 2002.

that there are actually two ways to describe a lunar revolution. The first, known as the *sidereal month*, is the actual time, 27⅓ Earth-days, required for the Moon to complete one revolution about Earth. During this time, however, Earth has been moving about the Sun, so the Moon is not yet aligned between Earth and the Sun. To reach an alignment of Earth, the Sun, and the Moon, about 2 more days, or a total of 29.5 Earth-days, are required. This is the *synodic month*. The new moon thus occurs every 29.5 days. The synodic month was the basis for the first Roman calendar and also for our secular calendar.

Lunar and Solar Eclipses

LUNAR ECLIPSES

A lunar eclipse occurs when the Moon passes behind Earth, so that Earth is between the Sun and the Moon. When this happens, on some occasions the Moon passes through Earth's shadow, thus producing a *lunar eclipse*. This can occur only during the full Moon phase, but does not occur during *every* full Moon. Why not? It turns out that the Moon orbits Earth on a slight (5°) tilt to the Earth–Sun plane. A lunar eclipse occurs only when the Sun, Earth, and the Moon are in line in the same plane (Figure 4.14). During most full Moons the Moon is slightly above or below the Earth–Sun plane, and Earth's shadow is cast off into space, not onto the Moon.

Total lunar eclipses occur when the Moon passes through the darkest part of Earth's shadow, called the *umbra*. Since Earth's umbra is fairly large compared to the size of the Moon, total lunar eclipses can last for over an hour. While the Moon is within Earth's *penumbra*, or more peripheral shadow, a partial lunar eclipse will occur (see Figure 4.15).

SOLAR ECLIPSES

Solar eclipses occur when the Moon passes between Earth and the Sun. Under certain circumstances, the Moon's shadow travels across part of Earth's surface. This can occur only during the new Moon phase and with the Sun, the Moon, and Earth all in the same plane.

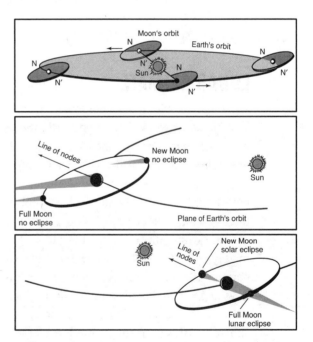

FIGURE 4.14 Eclipses of the Sun and Moon. Solar and lunar eclipses can occur when the Moon intersects the ecliptic in new-Moon or full-Moon position. If the Moon is above or below the ecliptic during these phases, no eclipse occurs. Source: *Let's Review: Earth Science—The Physical Setting*, 2nd Ed., Edward J. Denecke, Jr., Barron's Educational Series, Inc., 2002.

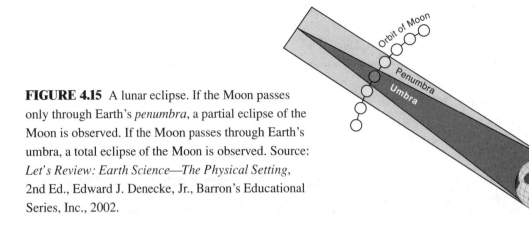

FIGURE 4.15 A lunar eclipse. If the Moon passes only through Earth's *penumbra*, a partial eclipse of the Moon is observed. If the Moon passes through Earth's umbra, a total eclipse of the Moon is observed. Source: *Let's Review: Earth Science—The Physical Setting*, 2nd Ed., Edward J. Denecke, Jr., Barron's Educational Series, Inc., 2002.

This event is less common than a lunar eclipse because the Moon's shadow is so small that only a limited portion of Earth experiences a total solar eclipse, whereas a much wider region can experience a lunar eclipse.

A total solar eclipse occurs when the Moon's umbra passes across Earth's surface. The period of totality usually lasts for only a couple of minutes before part of the Sun's disk becomes visible again. When the Moon's penumbra passes over a portion of Earth's surface, a partial eclipse occurs (see Figure 4.16).

LUNAR CLIMATE

Although the Moon does not experience weather as we know it here on Earth, it does have a climate. Days on the Moon, each of which lasts for 14 Earth-days because of the Moon's relatively slow period of rotation, are best described as hot, with surface temperatures reaching about 125°C (260°F). Nights are cold with temperatures falling to approximately −170°C (−280°F). The extreme temperatures occur in part as a result of the long days and nights, but also are testimony to the extremes that occur when there is no atmosphere. Earth's atmosphere moderates the temperatures here by absorbing heat lost by Earth's surface at night and by distributing heat absorbed during the day. Were it not for our atmosphere, the temperature range on Earth would be just as extreme as that on the Moon.

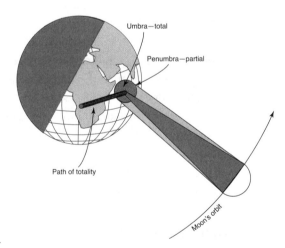

FIGURE 4.16 A solar eclipse. Observers in the umbra experience a total eclipse, and those in the penumbra experience a partial eclipse. Source: *Let's Review: Earth Science—The Physical Setting*, 2nd Ed., Edward J. Denecke, Jr., Barron's Educational Series, Inc., 2002.

Internet Resources to Lunar Phase and Tide Data

www.usno.navy.mil/usno/time/moon-phase-images/

"Virtual Reality Moon Phase Pictures" is an interactive Web site that allows the user to select a date between 1800 and 2199 and determine the lunar phase on that date.

http://aa.usno.navy.mil/data/docs/RS_OneYear.html

This site helps the user to determine the times of sun- or moonrise and of sun- or moonset for any given day.

www.usno.navy.mil/USNO/astronomical-applications/data-services/upcom-eclipses

Information about upcoming and recent eclipses of the Moon and Sun is available here.

www.fourmilab.ch/earthview/vplanet.html

This interesting site, called "Earth and Moon Viewer," offers many unique views of Earth and the Moon.

http://tidesonline.nos.noaa.gov/

"Tides Online" enables users to search by state to find current tide data as well as current coastal conditions, including coastal winds and sea surface temperatures.

http://tbone.biol.sc.edu/tide/sitesel.html

"The WWW Tide and Current Predictor" page allows the user to select a site and obtain tide information for a wide range of dates, both in the past and in the future. The interface enables the user to select from a wide variety of modes of presentation.

REVIEW EXERCISES FOR CHAPTER 4

WORD-STUDY CONNECTION

Arctic	gibbous	synodic
catastrophic collision hypothesis	gravity	tidal motion
circumference	lunar eclipse	tides
crescent	lunar phases	tropical
crust	mantle	waning
density	new Moon	waxing
eclipse	primordial	weathering
erosion	sidereal	
full Moon	solar eclipse	

SELF-TEST CONNECTION

PART A. Completion. *Write in the word or words that correctly complete the statement.*

1. The moon is composed of material similar to that found within Earth's _____ and mantle.

2. The formation of the moon is described by the _____.

3. Early Earth is also known as _____ Earth.

4. The Moon's gravity causes _____ here on Earth.

5. The layer of Earth beneath the crust is known as the _____.

6. The action of transporting weathered material on Earth is called _____.

7. The phase of the moon that precedes the full Moon is the _____.

8. When Earth's shadow is cast upon the Moon, a _____ occurs.

9. The period of time between new Moons is known as a _____.

10. The actual time for the Moon to complete one orbit about Earth is a _____.

11. As the moon orbits Earth, it appears in different _____.

12. Much of the lunar surface is over _____ years old.

13. A _____ occurs when the Moon's shadow passes over a portion of Earth's surface.

14. Of all the major systems found on Earth, the Moon has only a _____.

15. Lunar craters are well preserved on the Moon because of a lack of _____ and erosion.

PART B. Multiple Choice. *Write the letter of the item that correctly completes the statement or answers the question.*

1. Which of the following occur(s) on Earth but not on the Moon?
   (a) weathering
   (b) erosion
   (c) rain
   (d) wind
   (e) all of the above

2. The new Moon is followed by a
    (a) waxing crescent
    (b) waxing gibbous
    (c) full Moon
    (d) waning gibbous
    (e) waning crescent

3. A lunar eclipse can occur only during a
    (a) waxing crescent
    (b) waxing gibbous
    (c) full Moon
    (d) waning gibbous
    (e) new Moon

4. A solar eclipse can occur only during a
    (a) waxing crescent
    (b) waxing gibbous
    (c) full Moon
    (d) waning gibbous
    (e) new Moon

5. A synodic month best describes the period of time
    (a) between a full Moon and a new Moon
    (b) between full Moons
    (c) required for the Moon to complete one rotation
    (d) required for Earth to complete one rotation
    (e) required for Earth to complete one revolution

6. The most significant impact that the Moon has on our planet is
    (a) weather
    (b) erosion
    (c) tides
    (d) source of energy
    (e) ocean currents

7. The Moon may have formed from a catastrophic collision between a
    celestial object and Earth billions of years ago. In this collision, material
    from Earth's _____ may have been ejected and later formed
    the Moon.
    (a) crust
    (b) mantle
    (c) core
    (d) both a and b
    (e) both b and c

8. Earth and the Moon have a unique relationship when compared to other planets and their satellites in that
    (a) the Moon is unusually close to Earth
    (b) the Moon is one of the largest satellites
    (c) the Moon is relatively large as compared to Earth
    (d) both a and c
    (e) both b and c

9. Transport of materials may prove easier on the Moon than on Earth because
    (a) the Moon has more gravity than Earth and therefore objects weigh less
    (b) the Moon is farther from the Sun than is Earth
    (c) the Moon is closer to the Sun than is Earth
    (d) the Moon has less gravity than Earth and therefore objects weigh less
    (e) none of the above

10. The Moon experiences much greater temperature swings between day and night than occur on Earth. This can best be explained by the fact that the Moon
    (a) is farther from the Sun
    (b) is closer to the Sun
    (c) has no atmosphere to trap heat and distribute energy
    (d) is constantly bombarded by meteors
    (e) controls tides

PART C. Modified True/False. *If a statement is true, write "true" for your answer. If a statement is incorrect, change the* underlined *expression to one that will make the statement true.*

1. Of the five major systems, the geosphere is unique to our planet.

2. Even though Earth and its Moon formed at the same time, Earth's surface is much older.

3. Apollo astronauts gathered evidence supporting the catastrophic collision hypothesis, which explains the origin of the Moon.

4. An 80-kilogram person would weigh about 454 kilograms on the Moon.

5. The most pronounced impact the Moon has on Earth is the creation of waves in the ocean.

6. We always see the same side of the Moon because the rotational period of our satellite is equal to its period of revolution.

7. Lunar phases are connected to the synodic month.

8. A lunar eclipse can occur only during the new Moon phase.

9. Solar eclipses occur for a <u>shorter</u> time period than lunar eclipses.

10. Earth's <u>atmosphere</u> acts to moderate our climate, keeping temperature ranges much narrower than those experienced on the Moon.

CONNECTING TO CONCEPTS

1. Compare the natural forces that have acted to shape the surface of Earth versus that of the Moon.

2. Name the theory that best explains how Earth came to have a moon. Cite supporting evidence for this theory.

3. Describe the Moon's greatest effect upon Earth today.

4. Describe the difference between a sidereal and a synodic month.

5. Diagram the Earth–Moon–Sun alignment during a solar eclipse.

6. Compare the climates of the Moon and Earth. In what way(s) are they similar? Different? Why do these differences exist?

CONNECTING TO LIFE/JOB SKILLS

The Moon has been a subject of fascination to humans for millennia. This fascination eventually resulted in Project Apollo, during which six different missions brought astronauts to the Moon. It is likely that humans will return to the Moon in the near future, possibly to establish a colony in which research can be conducted. To accomplish such a task, hundreds or thousands of scientists and other professionals will need to work together. Today's students can become tomorrow's scientists. Most important are a dedication to study and the tenacity to stay the course.

ANSWERS
SELF-TEST CONNECTION

Part A

1. crust
2. catastrophic collision hypothesis
3. primordial
4. tides
5. mantle
6. erosion
7. waxing gibbous
8. lunar eclipse
9. synodic month
10. sidereal month
11. phases
12. 3 billion
13. solar eclipse
14. lithosphere (geosphere)
15. weathering

Part B

1. **(e)**	3. **(c)**	5. **(b)**	7. **(d)**	9. **(d)**
2. **(a)**	4. **(e)**	6. (c)	8. (c)	10. (c)

Part C

1. False; biosphere
2. False; the Moon's
3. True
4. False; 13 kilograms
5. False; tides
6. True
7. True
8. False; full moon
9. True
10. True

CONNECTING TO CONCEPTS

1. The Moon's surface has been shaped primarily by meteorite impacts and any subsequent lava flows resulting from those impacts. Earth's surface, in contrast, has been shaped by a number of natural forces, including moving water, glacial movement, and wind. Earth has also been impacted by meteoroids; however, other surface processes weather and erode impact craters over time.

2. The catastrophic collision hypothesis best explains how the Moon formed. The Moon's composition closely matches that of Earth's mantle and crust.

3. The Moon's greatest effect upon Earth is our satellite's gravitational attraction on the oceans. The gravitational pull creates tides.

4. A sidereal month is a period of $27\frac{1}{3}$ Earth-days. This is the actual time for the Moon to complete one revolution about Earth. During this time, however, Earth has been moving about the Sun, so the Moon is not yet aligned between Earth and the Sun. To reach an Earth–Sun alignment requires about 2 more days, or a total of 29.5 Earth-days. This is the synodic month.

5. During a solar eclipse, the Moon is between the Sun and Earth. The eclipse occurs as the Moon's shadow crosses over Earth's surface.

6. Since the Moon and Earth are virtually the same distance from the Sun, they receive the same energy from it. The Moon, lacking an atmosphere, responds differently to incoming solar energy than does Earth. Earth's atmosphere acts to trap incoming solar energy, thereby limiting the temperature range on the planet's surface. Therefore, although the average temperatures of the Moon and Earth are similar, the temperature range on Earth is much less than that observed on the Moon.

Minerals and Rocks– Evidence of Continual Change

5

WHAT YOU WILL LEARN

This chapter focuses on minerals and rocks and the evidence they provide of continual change. In this chapter you will learn

- the basic chemistry of minerals and how we group them;
- the importance of minerals as a natural resource;
- the difference between a mineral and a rock;
- the three major types of rocks, how they differ and how we classify them;
- what the rock cycle is.

SECTIONS IN THIS CHAPTER

- Minerals
- Igneous Rocks
- Sedimentary Rocks
- Metamorphic Rocks
- The Rock Cycle
- Related Internet Resources for Great Images and Information
- Review Exercises for Chapter 5

Minerals

BASIC CHEMISTRY

In any study of Earth's surface and interior, an understanding of minerals is essential. Minerals are naturally occurring solids formed through geologic processes. They have an identifiable chemical composition, a highly ordered atomic structure, and specific physical properties. There is even a specific branch of science devoted to the study of minerals known as mineralogy. Minerals are the building blocks of larger structures and are key components of all of Earth's major systems. Minerals are essential to virtually all life-forms, including humans. Humans need to regularly ingest minerals, sometimes referred to as electrolytes. Minerals are found as solids within the geosphere, as well as in great quantities throughout the world's oceans in dissolved form. Minerals readily flow through Earth's systems; for example, dissolved minerals in the ocean may be absorbed by a marine organism to help build its carbonate shell. When that organism dies, its shell is eventually incorporated into the sediment on the seafloor.

Only a few of the 100+ elements are commonly found in most minerals. In fact, in the 2000 known minerals, only eight elements are usually present. Table 5.1 lists the most common elements found in Earth's crust (an integral part of the geosphere), hydrosphere, and troposphere (an integral part of the atmosphere). The crust is the portion of the geosphere where minerals are often found in their solid forms.

From a geologic perspective, minerals are naturally occurring, inorganic, crystalline solids that possess definite chemical structures. They are generally composed of ionic

TABLE 5.1
AVERAGE CHEMICAL COMPOSITION OF EARTH'S CRUST, HYDROSPHERE, AND TROPOSPHERE

Element (symbol)	Crust		Hydrosphere	Troposphere
	Percent by Mass	Percent by Volume	Percent by Volume	Percent by Volume
Oxygen (O)	46.40	94.04	33.0	21.0
Silicon (Si)	28.15	0.88		
Aluminum (Al)	8.23	0.48		
Iron (Fe)	5.63	0.49		
Calcium (Ca)	4.15	1.18		
Sodium (Na)	2.36	1.11		
Magnesium (Mg)	2.33	0.33		
Potassium (K)	2.09	1.42		
Nitrogen (N)				78.0
Hydrogen (H)			66.0	
Other	0.66	0.07	1.0	1.0

Source: The State Education Department, *Earth Science Reference Tables*, 2001 ed. (Albany, New York: The University of the State of New York).

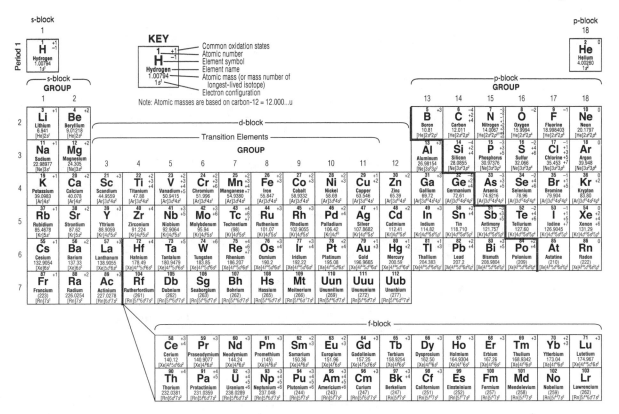

FIGURE 5.1 A basic periodic table. An interactive version of the periodic table is accessible on the Internet at: *http://pearl1.lanl.gov/periodic/default.htm.*

compounds. An ionic compound contains a metallic ion that is chemically bonded to a nonmetal ion or a polyatomic ion (charged group of atoms). This all sounds rather complex but in actuality is quite simple and understandable.

Let's begin with the periodic table (see Figure 5.1), which includes all known elements and helps us to see that they can be categorized into two basic types, metals and nonmetals.

Metals and nonmetals have a natural affinity to chemically bond together. Chemical bonding involves the gain, loss, or sharing of electrons. When a metallic atom bonds to a nonmetallic atom, the metallic atom tends to lose electrons, thus becoming a metallic ion, or *cation* (see Figure 5.2). The cation possesses a positive charge as the atom has lost one or more electrons, which carry negative charges. In other words, the metallic ion now has more protons, or positively charged particles, than electrons, or negatively charged particles; hence its net charge is positive.

The electrons lost by the metallic atom are transferred to, or gained by, the nonmetallic atom, making it an *anion*, or negatively charged ion. The nonmetal now has more electrons than protons. Sometimes a metal will bond to a simple nonmetallic ion, as when sodium bonds to a chloride ion to form sodium chloride. Other times a more complex ion is involved, for example, a carbonate, which consists of one carbon

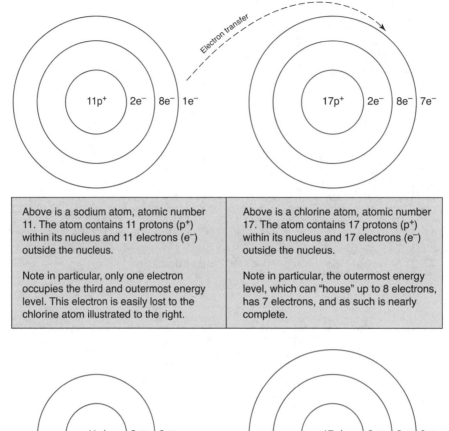

Above is a sodium atom, atomic number 11. The atom contains 11 protons (p^+) within its nucleus and 11 electrons (e^-) outside the nucleus. Note in particular, only one electron occupies the third and outermost energy level. This electron is easily lost to the chlorine atom illustrated to the right.	Above is a chlorine atom, atomic number 17. The atom contains 17 protons (p^+) within its nucleus and 17 electrons (e^-) outside the nucleus. Note in particular, the outermost energy level, which can "house" up to 8 electrons, has 7 electrons, and as such is nearly complete.

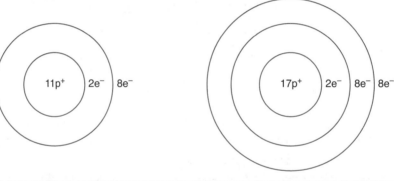

The sodium atom loses its outermost electron and now has more protons than electrons, leaving it with a net charge of +1. The newly formed sodium cation, which is positively charged, is now electrostatically attracted to the newly formed chloride anion (see box to the right.)	The chlorine atom gains an electron. By completing its outermost energy level, it becomes a negatively charged anion and is renamed as "chloride." Note that the anion now has more electrons than protons and as such carries a net negative charge.

FIGURE 5.2 Positively charged cations and negatively charged anions bond to each other as a result of their electrostatic attraction.

atom and three oxygen atoms bonded together. As a group, these atoms carry a charge and thus are known as a *polyatomic ion*, that is, a many-atom charged particle (see Figure 5.3). Since metallic ions (cations) and nonmetallic ions (anions) carry opposite charges, they are attracted to each other. The bond that forms between them, known as an *ionic bond*, is simply an electrostatic attraction of opposite charges. Ionic compounds, those that contain ionic bonds, are easily recognized as they contain both metals and nonmetals in their chemical formulas. Figure 5.2 depicts the formation of ions, which is a precursor to the formation of an ionic compound.

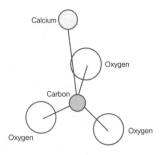

FIGURE 5.3 Molecular model of calcite ($CaCO_3$).

Ionic compounds tend to have relatively high melting points and consequently are solids at room temperature. (Recall that all minerals must be solid at room temperature.) While a chemist will refer to the compound with the formula NaCl as sodium chloride, a mineralogist will likely refer to it as halite. Halite is the naturally occurring form of sodium chloride found in Earth's crust. Many other minerals, in addition to halite, are primarily composed of one chemical compound. Some minerals, however, may contain more than one compound, so it would not be appropriate to say that all compounds are minerals and all minerals are compounds. Dolomite, for example, is a physical mixture of calcium carbonate and magnesium carbonate. To a chemist these are two distinctly different compounds, but to a mineralogist the distinction is not as critical because both compounds have similar physical and chemical properties. Halite and other minerals are shown in Figure 5.4. Properties of some minerals are given in Table 5.2.

In studying the periodic table, you will note that calcium and magnesium are in the same chemical family (vertical column). This relationship that chemists have identified tends to produce similar chemical and physical properties in elements.

(a)   (b)

PROPERTIES OF MINERALS

Minerals can be categorized by easily determined physical properties. One of the most recognizable is *hardness*, which describes a mineral's ability to resist scratching or to scratch other objects. Diamonds are

(c)   (d)   (e)

FIGURE 5.4 Photographs of a few common minerals.
(a) halite   (b) gypsum   (c) pink quartz   (d) olivine
(e) calcite

TABLE 5.2
CHEMICAL NAMES AND FORMULAS OF SOME COMMON MINERALS

Mineral	Chemical Name	Chemical Formula
Calcite	Calcium carbonate	$CaCO_3$
Galena	Lead sulfide	PbS
Gypsum	Calcium sulfate dihydrate	$CaSO_4 \cdot 2H_2O$
Olivine (fosterite)	Magnesium silicate	Mg_2SiO_4
Potassium Feldspar	Potassium aluminum silicate	$KAISi_3O_8$
Pyrite	Iron sulfide	FeS_2
Quartz	Silicon dioxide	SiO_2

known to be the hardest minerals; a diamond will scratch any other mineral. The softest mineral is talc. Talc, which can be refined into talcum powder, is so soft that even your fingernail will scratch it! Moh's hardness scale (Table 5.3) measures the property of hardness. The scale consists of ten minerals arranged from hardest (10) to softest (1).

Another test often conducted to identify a mineral is a streak test. Rubbing the mineral against a hard, unglazed porcelain surface identifies the mineral's *streak color*. Some minerals leave behind a colored streak when they are reduced to powder by this test. The color of the mineral may vary from sample to sample, but its streak color usually does not. This provides a reliable test to identify a mineral.

Minerals are crystalline solids. A crystalline solid has an orderly internal arrangement of atoms. This order produces a solid with a recognizable *geometry*. When a mineral forms without space restrictions inside Earth, a crystalline pattern may be produced. For example, pyrite, or fool's gold, has highly recognizable cubic crystals, and quartz tends to form hexagonal crystals. Thus the geometry of a crystalline solid is helpful in identifying it.

Luster is often used in the process of identifying a mineral. *Luster* describes the mineral's appearance, or the way in which light is reflected from its surface. If a mineral appears to be shiny like a metal, it is said to have a metallic luster. If the luster is nonmetallic, the mineral may appear vitreous (glassy) like quartz crystals, earthy (dull), or pearly, silky, and so on.

TABLE 5.3
MOH'S SCALE

Mineral	Hardness	
Talc	1	SOFTEST
Gypsum	2	
Calcite	3	
Fluorite	4	
Apatite	5	
Orthoclase	6	
Quartz	7	
Topaz	8	
Corundum	9	
Diamond	10	HARDEST

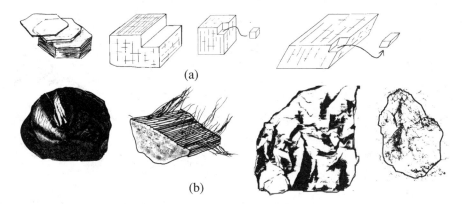

FIGURE 5.5 A few common minerals that display cleavage (a) and fracture (b).
Source: *Let's Review: Earth Science—The Physical Setting,* 2nd Ed., Edward J. Denecke, Jr.,
Barron's Educational Series, Inc., 2002.

A mineral may also be identified by determining whether it exhibits cleavage or fracture (see Figure 5.5). *Cleavage* is the tendency to cleave or break along regular planes, while *fracture* is the tendency to break irregularly.

Specific gravity, or a mineral's density relative to the density of water, is easily determined with basic laboratory equipment and varies significantly from mineral to mineral. The specific gravity of a mineral depends to a great extent upon the elements of which it is composed. Minerals rich in iron tend to have higher specific gravity (or density) than those rich in silicon and oxygen. Classic examples of iron-rich minerals include magnetite and hematite; both are forms of iron oxide.

Certain minerals have special, highly identifiable properties. One example is magnetite, whose name suggests that it is a magnetic mineral. Other minerals are known to fluoresce, that is, give off visible light in the presence of ultraviolet light. Northwestern New Jersey is rich in these minerals, which include calcite and willemite.

In recent years, despite a relative abundance of some minerals, there has been a move toward the conservation and recycling of existing minerals. Aluminum, for example, which is obtained from bauxite, is environmentally polluting to produce. When refined aluminum is recycled and reused, the process of converting bauxite or other aluminum ores into aluminum is avoided.

MINERAL GROUPS

Minerals can be classified into common chemical groupings. It is no surprise, since much of Earth's crust is composed of silicon and oxygen, that silicates are the most common group of minerals. A silicate mineral contains a group of silicon and oxygen atoms bonded together, often in a tetrahedral structure (see Figure 5.6). This structure often produces a rather hard mineral, though not always, as talc contains the silicon-oxygen tetrahedron.

Other common mineral groups are listed in Table 5.4. Note the relationship between the group name and the similarity in the members' chemical formulas. Can you identify, for example, what all oxides have in common, and what all sulfates have in common?

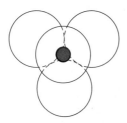

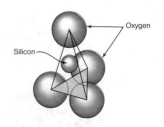

An overhead view, as though looking through the large oxygen ions at the smaller silicon ion between them.

The natural position of the silicon tetrahedron. Notice that the silicon ion is nestled among the oxygen ions and is hidden from view.

An exploded view, showing the positions of the silicon ion and oxygen ions.

FIGURE 5.6 Three views of the silicon tetrahedron. The silicon ion carries a +4 charge (Si^{4+}), and each oxygen ion carries a –2 charge (O^{2-}), so the net charge of the entire silicon tetrahedron is –4: $(SiO4)^{4-}$. Source: *Let's Review: Earth Science—The Physical Setting,* 2nd Ed., Edward J. Denecke, Jr., Barron's Educational Series, Inc., 2002.

TABLE 5.4
MAJOR ROCK FORMING MINERAL GROUPS

Mineral Group	Representative Mineral	Chemical Composition
Silicates	Olivine	$(Mg,Fe)_2SiO_4$
	Feldspar – Orhtoclase	$KAlSi_3O_8$
	Quartz	SiO_2
	Mica – Biotite	$K(Mg,Fe)_3AlSi_3O_{10}(OH)_2$
Sulfides	Pyrite	FeS_2
	Galena	PbS
	Chalcopyrite	$CuFeS_2$
Oxides	Hematite	Fe_2O_3
	Magnetite	Fe_3O_4
	Zincite	ZnO
Sulfates	Gypsum	$CaSO_4$
	Barite	$BaSO_4$
Carbonates	Dolomite	$CaMg(CO_3)_2$
	Calcite	$CaCO_3$
	Siderite	$FeCO_3$
Elements	Silver	Ag
	Gold	Au
	Mercury	Hg
Halides	Halite	$NaCl$
	Fluorite	CaF_2
	Villiaumite	NaF

MINERALS AS IMPORTANT RESOURCES

Minerals supply our society with many of the raw materials that we require to maintain our current standard of living. A mineral that contains commercially useful quantities of an important resource is called an *ore*. Bauxite, as noted earlier, is an ore of aluminum. The aluminum required to build a Boeing 747 plane originated as the rather indistinct mineral bauxite. All copper, iron, and silicon, to name just a few common and important minerals, must be mined from Earth as ores before copper wiring, steel girders, and silicon chips can be manufactured.

Minerals that are relatively rare and have aesthetic qualities are called *gems*. Many gems, as a result of their rarity and beauty, have been regarded as prize possessions since the beginning of recorded history. Precious, that is, rarest, most beautiful, and most durable, gems include diamonds, emeralds, opals, rubies, and sapphires. Semiprecious gems include jade, garnet, quartz, and topaz.

> **REMEMBER**
> Chemistry is at the core of any study of minerals.

MINERAL OR ROCK?

Minerals and rocks are often referred to interchangeably. In fact, however, they are not the same. A *mineral* is a chemical compound or a mixture of compounds, whereas a *rock* is an aggregate of minerals. An aggregate is a mixture that is cemented or mechanically joined together. In essence, rocks are composed of minerals. When we discuss rocks, we talk about them in terms of their modes of formation as well as the minerals of which they are composed.

Igneous Rocks

MAGMA VERSUS LAVA

A rock is classified according to its mode of formation. Rocks are an important area of study because they provide clues to past environments on Earth as well as containing valuable resources needed by our society and others. Rocks tell a fascinating story of a dynamic and at times volatile planet! Through their study we can decipher the locations of ancient oceans, life-forms, and environments. With this knowledge, we can understand how our planet responded to ancient stresses and thus gain insight into how it may respond to future ones.

Igneous rocks, meaning "rocks from fire," form from a *melt*. This melt is a complex mixture of liquids and gases and is known as *magma* or *lava*. The melt is initially quite hot, usually over 2000°C, and always forms deep within Earth. Radioactive decay of unstable elements provides the heat required to melt solid rock and form this complex mixture. Sometimes the material known as magma never reaches the surface but cools beneath the surface to form various structures composed of igneous rock within Earth's crust. When magma does reach the surface, it is chemically changed as it interacts with oxygen in Earth's atmosphere, and is then referred to as lava. This is the essential difference between magma and lava. Lava, because of its location on Earth's surface, tends to cool much more rapidly than its relative, magma, which cools deep within the crust.

When studying an outcrop (exposure) of igneous rock, geologists can generally determine whether the rock was *intrusive* (formed as magma without reaching the surface) or *extrusive* (formed as lava on the surface) by analyzing the crystal size. Intrusive rock tends to cool more slowly than extrusive rock, thus allowing crystals to grow larger. Crystals in granite, a common intrusive igneous rock, can grow to the size of pebbles or even larger, whereas crystals in basalt, a common extrusive igneous rock, are often too small to be seen without a magnifying glass. Figures 5.7 and 5.8 illustrate the cooling process of magma into a crystalline solid and the effect of cooling rate.

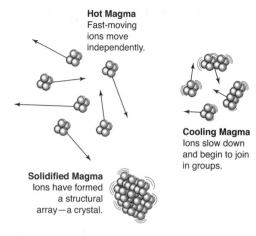

FIGURE 5.7 Cooling and crystallization. Source: *Let's Review: Earth Science— The Physical Setting,* 2nd Ed., Edward J. Denecke, Jr., Barron's Educational Series, Inc., 2002.

Lava is often associated with volcanic activity. Famous volcanoes, such as those in the Hawaiian Islands, erupt from time to time, with hot, glowing lava flowing down the mountainside toward the ocean. These spectacular lava flows produce igneous rock that shows signs of rapid cooling. Not all lava, however, is associated with volcanic activity. Magma has been known to penetrate the crust, reaching

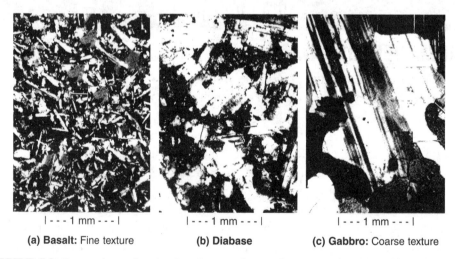

|- - - 1 mm - - -| |- - - 1 mm - - -| |- - - 1 mm - - -|

(a) Basalt: Fine texture   **(b) Diabase**   **(c) Gabbro:** Coarse texture

FIGURE 5.8 Comparison of grain sizes in a continuum from extrusive (a) and intrusive (b and c) rocks. Basalt (a) forms at the surface. Diabase (b) may form as an intrusion near the surface as in a dike or sill. Gabbro (c) cools very slowly deep underground. Source: *Physical Geology,* 2nd Ed., Richard Foster Flint and Brian J. Skinner, John Wiley, 1974, 1977.

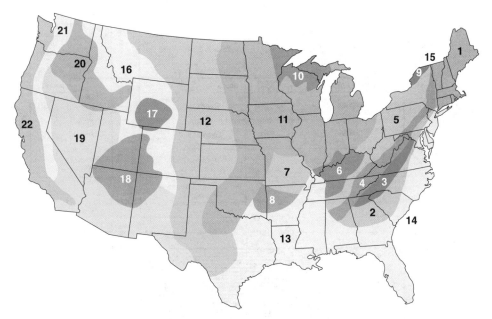

FIGURE 5.9 The physiographic provinces are coded by number: 1. New England,
2. Piedmont, 3. Highlands (Blue Ridge), 4. Ridge and Valley, 5. Appalachian Plateau,
6. Interior Low Plateaus, 7. Ozark Plateau, 8. Ouachita Province, 9. Adirondacks, 10. Superior
Uplands, 11. Central Lowlands, 12. Great Plains, 13. Atlantic Coastal Plain, 14. Atlantic
Continental Margin, 15. Saint Lawrence Lowlands, 16. Rocky Mountain, 17. Wyoming Basin,
18. Colorado Plateau, 19. Basin and Range, 20. Columbia Plateau, 21. Cascade-Sierra, 22.
Pacific Coastal. Source: L. Hanson, Salem State College, Salem, MA.

the surface as lava visible through fissures. This lava does not erupt; instead, it may
form large lava lakes that cool and harden over time into extrusive igneous structures. Portions of the Piedmont Physiographic Province (see Figure 5.9) along
the U.S. eastern seaboard contain igneous rock that formed in this manner. Also,
as North America and Africa separated over 200 million years ago, large fissures
opened, allowing lava lakes and near-surface intrusions called *sills* to form. Evidence
can be found along the Palisades (sill) and other portions of northern New Jersey,
specifically the Watchung Mountains, situated just west of New York City, which
are classic examples of these ancient lava flows.

CLASSIFYING IGNEOUS ROCKS

As stated above, igneous rocks can be classified (see Figure 5.10) as either *intrusive*
(forming within the interior as magma slowly cools and hardens into solid rock)
or *extrusive* (forming as lava on the surface quickly cools and hardens). Significant
physical differences are caused by the different rates of cooling.

The slower cooling intrusive rocks have much larger crystalline structures. A classic
example is granite. In granite, large crystals of quartz, mica, and feldspar, all common
minerals, can be seen. Granite is classified as a coarse-grained intrusive rock.

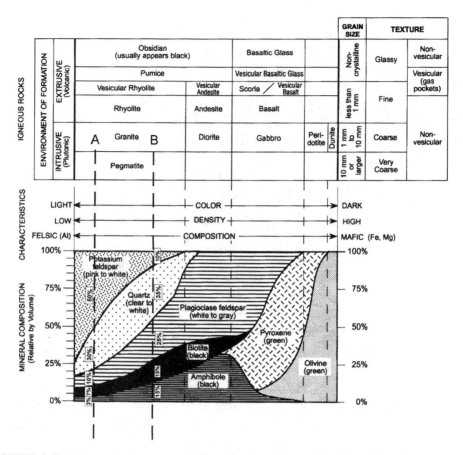

FIGURE 5.10   Scheme for igneous rock identification. Source: The State Education Department, *Earth Science Reference Tables,* 2001 ed. (Albany, New York: The University of the State of New York).

Rhyolite is similar in chemical composition to granite, but has a very different appearance because of its mode of formation. Rhyolite is an extrusive igneous rock. Rapid cooling does not allow for large crystallization, and therefore, the rock has a uniform gray appearance with individual crystals too small to identify. Basalt is classified as a fine-grained extrusive rock.

Igneous rocks are secondarily identified by their colors. The color categories into which most igneous rocks fit are light, dark, and ultra-dark. Although rhyolite and basalt are both extrusive rocks with the characteristic small crystal structure, rhyolite is much lighter in color than basalt. The difference in color results from the relative abundances of iron- and magnesium-containing minerals found in the rocks. Iron and magnesium, when present in significant quantity, tend to make a mineral dark (and dense). Basalt is rich in iron and magnesium—hence its dark color. Rhyolite, on the other hand, has little iron and magnesium; instead it is rich in the elements silicon and aluminum, which tend to produce a lighter colored rock.

> **REMEMBER**
> Igneous rocks are initially classified as to whether they are intrusive or extrusive.

COMMON IGNEOUS ROCKS AND THEIR USES

Igneous rocks are much more than just artifacts for study by geologists. They have significant value to our society. As noted earlier, rocks are made of minerals, and minerals supply the raw materials required by our technological and industrial society. Rocks are often cut and used for headstones, sculptures, and material for buildings. Crushed rock is used for road construction. Granite is highly desirable, as it is a hard rock that is quite resistant to weathering; structures made of granite tend to last for years with little impact by rain or wind.

Sedimentary Rocks

FORMATION OF SEDIMENTARY ROCKS

Water is the factor common to the formation of all sedimentary rocks. Regardless of individual variety, all sedimentary rocks form in an aqueous environment. The term *sedimentary* is derived from the Latin *sedimentum*, which means "settling." One common scenario involves a river that first weathers and erodes rock along its banks. The eroded material, or sediment, is carried downstream until the flow of the river weakens, allowing the sediment to settle out from the water. Over the years, more and more sediment builds up; then, if conditions change, for example, the river alters its course, the sediment dries and is cemented together. This stage, called *lithification*, is the final stage involved in the formation of what is known as "clastic" sedimentary rock, or rock that has formed from particles of various sizes (see Figure 5.11). The type of sedimentary rock that forms depends largely upon the size of the particles that collected in the sediment.

Sedimentary rock can also form by chemical and organic means. *Chemical sedimentary rock* forms when chemical reactions occur in water to produce precipitates, or solids, that then settle out of solution. Evaporation can also produce precipitates

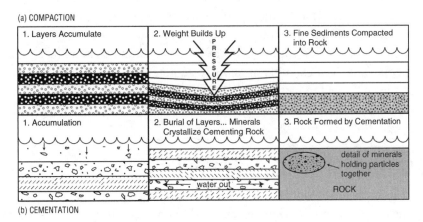

FIGURE 5.11 Compaction and cementation.

FIGURE 5.12 Grand Canyon National Park, Arizona.

as various ions become saturated in solution and bond to other ions. Depending upon the chemical composition of the precipitate, halite, a mineral rich in sodium chloride, or gypsum, a mineral rich in calcium sulfate, may form.

Organic sedimentary rock forms from the remains of plants and animals, including carbonate shells or fragments of shells that lithify to form solid rock. Two rocks that form under these conditions are limestone and coquina.

In summary, water plays a tremendous role in breaking rock down into sediment, transporting the sediment, and helping it to form new sedimentary rock.

About 75 percent of all the exposed rock on the surface of Earth is sedimentary rock. This fact is testimony to the great interaction between the hydrosphere and the geosphere on the surface. Much of the interior western United States is composed of beautiful sedimentary rock formations, and several of our national parks showcase these formations within their borders. The Grand Canyon (see Figure 5.12) is perhaps the best example of exposed sedimentary rock. Sedimentary rock often has a classic layered pattern that is highly recognizable. These layers, called *strata* or *beds*, result from the gradual deposition of sediment over long periods of time.

CLASSIFYING SEDIMENTARY ROCKS

Sedimentary rocks (see Figure 5.13) are classified first as to whether they are of detrital (clastic) or chemical origin. Rocks of *detrital origin* originate as solid particles from weathered rock, as described above. Rocks of *chemical origin* form when soluble

Scheme for Sedimentary Rock Identification

INORGANIC LAND-DERIVED SEDIMENTARY ROCKS					
TEXTURE	**GRAIN SIZE**	**COMPOSITION**	**COMMENTS**	**ROCK NAME**	**MAP SYMBOL**
Clastic (fragmental)	Pebbles, cobbles, and/or boulders embedded in sand, silt, and/or clay	Mostly quartz, feldspar, and clay minerals; may contain fragments of other rocks and minerals	Rounded fragments	Conglomerate	
			Angular fragments	Breccia	
	Sand (0.2 to 0.006 cm)		Fine to coarse	Sandstone	
	Silt (0.006 to 0.0004 cm)		Very fine grain	Siltstone	
	Clay (less than 0.0004 cm)		Compact; may split easily	Shale	
CHEMICALLY AND/OR ORGANICALLY FORMED SEDIMENTARY ROCKS					
TEXTURE	**GRAIN SIZE**	**COMPOSITION**	**COMMENTS**	**ROCK NAME**	**MAP SYMBOL**
Crystalline	Varied	Halite	Crystals from chemical precipitates and evaporites	Rock Salt	
	Varied	Gypsum		Rock Gypsum	
	Varied	Dolomite		Dolostone	
Bioclastic	Microscopic to coarse	Calcite	Cemented shell fragments or precipitates of biologic origin	Limestone	
	Varied	Carbon	From plant remains	Coal	

FIGURE 5.13 Scheme for sedimentary rock identification. Source: The State Education Department. *Earth Science Reference Tables,* 2001 ed. (Albany, New York: The University of the State of New York).

material, produced by chemical weathering, precipitates or becomes solid again. This usually occurs when the water that kept the chemical sediments soluble evaporates, leaving behind a chemical precipitate or sedimentary rock. Such rocks are often called *evaporites* because the process whereby they formed can be likened to the evaporation of salt water in a glass. The residue that remains in the glass is the salt that precipitated out of solution as the water evaporated.

Limestone, which is mostly calcite, is the most common sedimentary rock of chemical origin. Actually, its origin is biochemical since the chemical sedimentation involves marine life. Marine organisms absorb the dissolved calcium carbonate to form shells and other hard parts. When these organisms die, their skeletons accumulate on the sea or lake floor and then lithify into limestone over time.

Once sedimentary rocks are divided on the basis of detrital or chemical origin, the detrital rocks are further classified as to particle size. Clay-size particles, the smallest particles, form as mud lithifies into shale. Clay- and silt-size particles are less than $1/16$ millimeter in diameter. As a result of the small size, shale feels quite smooth if you rub your finger across it. As particle size increases, sandstone tends to form. With its

TABLE 5.5
THE WENTWORTH SCALE

Particle-Size Range (diameter, cm)	Particle Name	Sediment Composed of That Particle
Greater than 25.6	Boulder	Boulder gravel
6.4–25.6	Cobble	Cobble gravel
0.2–6.4	Pebble	Pebble gravel
0.006–0.2	Sand	Sand
0.0004–0.006	Silt	Silt
Less than 0.0004	Clay	Clay

characteristically rough surface, particle diameters can range up to 2 millimeters. When the diameter is greater than 2 millimeters, the particles are classified as gravel; they cement together to form a rock called *conglomerate*. The Wentworth scale (see Table 5.5) classifies sediments according to size.

> **REMEMBER**
> Clastic sedimentary rocks are classified according to particle size.

IMPORTANCE AND VALUE OF SEDIMENTARY ROCKS

Sedimentary rocks often contain important clues regarding Earth's history and the kinds of life-forms that could be found here in the past. The gradual process of sedimentation, compaction, and lithification tends to preserve fossils, the trace remnants of ancient life or imprints from ancient organisms. Many assessments can be made based upon the kinds of fossils and imprints found in rock. Ancient climates and land and ocean areas, as well as entire ecosystems, can be reconstructed by studying clues found in sedimentary rock. The kinds of sediments present in the rock provide clues as to the surface conditions while the sediments were being deposited. For example, coarse sediment settles only in fast-moving water, such as a youthful river, whereas silts will collect in a shallow ocean.

Sedimentary rocks are used throughout our society. Two important evaporites, halite and gypsum, have great commercial value. Halite is better known as table salt and is used, of course, as a food seasoning. Gypsum is the main ingredient in plaster of Paris.

Many a visitor to New York City has walked the streets of the various neighborhoods and admired the brownstones, that is, houses made of sandstone that were constructed largely in the late nineteenth century. Limestone is also used in construction.

Sedimentary rocks are often associated with aquifers, which are essentially underground locations where potable water can be found. Many sedimentary rocks are permeable; that is, water can filter through them. Sedimentary rocks are also somewhat soluble; water acts to dissolve the rock. These physical attributes allow water to collect in underwater cavities (aquifers). Aquifers are discussed in somewhat greater detail in Chapter 10.

Certain forms of coal are sedimentary rocks. The United States and other areas of the world still rely upon coal as a source of energy. Coal differs from most sedimentary rocks, however, as it is composed primarily of organic matter. Coal forms as the end product after the burial of large amounts of plant material over very long periods of time.

Sedimentary rock can also hold oil deposits. Shale, a highly impermeable sedimentary rock, often caps oil deposits. When the presence of a less dense material is suspected between layers of shale, that substance is likely to be oil or natural gas.

Metamorphic Rocks

METAMORPHISM

Metamorphism, a change in rock caused by heat and pressure, affects preexisting rocks. In other words, all rocks, whether sedimentary, igneous, or metamorphic, under the right conditions can turn into metamorphic rock. The right conditions include tremendous amounts of heat and pressure, so metamorphism occurs only deep within Earth's crust or, as "contact" metamorphism, near magma flows.

Metamorphism can occur when rocks are buried beneath thousands of feet of younger sediments, or when entire continental plates move and compress existing rock. Metamorphism can change the shape, texture, and mineral composition of an existing rock. Sometimes scientists can determine what the metamorphic rock was prior to the metamorphism; at other times, this is impossible.

Easily identified sequences include the metamorphism of shale into slate, limestone into marble, granite into gneiss, and sandstone into quartzite. In these cases, when samples of each metamorphic rock and its nonmetamorphosed relative are studied, the relationship is clearly identifiable. For example, both shale and slate are layered and fine grained, and granite and gneiss contain similar minerals. Gneiss, however, is *foliated*, that is, has a banded appearance (see Figure 5.14) because of the partial melting that occurs during metamorphism, allowing the minerals to rearrange and form the bands that give gneiss a layered appearance.

More relevant details about metamorphism are given in Table 5.6.

Compression causes randomly dispersed crystals to become oriented in a direction perpendicular to the pressure. The oriented crystals form thin sheets or layers called *foliation*.

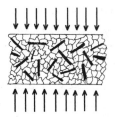
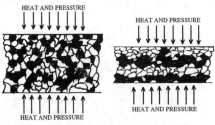

Migration of mineral crystals caused by density differences produces *banding*—regions of light- and dark-colored minerals.

FIGURE 5.14 Compression related to metamorphism causes randomly oriented grains to begin to organize into bands. Source: *Let's Review: Earth Science—The Physical Setting,* 2nd Ed., Edward J. Denecke, Jr., Barron's Educational Series, Inc., 2002.

TABLE 5.6

DERIVATIONS OF COMMON METAMORPHIC ROCKS

Parent Rock	Metamorphic Rock
Shale	Compaction and reorientation → slate
	Slate → micas and chlorite form → phyllite
	Phyllite → grains recrystallize and get coarser → schist
	Schist → banding occurs→ gneiss
Shale	Contact with magma → granofels
Sandstone	Quartzite
Limestone or dolostone	Marble
Basalt	Hydration reactions → greenschist
	Greenschist → dehydration → amphibolite
	Amphibolite→ banding occurs → gneiss

Source: *Let's Review: Earth Science—The Physical Setting*, 2nd Ed., Edward J. Denecke, Jr., Hauppauge, NY: Barron's Educational Series, Inc., 2002.

CLASSIFYING METAMORPHIC ROCKS

A metamorphic rock is classified (see Figure 5.15) as having a foliated or a nonfoliated texture. Foliation, as explained above, is recognized as the layering or banding of minerals. Gneiss is a classic example of a foliated, metamorphic rock. Marble, which often appears as a single-colored rock that shows no evidence of layering, is a classic example of a nonfoliated rock.

> **REMEMBER**
> Metamorphic rocks are initially classified as to whether they are foliated (layered) or nonfoliated.

Texture	Grain Size	Composition	Type of Metamorphism	Comments	Rock Name	Map Symbol
FOLIATED (MINERAL ALIGNMENT)	Fine	MICA / QUARTZ / FELDSPAR / AMPHIBOLE / GARNET / PYROXENE	Regional	Low-grade metamorphism of shale	Slate	
	Fine to medium		(Heat and pressure increase with depth) ↓	Folation surfaces shiny microscopic mica crystals	Phyllite	
				platy mica crystals visible from metamorphism of clay or feldspars	Schist	
(BANDING)	Medium to Coarse			High-grade metamorphism; some mica changed to feldspar; segregated by mineral type into bands	Gneiss	
NONFOLIATED	Fine	Variable	Contact (Heat)	Various rocks changed by heat from nearby magma/lava	Hornfels	
	Fine to coarse	Quartz	Regional or Contact	Metamorphism of quartz sandstone	Quartzite	
		Calcite and/or dolomite		Metamorphism of limestone or dolomite	Marble	
	Coarse	Various minerals in particles and matrix		Pebbles may be distorted or stretched	Metaconglomerate	

FIGURE 5.15 Scheme for metamorphic rock identification. Source: The State Education Department, *Earth Science Reference Tables,* 2001 Ed. (Albany, New York: The University of the State of New York).

COMMON METAMORPHIC ROCKS AND THEIR USES

Metamorphic rocks are quite useful to society as they are often hard and thus resistant to weathering. After only a century or so, the brownstones mentioned on page 128 show evidence of weathering of the soft sandstone they are made from, but gneiss or marble structures resist change. Change will eventually occur with metamorphic rocks, but more slowly than with a "soft" rock such as sandstone.

Marble is sufficiently weather resistant to be used in sink basins. Limestone would never suffice for such a purpose! Many statues and headstones are also made of marble. Slate is found on the roofs of old homes in the northeastern United States and is also used to make billiards (pool) tables, chalkboards, and floor tiles.

The Rock Cycle

The discussions in this chapter clearly suggest that no rock lasts forever and that existing rocks can be converted into other types of rocks by outside forces. Sandstone, for example, may be weathered and eroded over time. The sediments are carried downstream and deposited elsewhere to form newer sedimentary rock. That same sandstone, if buried over millions of years of sedimentation, can be metamorphosed into quartzite; and the quartzite, if additional heat is applied, can melt, forming magma that will someday solidify into igneous rock.

This continuity of change is called the *rock cycle* (see Figure 5.16). As you study the diagram, note that any rock can be converted into any other kind of rock, with sediment and magma/lava as intermediate steps along the way.

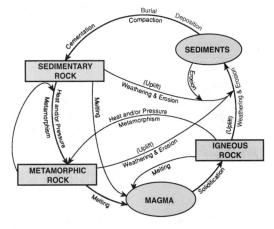

FIGURE 5.16 Rock cycle in Earth's crust. Source: The State Education Department, *Earth Science Reference Tables,* 2001 Ed. (Albany, New York: The University of the State of New York).

Using the rock cycle, we can see that, if we start with magma (a convenient, but not mandatory, starting point), igneous rocks can form that over time weather and erode into sediment. The sediment compacts and lithifies into sedimentary rock, which eventually is buried to sufficient depth that heat and pressure metamorphose the rock. The metamorphic rock, if subjected to additional heat and pressure, will melt, producing magma, the material with which we started.

The time required for these changes is well beyond the life span of a typical human; thus we tend to see rocks as permanent features on Earth's surface. When contemplating events over the course of geologic time, however, it is important to remember that nothing is permanent and unchanging!

Related Internet Resources for Great Images and Information

www.geolab.unc.edu/Petunia/IgMetAtlas/mainmenu.html

The University of North Carolina offers "Atlas of Metamorphic and Igneous Rocks, Minerals, and Textures."

http://volcano.oregonstate.edu/education/vwlessons/lessons/Slideshow/Slideindex.html

Oregon State University's "Rocks and Minerals Slide Show" is a part of the "Volcano World" Web site.

www.galleries.com

"Amethyst Galleries" presents a collection of pictures and information about minerals.

REVIEW EXERCISES FOR CHAPTER 5

WORD-STUDY CONNECTION

aggregate	electron	lava
aluminum	evaporites	lithification
anion	extrusive	luster
aqueous	fissure	magma
atom	foliated	magnesium
beds	formula	melting point
brownstones	fossil	metallic
calcium	fracture	metals
cation	gem	metamorphic
chemical bond	glassy	metamorphism
chemical group	halite	minerals
cleavage	hardness	Moh's hardness scale
compaction	heat	nonfoliated
copper	igneous	nonmetals
crystalline solid	intrusive	ore
detrital	ion	organic
diamond	ionic compound	oxygen
dull	iron	pearly

periodic table	rocks	specific gravity
polyatomic ion	sedimentary	strata
precipitates	sedimentation	streak
proton	silicate	talc
radioactive decay	silicon	
rock cycle	silky	

SELF-TEST CONNECTION

PART A. Completion. *Write in the word that correctly completes the statement.*

1. When an ionic compound forms, a cation bonds to an _____.

2. _____ is a property that describes a mineral's ability to resist scratching.

3. Minerals that tend to break along irregular surfaces are said to exhibit _____.

4. _____ is a common element that tends to produce a high specific gravity in minerals.

5. Minerals that are rich in silicon and oxygen are part of the _____ group.

6. Diamonds and emeralds are examples of _____.

7. Magma that reaches the surface is chemically changed and is then referred to as _____.

8. Igneous rocks that form within Earth's interior are classified as _____.

9. Plant and animal remains can form _____ sedimentary rock.

10. Gneiss is a rock that may have originally been _____ and was then metamorphosed.

11. Marble is an example of a _____ metamorphic rock.

12. Table salt is obtained from a mineral called _____.

13. Magma cools slowly to produce _____ rocks.

14. Metamorphism occurs as a result of _____ and pressure.

15. _____ rocks are most resistant to weathering and erosional forces.

PART B. Multiple Choice. Circle the letter of the item that correctly completes the statement.

1. A positively charged atom is called
    (a) an ion
    (b) a cation
    (c) an anion
    (d) a proton
    (e) an electron

2. When a mineral breaks along regular lines or planes, it is exhibiting
    (a) hardness
    (b) cleavage
    (c) fracture
    (d) softness
    (e) compaction properties

3. Minerals that form when aqueous solutions evaporate, leaving behind a precipitate, are called
    (a) detrital
    (b) silicates
    (c) evaporites
    (d) foliated
    (e) nonfoliated

4. A metamorphic rock that exhibits layering, such as gneiss, is said to be
    (a) intrusive
    (b) extrusive
    (c) basaltic
    (d) foliated
    (e) nonfoliated

5. Chemical bonding results from the gain, loss, or sharing of
    (a) electrons
    (b) protons
    (c) neutrons
    (d) ionic bonds
    (e) cations

6. The process by which a rock is changed by extreme heat and pressure deep within Earth is called
    (a) sedimentation
    (b) lithification
    (c) compaction
    (d) metamorphism
    (e) diversification

7. A mineral that has commercial, extractable quantities of a desirable element
   is called
   (a) a gem
   (b) an ore
   (c) semiprecious
   (d) precious
   (e) pearly

8. The process by which sediment is converted into solid rock is
   (a) sedimentation
   (b) compaction
   (c) lithification
   (d) metamorphism
   (e) radioactive decay

9. The term _____ denotes a mineral's resistance to being scratched or its ability
   to scratch other substances.
   (a) hardness
   (b) streak
   (c) color
   (d) cleavage
   (e) fracture

10. When a molten mixture of liquids and gases reaches Earth's surface it is called
    (a) magma
    (b) lava
    (c) a melt
    (d) intrusive
    (e) extrusive

PART C. Modified True/False. *If a statement is true, write "true" for your answer.*
If a statement is incorrect, change the <u>*underlined*</u> *expression to one that will make the*
statement true.

1. <u>Amorphous</u> is the term used to describe a mineral that exhibits a regular geomet-
   ric pattern.

2. Chemical bonding occurs with the gain, loss, or sharing of <u>electrons</u>.

3. Silicates are generally composed of silicon and <u>iron</u>.

4. Igneous rocks that form from the cooling and solidifying of lava are <u>intrusive</u>.

5. Metamorphic rocks that do not exhibit layering are classified as <u>nonfoliated</u>.

6. <u>Talc</u> is the hardest known mineral.

7. Rocks that form from molten material within Earth are called <u>igneous</u>.

8. Iron will tend to become <u>an anion</u> when bonding with a nonmetallic element.

9. When a metal bonds to a nonmetal the result is <u>an ionic compound</u>.

10. Houses in New York City that are made of sandstone are often called <u>brownstones</u>.

CONNECTING TO CONCEPTS

1. Is ice a mineral? If not, on what basis does it fail to meet the definition of a mineral?

2. Describe what transpires when a metallic atom chemically bonds to a nonmetallic atom. Be sure to apply the terms *cation* and *anion* in their proper contexts.

3. Imagine that you have a kit containing minerals of known hardness. One mineral has a hardness of 1, the second a hardness of 2, and so on. Describe how you can determine the hardness of a mineral not contained in the kit.

4. Describe the relationship between minerals and rocks.

5. How can one determine whether an igneous rock is of intrusive or extrusive origin?

6. Base your answers to this question on the data table below, which shows the relationship between the amount of aluminum in a type of rock and the energy needed to extract the aluminum from that rock.

    (a) On the grid provided, label the axes and select an appropriate scale for the data.

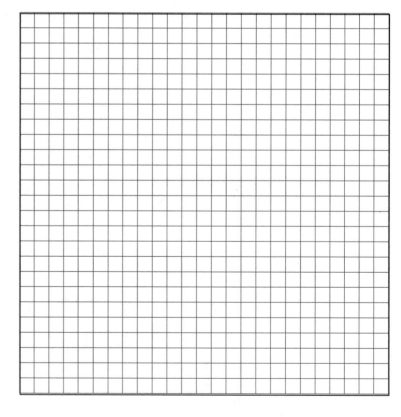

Sample	Aluminum Content of Ore (percent)	Energy Needed to Extract 1 Ton of Aluminum (thousands of kilowatt-hours)
1	3	220
2	5	140
3	10	90
4	20	65
5	30	54
6	40	49
7	50	48

(b) Plot the energy needed to extract 1 ton of aluminum versus the aluminum content of the ore from which it is extracted for each of the seven samples. Connect the seven points with a line.

(c) According to the graph you constructed, how much energy is needed to produce aluminum from rock that is 15 percent aluminum? Express your answer in thousands of kilowatt-hours.

7. Why might minerals be considered a nonrenewable resource?

8. Is gold a mineral? Why is it valuable to society? (Name at least three uses.)

CONNECTING TO LIFE

Geologists are often engaged in the study of rocks and minerals, which provide the raw materials that our society requires to function. As noted in this chapter, without bauxite, an ore of aluminum, we would not have airplanes. Steel, an alloy rich in iron, is a material needed in the construction of buildings. Iron is obtained from several ores, including magnetite and hematite. And what would a computer be without a silicon chip?

Geologists are also actively engaged in the search for oil, a vital resource for our society. Identification of certain types of rocks provides geologists with clues as to where to drill to find this important substance.

Geologists also study the planet in an effort to identify regions of high risk. This research includes the study of landslides, earthquakes, and volcanoes. Recent eruptions, including Mount Etna in Italy, have forced the evacuation of communities, and recent earthquakes in California have caused many casualties and destroyed millions of dollars' worth of public and private property.

ANSWERS
SELF-TEST CONNECTION

Part A

1. anion	6. gems	11. nonfoliated
2. Hardness	7. lava	12. halite
3. fracture	8. intrusive	13. igneous
4. Iron	9. organic	14. heat
5. silicate	10. granite	15. Metamorphic

Part B

1. **(b)**	3. **(c)**	5. **(a)**	7. **(b)**	9. **(a)**
2. **(b)**	4. **(d)**	6. **(d)**	8. **(c)**	10. **(b)**

Part C

1. False; Crystalline	5. True	8. False; a cation
2. True	6. False; Diamond	9. True
3. False; oxygen	7. True	10. True
4. False; extrusive		

CONNECTING TO CONCEPTS

1. Ice is not considered to be a mineral. It fails on the basis that all minerals are solids at room temperature. Ice, of course, melts at 0°C; since room temperature is usually defined as 20°C, ice will not exist at that temperature.

2. When a metallic atom bonds to a nonmetallic atom, the metal loses one or more electrons, which the nonmetal gains. By gaining electrons, the nonmetal takes on a negative charge and is then referred to as an *anion*. By losing electrons, the metal takes on a positive charge and is then called a *cation*.

3. A harder mineral will always scratch a softer mineral. Try scratching the unknown mineral with minerals of known hardness, starting with a hardness of 1. Eventually, one of the known minerals will scratch the unknown mineral and therefore must be harder than the unknown mineral. All of the other minerals used to that point are softer.

4. A mineral is a chemical compound or a mixture of compounds. A rock is an aggregate of minerals.

5. Intrusive igneous rocks cool more slowly than extrusive rocks. The slower a melt cools and hardens, the larger the mineral crystals will grow.

6. c. 77.5 kilowatt-hours

7. A mineral may be considered to be a nonrenewable resource if the energy required to recycle the mineral is excessive or if the process of recycling produces unwanted pollutants.

8. Yes, gold is a mineral. It meets all the requirements as stated in this chapter. Gold is used as currency, as an excellent conductor for electrical connectors, and as a component of jewelry. Gold is also used in an astronaut's visor as a filter for the Sun's high-energy radiation.

Geologic Time and Dating Rocks

WHAT YOU WILL LEARN

This chapter focuses on geologic time and dating rocks. In this chapter you will learn

- why we need a time scale and how that time scale is organized;
- the difference between the major "eras" of time;
- basic principles of dating rock sequences;
- radio-dating methods to establish actual ages;
- how Earth's age can be determined.

SECTIONS IN THIS CHAPTER

- The Geologic Time Scale
- Relative Dating of Rocks
- Absolute Dating of Rocks
- The Age of the Earth
- Related Internet Resources to Great Images and Information
- Review Exercises for Chapter 6

The Geologic Time Scale

RECOGNIZING THE NEED FOR A TIME SCALE

"We find no vestige of a beginning, no prospect of an end." With these words, written over 200 years ago by a Scottish physician named James Hutton, the era of modern geology began. Hutton, in his late-1700s publication titled *Theory of the Earth*, introduced the concept of *uniformitarianism* to the scientific world. This concept states that the physical, chemical, and biological laws that operate today also operated in the geologic past. This perspective allows us to observe geologic processes today and reason that these same processes have been occurring continuously and gradually for a very long time. Hutton is often referred to as the founder of modern geology.

Hutton believed Earth to be very old, perhaps more than 1 million years old. This was a radical departure from the dominant view of his day—that the age of Earth was about 6000 years. This view had its origin in a chronological history of Earth and all humanity based on the Old Testament in the Bible and constructed in the mid-1600s by an archbishop from Ireland named James Ussher.

Hutton studied Hadrian's wall, an ancient structure built near his home 17 centuries earlier, and noted the slow rate of its weathering and erosion. He noted also the layering of sediment in nearby mountains and observed erosional agents, including rivers, carry sediment downstream from these mountains. His observations led him to the conclusion that Earth was much older than the popular view held and that change occurred gradually over long periods of time.

The nineteenth century saw the construction of a geologic time scale (see Figure 6.1), which sought to place events in chronological order from Earth's formation to recent times. In the twentieth century, with the advent of radiometric dating as a technique to determine absolute age, dates could be amended and added to the geologic time scale.

ORGANIZATION OF THE GEOLOGIC TIME SCALE

The geologic time scale is organized by eons, eras, periods, and epochs. Each is a finer division of time than the preceding one. In other words, eons contain one or more eras; eras contain periods; and, in the most recent era, periods contain epochs. Each era begins and ends generally with a significant change in the life-forms on this planet. Major geologic events also separate one geologic time period from another.

PRECAMBRIAN TIME

The Precambrian era, sometimes referred to as an eon, covers approximately the first 4 billion years of Earth's history. As you read the next few sections, a story of gradual change and evolution will emerge. Figure 6.2 visually supports the ensuing discussion, and you may wish to refer to this figure periodically to locate humankind on the progression from Earth's formation to today. The Precambrian begins with the formation of Earth 4.5 billion years ago and ends with the beginning of the Paleozoic era 540 million years ago. Because of its antiquity, we know very little about this vast span of time. Much of the rock from this time period has been buried beneath younger rocks, is severely bent and folded, lacks fossils, or has simply been weathered and eroded away, thus erasing any record of events that occurred. The Precambrian was a time when Earth was still forming into the planet that we recognize today. For example, the atmosphere began to resemble the modern atmosphere as excess carbon dioxide and water vapor were deposited into the geosphere and hydrosphere and free oxygen started to appear. The planet's surface solidified and stabilized, readying itself for life to take hold.

Any forms of life that existed during this time were probably quite primitive, and few left any fossil record behind. One of the few fossil records from the Precambrian was left by stromatolites. Stromatolites are actually calcified material deposited by algae, one of the simplest life-forms. These fossils, dating back 2 billion years, provide evidence of the first *photosynthetic organisms.*

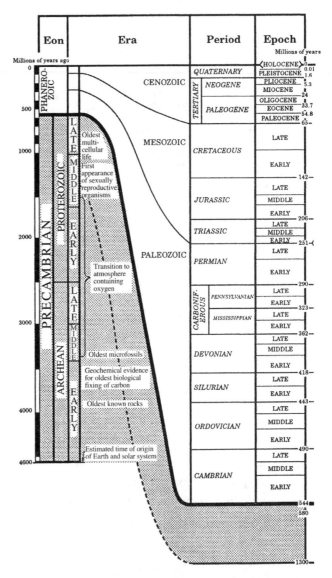

FIGURE 6.1 The geologic time scale. Source: The State Education Department, Earth Science Reference Tables, 2001 ed. (Albany, New York: The University of the State of New York).

These simple forms of life have been responsible for the production of atmospheric oxygen, which is vital for the survival of all members of the animal kingdom. It should come as no surprise, then, to learn that animal life evolved long after photosynthetic organisms, including algae and later plants, were established on Earth.

Chemosynthetic organisms probably appeared at this time also. Most plants obtain energy from photosynthesis, the process by which sunlight is converted into energy usable by the organism. Certain simple forms of life, however, survive by tapping

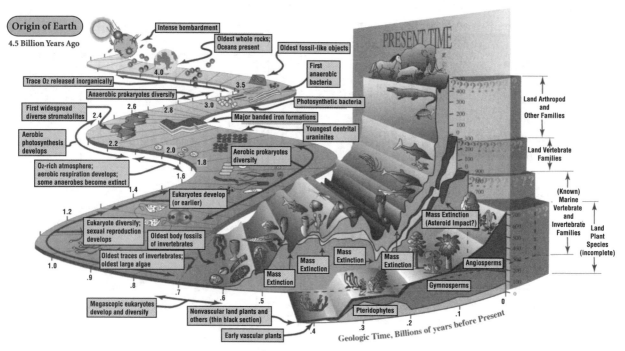

FIGURE 6.2 Schematic history of life on Earth. Source: *Evolution of a Habitable World*, Jonathan I. Lunine, Cambridge University Press, 1999. Reprinted with the permission of Cambridge University Press.

minerals released from deep ocean vents and produce energy through a series of chemical reactions. Photosynthesis is not an option for these deep-water organisms, as sunlight does not penetrate the great depths at which they live. Communities rich in biomass, living near deep ocean vents, seem to be the exception, not the rule, on the otherwise barren ocean floor.

PALEOZOIC ERA

The Paleozoic era, or the "age of ancient animal life," saw a wealth of marine life-forms develop and grow on this planet. The first period in the Paleozoic era is the Cambrian; this is one of two periods characterized as the "age of invertebrates." The two periods begin with the formation of the first shelled organisms and end with the appearance of the first fishes. There is an extensive fossil record of hard-shelled organisms from this time period, including an abundance of trilobites. A trilobite is basically a prehistoric crustacean or arthropod.

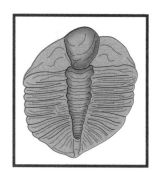

Geologically, during this time, the planet looked quite different from today. At the beginning of the Paleozoic, most of Earth's landmasses were centered about the South Pole. This explains the presence of glacier deposits in areas, including parts of South

America and Africa, that are tropical today. North America was oriented near the Equator, so fossils found in sedimentary strata dating to the Paleozoic contain tropical organisms.

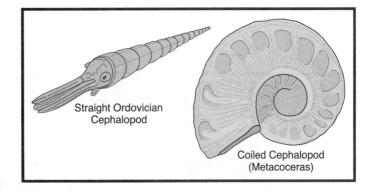

Straight Ordovician Cephalopod

Coiled Cephalopod (Metacoceras)

The Ordovician period followed the Cambrian period and witnessed the arrival of cephalopods, the first truly large organisms and predators of their day. Modern descendents include the squid and octopus. One species of these ancient sea dwellers reached a length of about 10 meters!

The middle Paleozoic is divided into the Silurian and Devonian periods. Collectively, these two periods are called the "age of fishes." This period of time, in addition to a dominance of fish life, saw the first land plants and first insects. By late in the Devonian period, amphibians with fishlike heads had evolved and were truly air-breathing organisms. Life was ready to crawl onto the land!

The late Paleozoic, in which amphibians throve and the first reptiles appeared, ended with a major extinction of trilobites and many marine species. Like most major extinctions, this one may have been caused by the impact of a large meteorite. Researchers have found that Earth has a history of major extinctions in which most life-forms were destroyed by cataclysmic events. It is thought that such events are often collisions of large meteorites with Earth.

Along with a wealth of animal life on the land, plants became abundant, as well as ancient varieties of trees, some of which grew to a height of over 30 meters. Geologically, this was a time of much tectonic activity as great landmasses moved about the surface to form Pangaea, an "all lands" supercontinent, toward the end of the Paleozoic.

As the continents moved and collided (see Figure 6.3), great mountain ranges formed, including the Ancient Appalachians in North America. The Ancient Appalachians were located in what today is the Highlands Physiographic Province (refer to Figure 5.9). Once Pangaea formed, this vast continent acted to modify the global climate system, causing extremes between seasons far greater than are experienced today. These extreme changes in climate from season to season may also have contributed to the mass extinction of life-forms that occurred toward the end of the Paleozoic. Over 75 percent of the amphibian families and many plant species disappeared as climatic extremes stressed these organisms into extinction.

MESOZOIC ERA

The Mesozoic era, or "age of middle animal life," is best characterized as the age of giant reptiles or dinosaurs. Dinosaurs ruled the land for over 100 million years. Major Hollywood movie releases have popularized at least one of the three periods into

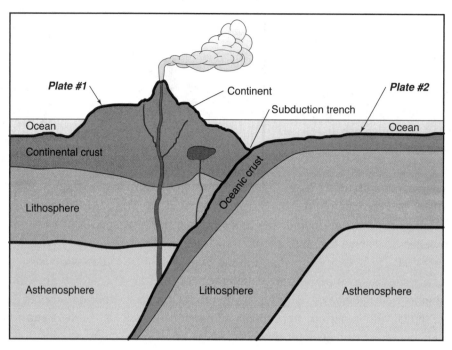

FIGURE 6.3 Mountain-building forces associated with the subduction of oceanic crustal plates. Source: Modified from Windows to the Universe, *www.windows.ucar.edu.*

which the Mesozoic era is divided. The earliest period is the Triassic, which is followed by the Jurassic and the most recent, the Cretaceous. The Mesozoic era also saw the first mammals, birds, and flowering plants. After the late Paleozoic extinction of many life-forms, new and more diverse species developed during the Mesozoic era. There is a significant difference in the fossils found in rock strata dating to the late Paleozoic as compared to fossils from the early Mesozoic. This difference is the basis on which the Paleozoic-Mesozoic boundary was established. Many species found in Paleozoic-age strata are missing from Mesozoic-age strata.

Geologically, the Mesozoic began with much of North America above sea level. Evidence for this is the lack of sedimentary strata from the Triassic period. During the Jurassic period, an inland sea formed in what is now the Colorado Plateau. In subsequent periods uplift and erosion exposed the sandstone sediments collected during the Jurassic. Swamps that formed later, during the Cretaceous period, invaded much of western North America and the Gulf Coast region. These swamps collected much organic material that later formed coal deposits that today are of great economic value to various parts of the United States and Canada. An interactive map of coal deposits across the United States can be found at this Web site: *http://pubs.usgs.gov/of/1996/of96-092/.*

Pangaea began to break apart during the Triassic and Jurassic periods. The breakup started as a rift between what is now the eastern United States and Africa and marked the origin of the Atlantic Ocean. This break-up process has continued for the past 200 million years. Also during the Mesozoic, mountains began to form in western North America as the land was deformed by collisions with the Pacific Plate.

The shelled egg is just one of the reasons that dinosaurs were so successful during the Mesozoic era, a time when evidence suggests that the climate was drier than it now is. Dinosaurs laid hard-shell eggs, each of which was a self-contained aqueous environment. This eliminated the need for a water-dwelling stage, such as the tadpole stage of a frog, so that the egg could be laid on land. Thus dinosaurs were not as dependent upon the presence of water as were other animals. Some evidence even suggests that some dinosaurs were warm blooded, further facilitating their adaptability to the existing climate.

The dinosaurs ranged in size from quite small to extremely large. One of the largest was *Tyrannosaurus rex*. A giant carnivore, *T rex* needed to spend much of its day eating just to sustain its incredible body mass! Other dinosaurs had special adaptations, including the long neck of *Apatosaurus*, which enabled it to feed on conifer trees, and the wings of *Archaeopteryx*, which perhaps evolved into the modern bird.

As the Mesozoic drew to a close, most reptiles became extinct. A few, however, escaped this fate, particularly the smaller species, including turtles, lizards, and snakes. This great extinction may have been caused by a tremendous collision between Earth and a meteorite that was probably about 10 kilometers in diameter. This event, thought to have occurred about 65 million years ago, caused a rapid and major change in the climate by spreading dust into the atmosphere worldwide and cooling the planet. Many species were unable to adapt in such a short span of time and thus became extinct.

Mammals first appeared during the Mesozoic era and were among the more successful species, ultimately surviving the mass extinction at the end of that era. Most species of mammals were quite small physically during the Mesozoic.

For evidence of such an Earth-meteorite collision, geologists search the rock record for the K-T boundary. The K refers to the Cretaceous period in the Mesozoic era, and the T to the Tertiary period in the Cenozoic era. Evidence found along the K-T boundary includes an unusual concentration of the element iridium. This element, while rare on Earth, is common in meteorites. A collision of the magnitude that is envisioned would have spread iridium-containing remnants of the meteorite across a wide portion of the planet. As the dust settled, and sediment collected on top of this thin layer of iridium, the layer was "permanently" preserved in the rock record for our study today. Satellite images suggest the meteorite impact occurred in the northern Yucatan Peninsula and the Gulf of Mexico.

An interesting article about this event appears at this Web site: *http://spaceflightnow. com/news/n0011/20crater/*.

CENOZOIC ERA

The Cenozoic era, or "age of recent animal life," is the era in which we live today. It began 65 million years ago with the mass extinction of the dinosaurs and many other species. As would be expected, scientists know more about the Cenozoic era than about earlier time periods. Evidence of geologic and biologic events in the recent past is clearer and more abundant, so that these events are easier to study. The Cenozoic era is a time when mammals became dominant on Earth.

The Cenozoic era is subdivided into two periods: the Tertiary, which borders the Mesozoic, and the more recent Quaternary, during which humans developed. The Quaternary began less than 2 million years ago and includes two epochs: the Pleistocene and the most recent, the Holocene, which began about 10,000 years ago. Humans appeared late in the Pleistocene but became truly dominant in the Holocene.

The Cenozoic saw major increases in the diversity and size of mammal species. Some of the largest mammals, now extinct, included the hornless rhinoceros, which reached a height of 5 meters, saber-tooth cat, giant bison, and wooly mammoth. These large beasts failed to survive the most recent ice age, which ended about 10,000 years ago.

> **REMEMBER**
> The geologic time scale is organized so that each era has a different life-form dominating the planet.
> Paleozoic: Age of Invertebrates; Mesozoic: Age of Giant Reptiles; Cenozoic: Age of Mammals.

Flowering plants also took hold on the land during the Cenozoic. As a result of the seed dispersal of these plants, they spread rapidly across the land. The presence of flowering plants brought birds that feed on their seeds. Grazing animals were also drawn to the plants, and predators followed the grazing animals.

Geologically, the Cenozoic in North America has been a period of mountain building in the West and erosion in the East. The difference results from the active plate boundary along the west coast and the lack of significant tectonic activity along the eastern seaboard.

Relative Dating of Rocks

Geologists use a variety of techniques to establish dates such as those cited in the preceding sections. One may reasonably ask, "How do we know when certain events occurred, such as the mass extinction at the close of the Mesozoic era?" or "How has the sequence of events in geologic time been established?" To answer these questions, two methods designed to establish relative dates or a sequence of events will now be considered.

LAW OF SUPERPOSITION

The *law of superposition* states that, when studying a series of rock layers or strata, older sediments, those first deposited, are found at the base of the series and the youngest rocks are at the top. The law of superposition is quite useful in establishing the relative ages of rocks in a series and the sequence of events that produced that series. The objective in relative dating or establishing relative ages is simply to determine which rocks in an outcrop (exposure of rock) are oldest and which are youngest. Figure 6.4 provides an example of how the law of superposition is applied.

PRINCIPLE OF ORIGINAL HORIZONTALITY

The *principle of original horizontality* states that all sediments are initially deposited horizontally. When this principle is applied in conjunction with the law of super-position, the relative ages of the layers in a rock outcrop and the geologic series of events that occurred may be established. In Figure 6.5, for example, the lowest three rock layers were deposited first. Then an "event" occurred that forced the existing layers to tilt, and the last event was the deposition of the top three rock layers. Using the

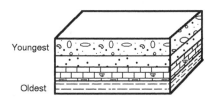

FIGURE 6.4 Law of superposition. Series of rock layers with the oldest at the base and the youngest at the top. Source: *Let's Review: Earth Science—The Physical Setting*, 2nd Ed., Edward J. Denecke, Jr., Barron's Educational Series, Inc., 2002.

principle of original horizontality, it is clear that at some time between the deposition of the lowest three rock layers, which, according to the law of superposition, are the oldest, and the deposition of the top three layers the lower layers were tilted. Notice that the top three layers were not exposed to the same forces as the lowest three layers.

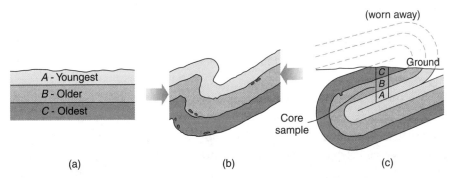

FIGURE 6.5 Principle of original horizontality. The original position of rock layers must be determined before applying the law of superposition. Originally horizontal rock layers (a) may be folded by compression (b), causing older rock layers to appear above younger rock layers (c). Source: *Macmillan Earth Science*, Eric Danielson and Edward J. Denecke Jr., Macmillan Publishing Co., 1989.

There may have been a period of erosion before the deposition of the top three layers. This would create an unconformity between the earlier and more recent layers. An *unconformity* is essentially a break in the rock record. Millions of years may have passed between the event that caused the tilting of the older rocks and the deposition of the more recent layers.

CORRELATION OF ROCK LAYERS

Correlation is the act of matching rocks of similar age in different regions. Correlation is best done when the rocks are not covered by soil and vegetation. Correlating rock layers can be as simple a task as identifying a layer that is unique in some way, perhaps in color or mineral composition, and then searching for it in two different areas.

Through the use of *index fossils* (see Figure 6.6), which are the remains of animals known to have lived and died within a particular period of time in Earth's history, geologists can correlate rock layers from different regions. For example, correlations using index fossils, can be made between Zion National Park in Utah and Grand Canyon National Park in Arizona.

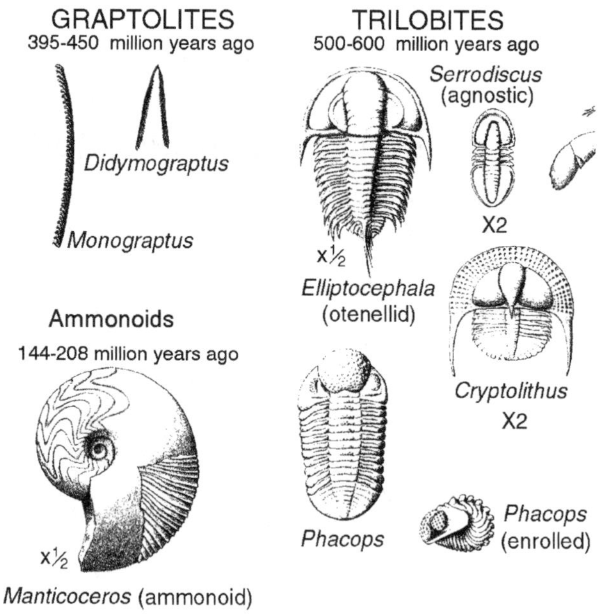

Index fossils are quite useful also in deciphering rock records. Organisms classified as index fossils are known to be found only within specific rock layers. To become an *index fossil*, an organism must have been easily recognizable and abundant, have been found over a wide geographical area, and have existed for only a brief period in order to meet the criterion that its

GRAPTOLITES
395-450 million years ago

Didymograptus

Monograptus

Ammonoids
144-208 million years ago

x½

Manticoceros (ammonoid)

TRILOBITES
500-600 million years ago

Serrodiscus (agnostic)

X2

x½

Elliptocephala (otenellid)

Cryptolithus X2

Phacops

Phacops (enrolled)

FIGURE 6.6 Various index fossils labeled with ages of existence. Source: The New York State Museum. Educational Leaflet #28, *Geology of New York, A Simplified Account*, Second Edition. Printed with permission of New York State Museum, Albany, N.Y.

remains are found only within a single rock layer. Very successful organisms, those that have survived as species for hundreds of millions of years, would not make very good index fossils, as they will be found in many rock layers of differing ages.

Correlation enables geologists to create a more comprehensive picture of the geologic history of a region. Figure 6.7 indicates how correlation of fossil strata from different locations can help in constructing a geologic history.

CROSS CUTTING

Cross cuts are created in a series of rock layers when a fault causes a linear break in the series, or when an intrusion of igneous material causes a dike to cut through the preexisting layers. The relative ages of rock layers can be established as the fault or dike can impact only layers in existence at the time of the event. Any rock

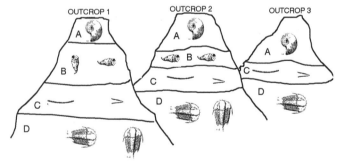

FIGURE 6.7 Correlation of fossils using rock strata (layers of rock) in several locations. Source: *Let's Review: Earth Science—The Physical Setting,* 2nd Ed., Edward J. Denecke, Jr., Barron's Educational Series, Inc., 2002.

layers deposited after the formation of the fault or the igneous intrusion will not show any effect.

Study Figure 6.8, which shows the effect of an igneous intrusion, and read the caption below the diagram. Note the intrusion of magma (layer 0), which later cools to become granite. The magma cross-cuts older, preexisting layers 3, 4, and 5. Layers 2 and 1 overlie the intrusion and are warped by the event. Contact metamorphism also occurs along all surfaces that come into contact with the intruding magma. Where the magma cross-cuts layers 3, 4, and 5, a dike has formed. Where layers 2 and 1 have warped, the pool of magma has formed a laccolith. Often all these features are revealed, millions of years after they formed, as erosion removes surface materials or as humans cut through mountains to build highways and other structures.

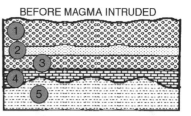

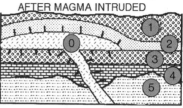

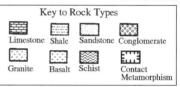

FIGURE 6.8 Five rock layers before and after the intrusion of magma. Layer 0, which formed from the cooling magma, is younger than the five (preexisting) rock layers into which it intruded. Source: *Let's Review: Earth Science—The Physical Setting*, 2nd Ed., Edward J. Denecke, Jr., Barron's Educational Series, Inc., 2002.

UNCONFORMITIES

Periods of deposition, resulting in the formation of rock layers or strata, are often followed by periods of erosion. These erosional events, or just periods of time when no deposition occurs, create unconformities, or breaks in the continuous or otherwise conformable rock record. No place on Earth has a complete rock record from all times in the past.

As defined earlier, an unconformity is a break in the geologic history of a rock series. Younger rocks do not always overlie older rocks in a continuous sequence of events. A break can occur when, after a period of deposition, a site experiences some kind of regional uplift, which, after a period of weathering and erosion, is followed by additional deposition. This deposition can occur several periods later in the overall geologic history of Earth. Figure 6.9 illustrates examples of unconformities.

FOSSILS

Fossils provide important clues in sedimentary rock. As mentioned above, index fossils help to correlate rock layers of similar age. Since different organisms existed at different times, the discovery of similar fossils in geographically separated rock layers can help to establish a relationship between those layers. This approach has limitations, however, as very old rocks or rocks of igneous or metamorphic origin will not contain fossils.

A *fossil* is defined as any evidence of earlier life preserved within a layer of rock, almost always sedimentary rock. Paleontologists study ancient (prehistoric) life and environments. Examples of fossils include tracks or animal footprints (see discussion of molds and casts below) preserved in what was soft sediment at the time; burrows or tubes made in sediment; coprolites or fossilized dung; or original or replaced remains, including bones and shells.

If an animal dies in an environment where it is covered by sediment (perhaps tar) before decaying, and if that region is left undisturbed as lithification (rock formation) occurs, the animal's remains may be found in their original form. For example, insects trapped in resin have been found as *original remains*.

More commonly, however, minerals replace the original bone, or hard parts, of an animal as decay occurs, resulting in the formation of *replaced remains*. The fossilized remains are an exact copy, but are not the original ones. Petrified wood is a good example of replaced remains.

Molds and casts can form when an organism is buried in mud and its hard body parts fossilize as the sediments lithify. If the fossil later dissolves, the hollow depression matching the form of the remains is called a *mold*. If new minerals fill the hollow region, a *cast* is said to have formed.

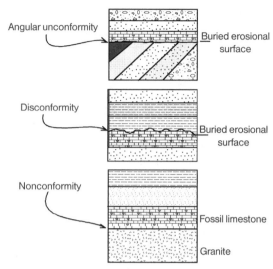

FIGURE 6.9 Illustration of an angular unconformity (deposition, followed by tilting of preexisting rock layers, then erosion, and a new period of deposition), a disconformity (a break in time between one period of deposition and another, with erosion typically occurring between the two), and a nonconformity (younger sedimentary rocks overlay older metamorphic or igneous rocks). Source: *Let's Review: Earth Science—The Physical Setting*, 2nd Ed., Edward J. Denecke, Jr., Barron's Educational Series, Inc., 2002.

Fossils also provide clues about the environments in which the organisms lived. Figure 6.10 illustrates some fossils and the environments that may have supported those life-forms.

Absolute Dating of Rocks

RADIOACTIVE DECAY

Relative dating is a good technique if one's objective is to organize into chronological order a sequence of geologic events as recorded in the rock record; however, if one wishes to establish how long ago a geologic event actually occurred or a biological species existed, absolute dates must be established. The ability to systematically and scientifically determine reasonably accurate dates is a relatively new skill practiced by geologists and other scientists.

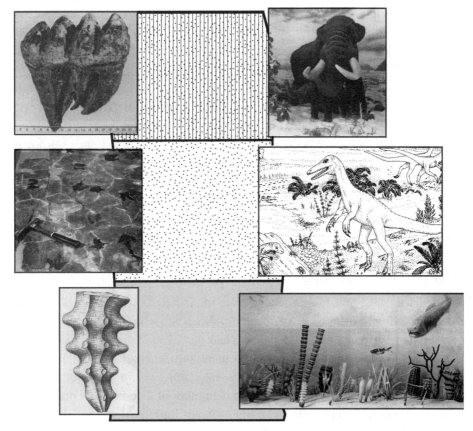

FIGURE 6.10 Fossil-bearing rock layers with illustrations of environments inferred from fossil evidence. Source: *Let's Review: Earth Science—The Physical Setting*, 2nd Ed., Edward J. Denecke, Jr., Barron's Educational Series, Inc., 2002.

This technique, called *radiometric dating*, involves the use of radioactive isotopes that help to determine the ages of rocks and other samples. By establishing the age of a rock that contains the fossilized remains of an organism, the age of the organism can be inferred. All the dates for the events recorded on the geologic time scale were determined by taking advantage of the fact that certain isotopes radioactively decay over time.

BASICS OF NUCLEAR CHEMISTRY

Nuclear chemistry involves the study of atomic nuclei. An atom contains protons and neutrons within its nucleus, as well as electrons outside the nucleus. (Electrons were of little significance in constructing the geologic time scale.)

From the time that Ernest Rutherford, a British physicist, first discovered that atoms have nuclei, scientists have refined their understanding of the atom through continuing research. The key to understanding the geologic time scale is radiometric dating, which, as stated above, is based upon the use of various isotopes.

The study of isotopes is less than a century old. The discovery of the neutron in the 1930s led to the realization that elements, while having a fixed number of protons in their nuclei, exhibit some variability in the number of neutrons. For example, every isotope of uranium has 92 protons in the nucleus; in fact, the number of protons, or the atomic number, determines the type of element. The difference between the various uranium isotopes lies in the number of neutrons in their nuclei. Considering just two of uranium's many isotopes, we find that the nucleus of one contains 143 neutrons, whereas the nucleus of the other has 146. Every isotope is identified by a mass number, which is the sum of its neutrons and protons. Thus, the first uranium isotope considered has 92 protons and 143 neutrons, for a mass number of 235. This isotope is referred to as uranium-235, sometimes written as ^{235}U. The other isotope, with 146 neutrons, has a mass number of 238 and is referred to as uranium-238.

RADIOMETRIC DATING

The importance of the preceding section lies in the fact that many isotopes, including both uranium isotopes discussed, are *unstable*. When an isotope is unstable, it will decay over time. To *decay* means to emit nuclear particles; in doing so, the isotope, the *parent element*, *transmutates* into another element, the *daughter element*. For example, over time, uranium-238 emits, at a predictable rate, *alpha particles*, which resemble helium nuclei. A helium nucleus has an atomic number of 2 and a mass number of 4. As uranium-238 emits alpha particles, it transmutates into thorium-234. The transmutation equation representing this process is as follows:

$$^{238}_{92}U \xrightarrow{\ \alpha\ } {}^{234}_{90}Th + {}^{4}_{2}He$$

Note that the alpha decay process converts uranium-238 into thorium-234 by releasing an alpha particle ($^{4}_{2}He^{+2}$).

Take note also that the process of transmutation illustrates both conservation of mass, as the mass of the reactant (238) is the same as the total mass of the products, and conservation of nuclear charge, as the nuclear charge, initially 92, totals 92 for the two products. Using these two principles, conservation of mass and conservation of charge, geologists can often predict the daughter element, or the element produced during the process of transmutation, if the parent element and the type of nuclear emission are known.

The key to the entire process is the realization that the amount of a radioactive isotope found in a particular sample will decrease over time. Specifically, in the case of uranium-238, the amount of this isotope present at the formation of a particular rock will decrease over time as the uranium-238 transmutes into thorium-234. It should be noted that thorium-234 is only one step in the process whereby uranium-238 decays into a stable isotope, that is, one that will no longer decay. Ultimately, uranium-238 will decay into lead-206. Thus, in a particular sample, the amount of uranium-238 is decreasing over time and the amount of lead-206 is increasing. In this example, lead-206 is called the *stable daughter* element and, once produced, remains as lead-206.

Thus, geologists will take a sample of rock and determine the ratio of lead-206 to uranium-238 in the rock. On the basis of this ratio and the known rate of decay of uranium-238, the age of the sample can be determined.

The rate of decay is measured by using the half-life of an isotope. The *half-life* is the time required for one-half of a radioactive isotope to decay. Let us consider two examples. The first is the radioactive isotope carbon-14, which decays over time into nitrogen-14. The half-life for this process is about 5570 years, far too brief a time for any significant amount of carbon-14 to remain in rock or mineral samples that are millions of years old. In fact, so little carbon-14 remains in a sample after 50,000 years or so that this isotope is useless for dating anything truly ancient. For this reason, carbon-14 is not discussed in the context of dating rocks. Carbon-14, is of great value, however, in dating archaeological artifacts that are hundreds or thousands of years old. This isotope has also proved useful in establishing the end of the last ice age, some 10,000 years ago.

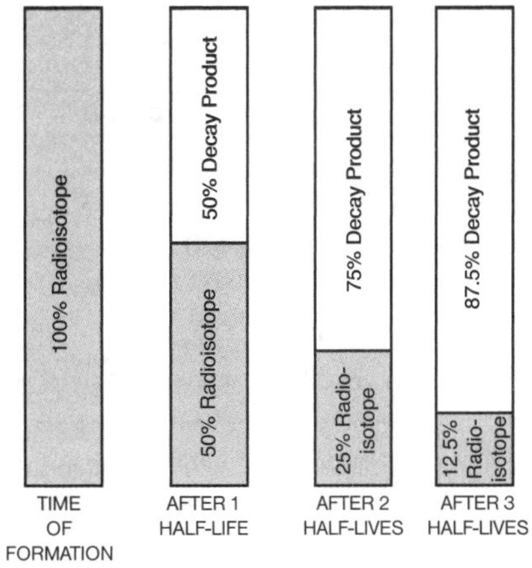

As a second example, let us consider uranium-235 with a half-life of 700 million years. We will assume that a rock contained some uranium-235 when it formed 700 million years ago, and that the rock has remained undisturbed ever since. One-half of the original uranium-235 will remain in the rock, as compared to the amount present when it formed, since one half-life has been completed. Figure 6.11 shows how the relative amounts of an isotope and its decay product change over time. As noted earlier, the original matter is not lost; rather, it is converted into other particles and energy, ultimately resulting in this example, in the production of lead-207.

It is interesting that lead, with atomic number 82, is the heaviest element to have any stable isotopes. All heavier elements are unstable, and many of the lighter elements, although having stable isotopes, also have unstable ones (see Table 6.1).

FIGURE 6.11 Ratios of isotope and decay product after one, two, and three half-lives. Source: *Let's Review: Earth Science—The Physical Setting*, 2nd Ed., Edward J. Denecke, Jr., Barron's Educational Series, Inc., 2002.

The Age of Earth

Prevailing views as to Earth's age have changed significantly since the seventeenth century, when Archbishop James Ussher proclaimed that Earth formed on October 26, 4004 B.C. This proclamation was based upon an interpretation, widely accepted at that time, of genealogy as described in the Bible. Modern methods for establishing Earth's age involve the use of radioactive isotopes.

TABLE 6.1
RADIOACTIVE DECAY DATA

Radioactive Isotope	Disintegration	Half-Life (years)
Carbon-14	$C^{14} \rightarrow N^{14}$	5.7×10^3
Potassium-40	$K^{40} \nearrow Ar^{40}$ $\searrow Ca^{40}$	1.3×10^9
Uranium-238	$U^{238} \rightarrow Pb^{206}$	4.5×10^9
Rubidium-87	$Rb^{87} \rightarrow Sr^{87}$	4.9×10^{10}

Source: The State Education Department, *Earth Science Reference Tables*, 2001 ed. (Albany, New York: The University of the State of New York).

Geologists have identified several isotopes with very long half-lives. These isotopes, including uranium-238, uranium-235, thorium-232, rubidium-87, and potassium-40, are the ones of choice for dating something as old as Earth. Table 6.2 gives the half-life of each of these isotopes and indicates its usefulness in radioactive dating.

The method, as described earlier, involves finding a sample of rock that has presumably been undisturbed for hundreds of millions of years. Such samples are generally located in continental cratons. *Cratons* are regions that have been tectonically stable for billions of years. In North America, the continental shield, stretching from the northern Great Plains to northern Canada, is a craton.

Meteoroids are thought to be among the original materials of the solar system. As meteors orbit the Sun, these bodies have remained essentially undisturbed for billions of years. On occasion, a meteorite impacts Earth's surface and is subsequently located. The meteorite can then be tested to determine its age.

TABLE 6.2
SIX RADIOISOTOPES USED IN DATING

Isotopes		Half-life (years)	Dating Range (years)	Minerals or Other Materials That Can Be Dated
Parent	Decay Product			
Uranium-238	Lead-206	4.5 billion	10 million–4.6 billion	Zircon
Uranium-235	Lead-207	710 billion	10 million–4.6 billion	Uraninite
Uranium-232	Lead-208	14 billion	10 million–4.6 billion	
Potassium-40	Argon-40 Calcium-40	1.3 billion	50,000–4.6 billion	Muscovite, biotite, hornblende
Rubidium-87	Strontium-87	47 billion	10 million–4.6 billion	Muscovite, biotite, potassium feldspar
Carbon-14	Nitrogen-14	$5,730 \pm 30$	100–70,000	Wood, peat, grain, charcoal, bone, tissue, cloth, shell, stalacites, glacier ice, ocean water

Source: *The Dynamic Earth: An Introduction to Physical Geology*, Brian J. Skinner and Stephen C. Porter, copyright © 1992 by John Wiley & Sons, Inc. Reprinted by permission of John Wiley & Sons, Inc.

The results of this test are significant, as modern theories regarding the solar system's formation suggest that all objects within our solar system formed at about the same time. The test allows the researcher to determine a ratio of the stable daughter isotope to the unstable parent isotope. Using the results of this test, the researcher can work backward in time to establish the date when the sample did not contain any of the stable daughter isotope. The youngest, or most recent, of the dates determined using each of the above radioactive parents is the oldest the Earth can be. An additional correction factor is built in because lead-204, which is stable and has no radioactive parent, yields clues as to how much lead was in primordial Earth. As determined by this method, the oldest known rocks are on the order of 4.28 billion years old. Scientists conclude that Earth and all other objects in the solar system are approximately 4.5 billion years old since Earth must have taken 1 billion years or so to cool enough to produce a stable solid surface.

Confirming evidence of the age of the solar system has been acquired from study of meteorites that have been recovered after impacting the planet. Through radiometric dating, samples of various meteorites are estimated to be approximately 4.5 billion years old. It is believed that these objects changed little after forming until they impacted Earth. Thus, when a geologist studies a meteorite, he or she is looking at material that formed when the entire solar system formed and has not changed since.

Additional supporting evidence for an age of this magnitude is based upon the apparent changes that have transpired on Earth over time. Consider, for example, a feature such as the Grand Canyon, where many layers of rock have been built over time and a river has carved the canyon through preexisting rock; many hundreds of millions of years were required for this process to produce what we marvel at today. Furthermore, it is likely that Earth's early atmosphere was nothing like what it is now. Literally billions of years were required for Earth's atmosphere to evolve from its primordial to its present composition.

Related Internet Resources to Great Images and Information

http://emuseum.mnsu.edu/archaeology/dating/

"Radioactive Dating Techniques" contains lots of information on the various methods used to determine the relative and absolute ages of strata. There is a link to an interactive Java-based site from this page.

http://pubs.usgs.gov/gip/geotime/

This contribution from the United States Geological Survey (USGS) is titled "Geologic Time." Topics include geologic time, relative time, and the age of Earth.

http://pubs.usgs.gov/gip/fossils/contents.html

This USGS Web site has the name "Fossils, Rocks, and Geologic Time."

www.nature.nps.gov/geology/tour/

"Tour of Park Geology" is a National Park Service Web site. Links are provided to a number of national parks and other sites that have geological themes or abundant fossils to explore.

REVIEW EXERCISES FOR CHAPTER 6

WORD-STUDY CONNECTION

absolute dating	extinct	photosynthetic
algae	extinction	Piedmont Physiographic Province
amphibians	fossil	Precambrian
Appalachians	fossil record	principle of original horizontality
archaeologist	geologic time scale	radioactive decay
Archaeopteryx	geology	radiometric dating
Cambrian	half-life	relative dating
Cenozoic	highlands	stable
conservation of mass	iridium	strata
coprolites	isotopes	stromatolites
correlation	Jurassic	thorium
craton	K-T boundary	transmutation
Cretaceous	law of superposition	Triassic
cross cutting	lead	trilobites
daughter element	Mesozoic	*Tyrannosaurus rex*
dinosaur	nuclear chemistry	unconformity
eon	Ordovician	uniformitarianism
epoch	Paleozoic	unstable
era	Pangaea	uranium
erosion	parent element	weathering
erosional agents	period	

SELF-TEST CONNECTION

PART A. Completion. *Write in the word or words that correctly complete the statement.*

1. The process of determining the actual ages of rock strata is called

    _____.

2. The _____ are an ancient mountain range that formed in the east-
    ern United States when Pangaea formed over 250 million years ago.

3. _____ is the heaviest element known with any stable isotopes.

4. _____ are ancient dinosaur dung.

5. Humans appeared during the _____ era.

6. The K-T boundary is identified by a layer with traces of the element

    _____.

7. Very little is known about the _____ Time.

8. The process of converting one element into another is called _____.

9. The largest of the dinosaurs was _____.

10. When a _____ decays, a daughter element is produced.

11. All birds may have descended from a flying dinosaur called _____.

12. When geologists study a series of rock layers, they apply the principle of original
    horizontality and then the _____ to determine the relative age of
    each layer.

13. The rate at which a radioactive isotope decays is called its _____.

14. Atoms that have the same number of protons in their nuclei but differ in the num-
    ber of neutrons are called _____.

15. A _____ isotope does not decay over time.

PART B. Multiple Choice. Circle the letter of the item that correctly completes the statement or answers the question.

1. The era known as the "age of dinosaurs" is the
   (a) Precambrian
   (b) Paleozoic
   (c) Mesozoic
   (d) Cenozoic
   (e) Jurassic

2. The era that is best referred to as the "age of recent life" is the
   (a) Precambrian
   (b) Paleozoic
   (c) Mesozoic
   (d) Cenozoic
   (e) Jurassic

3. Which represents the greatest span of time?
   (a) an eon
   (b) an epoch
   (c) an era
   (d) a period
   (e) a group

4. A break in a rock record is called
   (a) a disconformity
   (b) an unconformity
   (c) a cross cut
   (d) a coprolite
   (e) a craton

5. A portion of the continent that has been geologically stable for long periods of time is called
   (a) a disconformity
   (b) an unconformity
   (c) a cross cut
   (d) a coprolite
   (e) a craton

6. When a radioactive isotope decays, a _____ reaction occurs that produces a _____ element.
   (a) chemical, parent
   (b) chemical, daughter
   (c) transmutation, parent
   (d) transmutation, daughter
   (e) none of the above

7. Uranium, a radioactive element, eventually decays into the stable isotope
   (a) thorium
   (b) plutonium
   (c) lead
   (d) gold
   (e) hydrogen

8. The first birds and flowering plants appeared during the
   (a) Precambrian era
   (b) Cambrian period
   (c) Paleozoic era
   (d) Mesozoic era
   (e) Cenozoic era

9. Organisms that no longer exist are said to be
   (a) endangered
   (b) protected
   (c) extinct
   (d) hiding
   (e) both a and d

10. Stromatolites are ancient forms of
    (a) bacteria
    (b) viruses
    (c) plants
    (d) animals
    (e) algae

PART C. Modified True/False. *If a statement is true, write "true" for your answer. If a statement is incorrect, change the* <u>*underlined*</u> *expression to one that will make the statement true.*

1. Carbon-14 and carbon-12 are <u>isotopes</u> since they share the same nuclear charge (number of protons) but differ in the number of neutrons.

2. The <u>Paleozoic</u> represents the longest time period of any era.

3. Radiometric dating is a technique that is used to determine <u>relative age</u>.

4. A layer of <u>iridium</u> marks the time when a great meteoroid probably collided with Earth, resulting in the destruction of many species, including the dinosaurs.

5. An isotope that decays over time or is radioactive is said to be <u>stable</u>.

6. Eras are divided into <u>eons</u>.

7. The period of time required for one-half of a radioactive isotope to break down is called the <u>half-life</u>.

8. The geologic time scale is broken down into Precambrian time and three main eras: the Paleozoic, the Mesozoic, and the <u>Triassic</u>.

9. *Archaeopteryx* was a flying dinosaur.

10. <u>Uranium</u> is the heaviest element with any stable isotopes.

CONNECTING TO CONCEPTS

1. What major events mark the close of each era and the dawn of a new era?

2. If the dinosaurs had not become extinct, would we still be in the Mesozoic era?

3. Is it possible to bring a species back from extinction? Do some research and learn what, if anything, has been done in this area in recent years.

4. If a column of rock layers is found, which layer will be the oldest? What law or principle are you applying in answering this question?

5. Isotope *A* has a half-life of 100 million years. If there were 100 grams of isotope *A* 200 million years ago, what mass remains of the original isotope?

6. In the diagrams below, columns *A*, *B*, and *C* are layers of rock obtained from locations about 5 kilometers apart and have not been overturned.

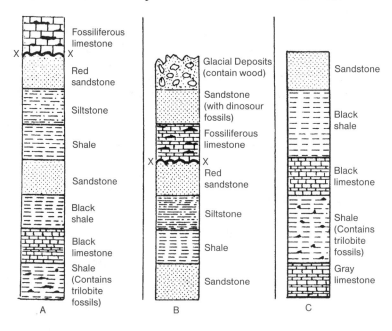

a. Determine the oldest rock layer in the series.
b. Which rock layer was deposited most recently?

7. Base your answers to the following question on the diagram (right) and your knowledge of Earth science. The diagram represents a profile view of a rock out-crop. The layers are labeled *A* through *H*.

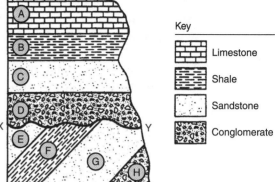

Key
Limestone
Shale
Sandstone
Conglomerate

a. Briefly describe the geologic process that resulted in the boundary represented by line *X-Y*.

b. Assume that none of the layers has been overturned. Layer *D* is 505 million years old, and layer *B* is 438 million years old. State the geologic period during which layer *C* could have formed.

ANSWERS
SELF-TEST CONNECTION

Part A

1. absolute dating
2. Appalachians
3. Lead
4. Coprolites
5. Cenozoic
6. iridium
7. Precambrian
8. transmutation
9. *Tyrannosaurus rex*
10. parent element
11. *Archaeopteryx*
12. law of superposition
13. half-life
14. isotopes
15. stable

Part B

1. **(c)**
2. **(d)**
3. **(a)**
4. **(b)**
5. **(e)**
6. **(d)**
7. **(c)**
8. **(d)**
9. **(c)**
10. **(e)**

Part C

1. True
2. False; Precambrian
3. False; absolute age
4. True
5. False; unstable
6. False; periods
7. True
8. False; Cenozoic
9. True
10. False; Lead

CONNECTING TO CONCEPTS

1. Each era generally ends with a mass extinction and a new life-form becoming dominant. For example, the Mesozoic era ended with the extinction of the dinosaurs (reptiles were dominant in the Mesozoic), and in the Cenozoic era mammals and birds began to thrive.

2. By definition, if dinosaurs and reptiles still dominated all other species, then yes, we would still be in the Mesozoic era.

3. No—and maybe. Once a species is extinct, it's extinct, or is it? Because of recent work with cloning and DNA, some scientists are hopeful of "bringing back" an extinct species by using remnants of its DNA. Most scientists, however, believe that the DNA of a member of an extinct species is too incomplete to serve this purpose.

4. The law of superposition states that the lowest layer is the oldest layer.

5. Isotope A has a half-life of 100 million years. Therefore, if the sample is 200 million years old, it has passed through two half-lives, and one-half of one-half (or one-quarter) of the original sample remains. Since there was originally 100 grams, 25 grams of the original radioactive "parent" remains after 200 million years.

6. a. Gray limestone (look at the base of column C) is the oldest. The oldest deposit should logically be at the base of any series of rock layers.
    b. Glacial deposits were deposited most recently. Clues include the presence of wood, the fact that glaciers deposited the materials (glaciers are relatively recent in geologic history), and the fact that the material is not solid rock, but sediment.

7. a. Line X-Y is most likely an unconformity. It resulted from a period of erosion followed by new deposits, beginning with layer D.
    b. Layer C could have formed anytime from the late Cambrian through the entire Ordovician to the early Silurian period.

Plate Tectonics– Our Dynamic Planet

7

WHAT YOU WILL LEARN

This chapter focuses on the dynamic nature of our planet and the study of plate tectonics. In this chapter you will learn

- early evidence that led to the continental drift hypothesis;
- the types of motions that can occur along plate boundaries;
- how plate tectonic theory explains the existence of the Hawaiian Islands;
- how earthquakes occur and how we measure their strength.

SECTIONS IN THIS CHAPTER

- Continental Drift
- Convection Cells and the Mantle
- Plate Boundaries and Tectonic Activity
- Plate Boundaries and Earthquakes
- Special Topics including Predicting Earthquakes, Tsunamis, and the Geology of Our National Parks
- Internet Resources to Study Current Earthquakes and Related Information
- Review Exercises for Chapter 7

Continental Drift

EARLY EVIDENCE PRESENTED

In a book published in 1912, Alfred Wegener, a German meteorologist and geophysicist, presented the *continental drift hypothesis*. Wegener noted the apparent jigsaw-puzzle fit of the Americas, Africa, and Europe and argued that, at some time in the past, perhaps 200 million years ago, the continents were all connected as one supercontinent, which he dubbed *Pangaea*, meaning "all land." Although Wegener was not the first to observe the shapes of the continents and their seeming to fit together, he was among the first to suggest that the continents actually drifted across the surface of our planet.

Wegener set out to collect additional evidence, including the correlation of fossils among continents. He learned that fossils of *Mesosaurus*, a genus of small, aquatic reptiles, had been found in regions of Africa and South America. These regions, thousands of kilometers apart today, would have been adjacent to each other when Pangaea existed (see Figure 7.1). Further support for the existence of Pangaea was the discovery of glacial deposits in southern South America and southern Africa. This finding suggested not only that these areas were once in a geographically similar location, but also that they were centered much closer to the South Pole than they are today, supporting the notion that the continents have drifted across the globe (see Figure 7.2). Mountain ranges also support the theory of continental drift. The Appalachians appear to match quite nicely, in many respects, the Caledonian Mountains of northern Europe (see Figure 7.3).

Despite the evidence presented, Wegener's ideas were not met with widespread acceptance; in fact, he was ridiculed by many of his contemporaries. Wegener faced two key problems: First, he was unable to propose a mechanism by which continents could drift, and second, he was said to be a rather dull speaker, unable to captivate and motivate his

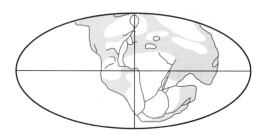

FIGURE 7.1 Pangaea as proposed by Wegener. Source: **Earth Science on File** by the Diagram Group. Copyright © 1988 by The Diagram Group. Reprinted by permission of Facts on File, Inc.

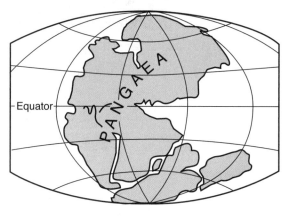

PERMIAN
225 million years ago

FIGURE 7.2 View of continents and their proximity to South Pole and the Equator during the Triassic Period (225 mya). Source: U.S. Geological Survey (USGS).

audience. Had he been a more persuasive speaker, his ideas might have gained acceptance decades before they finally did take hold and become the dominant view.

RECENT EVIDENCE PRESENTED

The International Geophysical Year of 1957–58 initiated a massive project to map the seafloor. Sonar and other instruments helped to identify structures, including the Mid-Atlantic Ridge, previously hidden in the depths of the oceans. Around the same time, an extensive international network of seismographs was established and much knowledge was gained about events in Earth's interior. A *seismograph* (see Figure 7.4) is an instrument designed to create a record, called a *seismogram* (see Figure 7.5), of the various kinds of movements caused by an earthquake. The data thus acquired helped geologists to formulate a modern theory

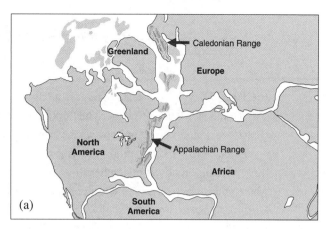

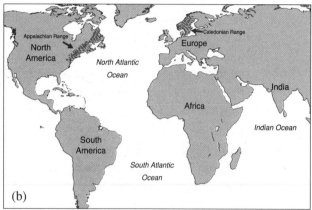

FIGURE 7.3 The Appalachian and Caledonian Mountains. (a) Appalachian and Caledonian ranges as they appeared shortly after their formation. (b) Appalachian and Caledonian ranges as they appear today.

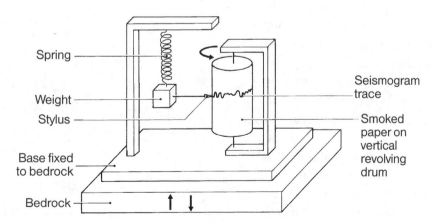

FIGURE 7.4 A simple seismograph. Source: **Earth Science on File** by the Diagram Group. Copyright © 1988 by The Diagram Group. Reprinted by permission of Facts on File, Inc.

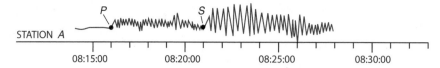

FIGURE 7.5 A typical seismogram. Note the arrival of the *P*-waves at 8:16:00; the *S*-waves arrive 5 minutes later at 8:21:00. The farther the seismograph station is from the epicenter, the greater the difference will be between the arrival of the *P*-waves and the slower moving *S*-waves. A method called "triangulation," described at greater length later in this chapter, uses data from three different seismograph stations to determine the actual epicenter. A good Web site where you can actually work through the actual process of identifying an epicenter is *http://vcourseware.sonoma.edu/*.

known as *plate tectonics* that accepts the notion of continental drift and provides a reasonable mechanism for its occurrence.

Additional support for the plate tectonics theory has come from studies of rocks that form the ocean floor. One of the most important observations is the youthful nature of these rocks (Figure 7.6a). The oldest rocks found along the ocean floor are about 200 million years old. According to the plate tectonics theory, new rock is formed along zones called *divergent plate boundaries* (Figure 7.6b). Mid-ocean ridges, formed largely by the new material flowing from deep within the crust and the mantle, are located along these divergent boundaries. As these ridges form along divergent boundaries, older material is transported away toward *convergent plate boundaries*, where it subducts (returns to deep within the crust) and is "recycled." The slow but continual motion described here ensures that igneous rock located in the seafloor will not exist in its present form for more than a few hundred million years.

When new rock forms along the divergent boundaries, Earth's magnetic field causes certain minerals, rich in iron, that are still molten to align

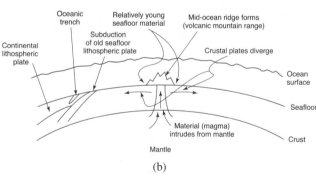

FIGURE 7.6 (a) Digital-age map of the ocean floor. The youngest rocks on the seafloor are located along the mid-ocean ridge. Source: NOAA/NGDC (National Geophysical Data Center). (b) Idealized ocean floor process.

with magnetic north. The planet's magnetic field has reversed itself several times in the past several million years, and the alignment of the magnetic minerals within the rock indicates these reversals. The evidence of the conveyor-belt-like mechanism described above is the mirror image of magnetic reversals found on either side of the mid-ocean ridges (see Figure 7.7).

Divergent and convergent plate boundaries are discussed more fully in the section headed "Plate Boundaries and Tectonic Activity."

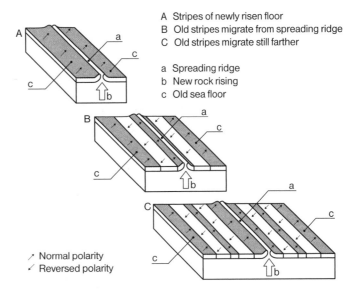

A Stripes of newly risen floor
B Old stripes migrate from spreading ridge
C Old stripes migrate still farther

a Spreading ridge
b New rock rising
c Old sea floor

↗ Normal polarity
↙ Reversed polarity

FIGURE 7.7 Paleomagnetism patterns on the ocean floor. Source: **Earth Science on File** by the Diagram Group. Copyright © 1988 by The Diagram Group. Reprinted by permission of Facts on File, Inc.

Convection Cells and the Mantle

RELATIONSHIP OF THE MANTLE TO THE CRUST

The plate tectonics theory presents a model of Earth's interior that has the uppermost mantle and the overlying crust making up a strong, rigid layer known as the *lithosphere*. This rigid layer is broken into several large segments called *tectonic plates*. The lithospheric plates ride atop the *asthenosphere*, a more plastic, or pliable, region of the mantle. Convection cells within the asthenosphere drive movements in the plates at the surface. Figure 7.8 illustrates this process. Figure 7.9 shows the relative positioning and thickness of each layer of Earth's interior.

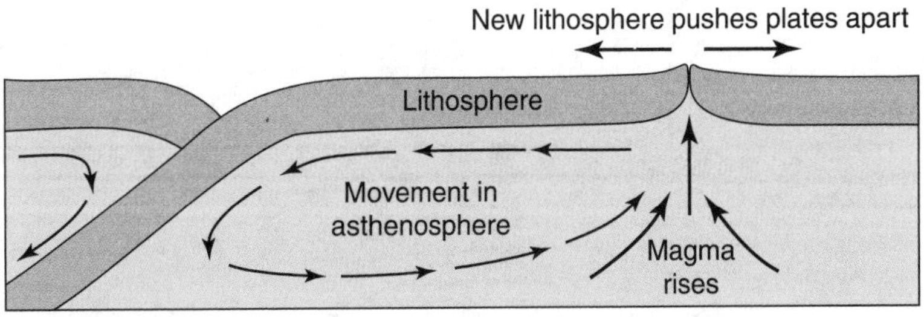

New lithosphere pushes plates apart
Lithosphere
Movement in asthenosphere
Magma rises

TECTONIC PLATES

As noted above, the tectonic plates are large pieces of the lithosphere

FIGURE 7.8 Ocean floor is destroyed at subduction zones as old crustal material plunges into the mantle. Trenches form along subduction zones and melting occurs at greater depths as friction between the plates creates tremendous heat and pressure. Source: *The Dynamic Earth*, Brian J. Skinner and Stephen C. Porter, John Wiley, 1989.

that ride on top of the asthenosphere. These large plates (see Figure 7.10) include the North American Plate, upon which most of the United States rests. Bordering the North American Plate to the west are the Pacific Plate and several other, smaller plates. Plate movements along the boundaries between plates can wreak havoc for residents of those regions.

Mount Saint Helens provides a perfect example of this kind of action. As subduction occurs to the west of this volcanic peak and to the Cascades in general, highly viscous magma is produced beneath the surface of the northwestern United States. On occasion, most recently in the early 1980s, Mount Saint Helens has erupted violently, sending ejecta composed of ash, lava, and volcanic "bombs" (rocks) all over the surrounding countryside. Damage and destruction from the latest eruption was extensive. More than 50 people died, though advance warnings caused most to evacuate.

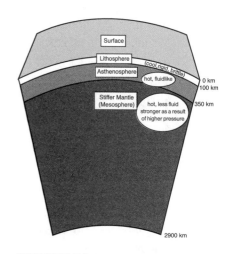

FIGURE 7.9 Layering beneath the surface. Note the relative thickness of the lithosphere, asthenosphere, and mantle. Source: *Let's Review: Earth Science—The Physical Setting*, 2nd Ed., Edward J. Denecke, Jr., Barron's Educational Series, Inc., 2002.

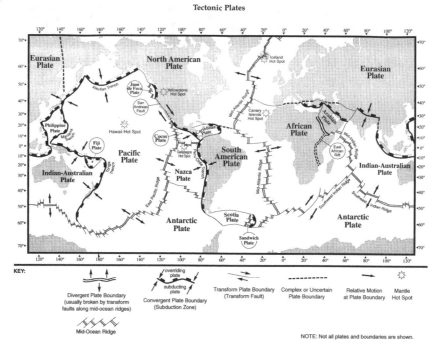

FIGURE 7.10 Diagram illustrating plate tectonics. Note the relative movement along each of the plate boundaries identified. All three major types of boundaries—convergent, divergent, and transform—are illustrated; refer to the key beneath the map. Source: *Let's Review: Earth Science—The Physical Setting*, 2nd Ed., Edward J. Denecke, Jr., Barron's Educational Series, Inc., 2002.

PANGAEA–THEN VERSUS NOW

Geologists have constructed a series of views of how the orientations and geographic locations of Earth's land-masses probably changed from the time when Pangaea existed 200 million years ago until today. Careful study of the images in Figure 7.11 reveals that North America once straddled the Equator and that South America was once near the South Pole.

> **REMEMBER**
> Tectonic activity explains many of today's observed surface features as well as seemingly unrelated colonies of ancient life-forms and glacial remnants in now tropical locales.

Plate Boundaries and Tectonic Activity

DIVERGENT BOUNDARIES

As stated above, *divergent boundaries* are zones where lithospheric plates are being separated over time as new material forces its way through cracks or rifts. A divergent boundary is, therefore, a site where new rock is formed. Since Earth's volume remains relatively constant, the new material produced at divergent boundaries must later be destroyed in other regions, known as convergent boundaries, which are addressed in the next section.

Divergent boundaries are typically located along the ocean floor. One of the most notable divergent boundaries is the Mid-Atlantic Ridge. As its name suggests, the Mid-Atlantic Ridge is actually a submerged mountain range oriented on a roughly north-south line in the middle of the Atlantic Ocean.

Plate tectonics theory suggests that Pangaea, the supercontinent, began to break apart as molten material forced its way to the surface along the North American–Eurasian plate boundary. This action ultimately resulted in the separation of North America and Africa and the formation of the Atlantic Ocean. Today the Atlantic Ocean completely covers this submerged chain of volcanic mountains, with the exception of a few islands that penetrate the ocean surface in the North Atlantic. These islands, which include Iceland, Surtsey (adjacent to Iceland), and the Azores, are the only part of the entire Mid-Atlantic Ridge that is above sea level.

The Mid-Atlantic Ridge, like most mid-ocean ridges, has a characteristic rift valley along the zone where the plates are spreading apart. The youngest bedrock along the seafloor can be found near the rift zone; the seafloor gets progressively older as one proceeds away from the rift zone and toward the coastline. One of the most interesting discoveries of recent geologic studies of the ocean was the extreme youth of the seafloor. The process occurring along the mid-ocean ridges helps to explain how new seafloor is produced and why the observed pattern occurs (see Figure 7.12).

CONVERGENT BOUNDARIES

Convergent boundaries are zones where lithospheric plates are converging or colliding. In these locations old lithospheric crust is recycled deep into Earth's crust and

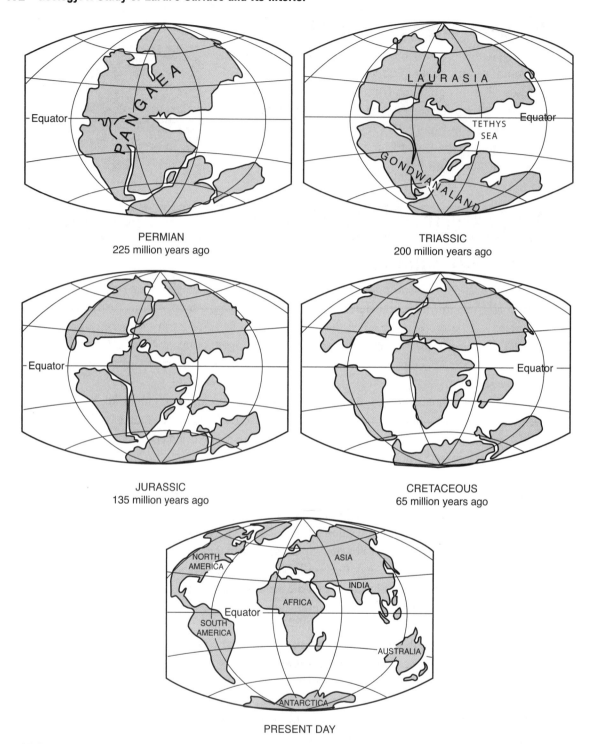

FIGURE 7.11 Change in continental position over time because of continental drift. Source: U.S. Geological Survey (USGS).

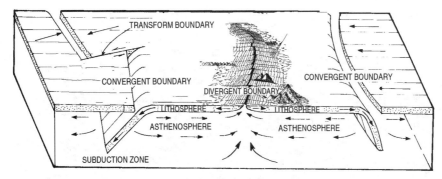

FIGURE 7.12 Convergent, divergent, and transform boundaries illustrated in 3D. Source: *Let's Review: Earth Science—The Physical Setting,* 2nd Ed., Edward J. Denecke, Jr., Barron's Educational Series, Inc., 2002.

mantle. The material subducted into the mantle thus begins a long, slow journey that may eventually take it through a complete cycle, at the end of which (after nearly a billion years) it is extruded back to the seafloor through a mid-ocean ridge along a divergent plate boundary.

Much research on convection within the mantle is ongoing. Several Web sites may be helpful in exploring this topic further. Here are two of them:

www.npaci.edu/successes/1999_mantle.html

"Mantle Convection in Three-Dimensions" by Paul Tackley, researcher at UCLA

www.lpi.usra.edu/science/kiefer/Research/convection.html

"Mantle Convection Research" by Walter Kiefer of the Lunar and Planetary Institute

A closer look at convergent plate boundaries reveals three types. One type occurs when a plate composed of oceanic crust collides with a plate composed of continental crust; a second, when an oceanic plate collides with another oceanic plate; and a third, when a continental plate collides with another continental plate.

In the first case, the collision of an oceanic plate with a continental plate, the oceanic plate *subducts* under the continental plate. This occurs because oceanic crust is composed of denser material than continental crust. In classic example of this kind of action, the Juan de Fuca Plate, located along the west coast of North America from northern California north to British Columbia, is subducting under the North American Plate. (Figure 7.13 is a general illustration of the process of subduction.) The result has been the formation of the Cascades Mountain Range. Individual peaks include Mounts Rainier, Hood, Bachelor, and, of course, Saint Helens.

Locate the Nazca Plate on Figure 7.10, and take note of its subduction zone, which spans virtually the entire western coastline of South America. The Nazca Plate is another site where subduction occurs; in fact, it impacts a much larger geographic region than does the Juan de Fuca Plate. As the Nazca Plate subducts under the South American Plate, deep earthquakes are observed a few hundred kilometers to the west

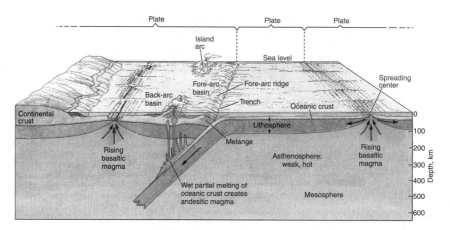

FIGURE 7.13 A cross section of the crust and upper mantle illustrating the structures created by plate movement. Source: *The Dynamic Earth*, 2nd Ed., Brian J. Skinner and Stephen C. Porter, John Wiley, 1992.

in parts of South America. The Andes, an impressive mountain range that forms the "backbone" of western South America, formed largely as a result of the subduction taking place along and near the Pacific coast of South America.

This action often produces a deep-sea trench at the site where the oceanic plate is subducting under the continental plate. As the oceanic plate subducts, often at a 45° angle under the continental plate (refer to Figure 7.13), friction causes the oceanic plate to heat up and eventually melt, forming a pool of magma several hundred kilometers beneath the continental crust. The magma then works its way back to the surface, forming a chain of volcanic mountains, typically located about 300 kilometers from the actual subduction zone. Figure 7.14 shows the relative positions of the volcanic mountain chain and the subduction zone. The best example of this in North America is the Cascade Mountain Range, which parallels the Juan de Fuca Plate. The mountains, as expected, are located approximately 300 kilometers to the east of the site where the Juan de Fuca Plate subducts under the North American Plate.

The deep-sea trench formed along the subduction zone produces the deepest regions in the ocean, with depths exceeding 10,000 meters (30,000 ft). The deepest ocean trench,

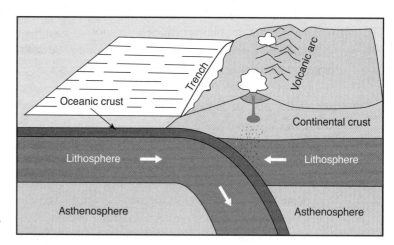

FIGURE 7.14 This diagram depicts a volcanic mountain chain and its proximity to subduction zones. Source: National Park Service.

the Marianas Trench, located just east of the Philippines in the Pacific Ocean, reaches a depth of nearly 11 kilometers (36,000 ft). An excellent Web site that describes the Marianas Trench in detail is *www.geocities.com/thesciencefiles/marianas/trench.html*.

In the second case, the collision of an oceanic plate with another oceanic plate, density differences in the plates' crusts, although minimal, are sufficient to produce subduction of the denser plate beneath the less dense plate. The resulting scenario is similar to the first case, described above, the most significant difference being the production of a volcanic island arc rather than a volcanic mountain chain on a continent. The Aleutian Islands in Alaska are an example of an island arc produced by this kind of action. Japan and the Philippines, which rank among the most populated island chains in the world, were also produced in this manner.

In the third case, convergence occurs between continental plates. The result is the welding of the continents and the formation of major mountain chains on land. Millions of years ago, Europe and Asia collided, forming the "Eurasian continent" and the Ural Mountains. Continental convergence can be observed today in the Himalayas. The Indian subcontinent is moving northward at a rate of 5 centimeters per year into China, which sits atop the Eurasian Plate. The plate India rests on has actually been partially forced under Asia. This has created an unusually high density in the continental crust of southern Asia. The end result is the anomalously high Tibetan Plateau, complete with a string of mountains that reach the highest altitudes on Earth.

North America has not been spared this kind of geologic activity. The Appalachians formed when Africa collided with North America around 250 million years ago. Today, the Appalachians are only a small fraction of their original height; time has extensively weathered and eroded these once-majestic mountains.

The middle Rocky Mountains, situated primarily in northern Utah and western Wyoming, are also folded mountains, resulting from the gradual movement of the North American Plate as it rotates northwestward and interacts with other plates to its west.

TRANSFORM FAULT BOUNDARIES

Transform fault boundaries are zones where two plates are ever so slowly grinding past each other. Crust is neither produced nor destroyed along these boundaries, in comparison to the action along divergent and convergent boundaries. Transform fault boundaries are part of a global conveyor-belt system that transports the rigid plates across the surface of our planet.

A famous, and classic, example of a transform fault boundary is the San Andreas Fault, which is particularly important to residents of California. The San Andreas Fault is a 1200-kilometer boundary, oriented roughly on a north-south line, that separates the Pacific Plate from the North American Plate (see Figure 7.15).

A small portion of the extreme western United States, including Los Angeles, is situated on the Pacific Plate. This portion of coastal southern California is slowly pivoting northwestward as the rest of California is pivoting southeastward. At the rate that these plates are moving, the Los Angeles Dodgers and San Francisco Giants baseball teams, both of which were once located in New York City, could again find themselves cross-town rivals in about 15 million years!

HOT SPOTS AND THE HAWAIIAN ISLANDS

Hot plumes deserve special mention because they have resulted in the formation of one of the most beautiful states in our nation—Hawaii. In fact, as you read this, the Hawaiian Islands are actually growing! Read on to discover why.

Most earthquake and volcanic activity occurs along or near plate boundaries.

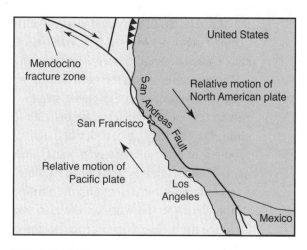

FIGURE 7.15 The San Andreas Fault. Source: National Park Service.

Casual inspection of global volcanic activity (see Figure 7.16) reveals this to be a pattern well adhered to by nature. There are, however, a few areas where volcanic activity seems to occur outside the usual pattern. The Hawaiian Islands are among these regions.

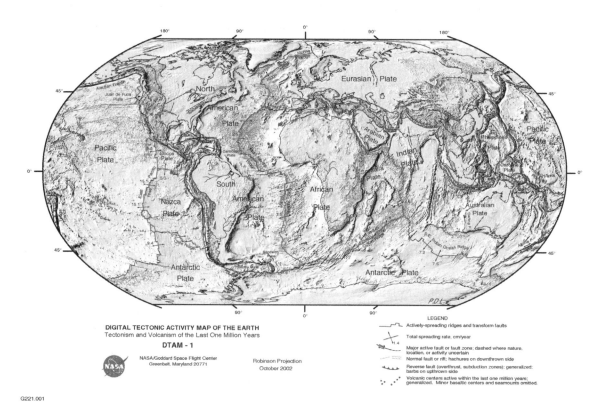

FIGURE 7.16 Global volcanic and earthquake activity. Source: NASA/Goddard Space Flight Center. Source: http://core2.gsfc.nasa.gov/dtam.

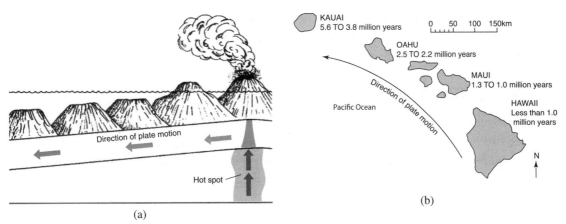

FIGURE 7.17 (a) Cross section showing a series of volcanoes formed by a plate moving over a hot spot. The extinct volcanoes are eroded to nearly flat surfaces at sea level and are submerged, forming flat-topped seamounts called *guyots*. Source: *Mountains of Fire*, Robert W. Decker and Barbara B. Decker, Cambridge University Press, 1991. Reprinted with the permission of Cambridge University Press. (b) The progressive increase in age of the Hawaiian Islands with distance away from currently active Hawaiian volcanoes. This supports the hot-spot hypothesis and indicates plate movement toward the northwest. Source: *Let's Review: Earth Science—The Physical Setting*, 2nd Ed., Edward J. Denecke, Jr., Barron's Educational Series, Inc., 2002.

Detailed analysis of the lithosphere and mantle reveals hot spots within the mantle that can break through and create volcanic islands in the middle of a plate. A hot spot itself is relatively stationary; however, the crustal plate above it moves over time. Studies of the seafloor have revealed a string of submerged volcanic mountains stretching from the northern Pacific, near Siberia, southward and then eastward toward the current location of the Hawaiian Islands (see Figure 7.17). Observations confirm the theory that the submerged volcanoes near Siberia are dormant and are the oldest of the chain, whereas Mauna Loa is still active and is nearly the youngest of the chain. Actually, a new Hawaiian island is being built to the east of the Big Island of Hawaii as the Pacific Plate continues its slow drift westward. A new section of the plate is now over the hot spot, and therefore the volcanic activity has largely shifted eastward from the Big Island.

Volcanoes are discussed in greater detail in Chapter 8.

Plate Boundaries and Earthquakes

WHAT IS AN EARTHQUAKE?

Earthquakes tend to occur along plate boundaries. An *earthquake* is a shaking of the crust caused by the release of energy as lithospheric plates shift in a rapid and violent manner. Great frictional stresses build up along plate boundaries until the stress at a particular location exceeds the ability of the rocks to resist movement. At that time, a movement of several meters can occur along portions of a fault. The resulting shock waves are associated with an earthquake.

P-, S-, AND *L*-WAVES

Earthquakes produce various types of waves. *P*-waves, or primary waves, are one of the two types of body waves (see Figure 7.18), which are capable of traveling through Earth. *P*-waves are compression waves. Body waves cause objects in their path (buildings, bridges, etc.) to vibrate back and forth in the same direction as the shock wave caused by the earthquake is moving.

S-waves, also known as secondary waves, are the other type of body wave created by an earthquake. *S*-waves are also affected by the material they travel through, though at

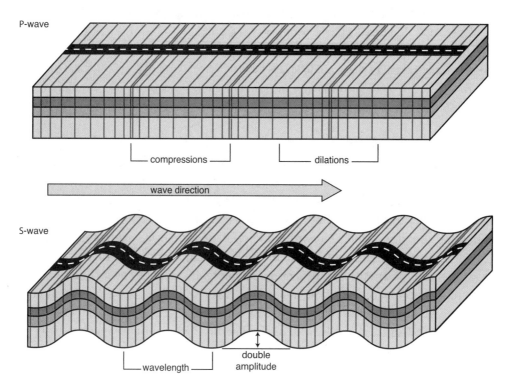

FIGURE 7.18 Earthquakes produce movements that originate at their focus, or the point of origin of the event. These movements "radiate" from the focus as seismic waves. Some seismic waves produce stretching and compressional movements; others, side-to-side or up-and-down movements. The *P*-wave (primary wave) is a compressional wave and produces stretching and compressional movements. *P*-waves move rapidly and can pass through both solid and liquid material; hence, seismographs can detect their occurrence from all points on the planet. *P*-waves are known as longitudinal waves; that is, they cause movement parallel to the direction in which the wave is traveling. *S*-waves, or transverse waves, produce side-to-side or up-and-down motion; that is, the movement created is perpendicular to the motion of the wave.

all times they move with less velocity than *P*-waves, a fact that helps geologists to locate the epicenter of an earthquake. The *epicenter* is the actual site on the surface where the earthquake occurs. This topic is discussed in greater detail in the section headed "Epicenter Versus Focus." Another interesting difference between *S*-waves and *P*-waves is that *S*-waves are not able to penetrate Earth's outer core; they can travel through solids, but not through liquids. *S*-waves cause oscillations at right angles to the primary wave. A shaking motion is produced by *S*-waves as they temporarily change the shape of the material transmitting them.

Surface waves, also known as *L*-waves or Love waves, are generated along the surface when an earthquake occurs. They cause a groundswell, similar to an ocean swell that can rock a ship. These waves are particularly damaging to buildings and other structures. Fortunately, surface waves travel relatively slowly and dampen out rather quickly; they are typically observed only within 500 kilometers of the earthquake's epicenter.

AFTERSHOCKS AND FORESHOCKS

Most of the plate movement takes place during the main earthquake; however, after the event, more adjustments in the positioning of the plates, known as *aftershocks*, typically occur. There may be hundreds of these additional earthquakes, or aftershocks. Typically they are much weaker than the original earthquake, but significant damage may still result as structures already weakened from the original quake are brought down by aftershocks.

There is also a phenomenon known as a *foreshock*. Foreshocks may precede a major earthquake by a period of time that can range from days to years. Although foreshocks are monitored as potential harbingers of earthquake activity, results indicate, unfortunately, that foreshocks are not highly reliable indicators of impending quakes. In fact, there is as yet no effective means by which to forecast the occurrence of a major earthquake.

READING A SEISMOGRAPH

Refer to Figure 7.4, which shows a typical seismograph record from a reporting station located about 400 kilometers from the earthquake's epicenter. Note the arrival of the first *P*-wave, followed by the first *S*-wave, and finally the surface waves. The greater the interval between the arrival of the first *P*-wave and that of the first *S*-wave, the greater the distance of the reporting station from the actual epicenter.

EPICENTER VERSUS FOCUS

The *epicenter*, as stated above, is the site on the surface below which the earthquake occurs. It is typically described in terms of latitude and longitude. The *focus*, in comparison, is the actual site of the earthquake or the point where the rock begins to move or break. The focus is located directly beneath the epicenter; thus the same latitude and longitude apply, and the depth is provided as additional information. By identifying the depth of an earthquake, geologists gain greater understanding of what is happening deep within the crust or the mantle of Earth. Deep earthquakes, with foci of up to 700 kilometers beneath the surface, have been known to occur in subduction zones where an oceanic plate is subducting under a continental plate. Some of the deepest earthquakes in the world have been recorded in Bolivia as the Nazca Plate subducts under the South American Plate.

LOCATING AN EPICENTER

When an earthquake occurs, the first question asked by many is, "Where was the epicenter?" To locate the epicenter of an earthquake, a method called *triangulation* is employed. This method requires three seismographic records from three different seismograph stations. As stated previously, *S*-waves travel more slowly than *P*-waves, so that, the farther from the epicenter, the greater the difference in time between the arrivals of *S*-waves and *P*-waves. The seismograms in Figure 7.19 can be used to determine the distances of the three stations from the epicenter. Circles representing these distances are then drawn on a world map around the three stations. Figure 7.19

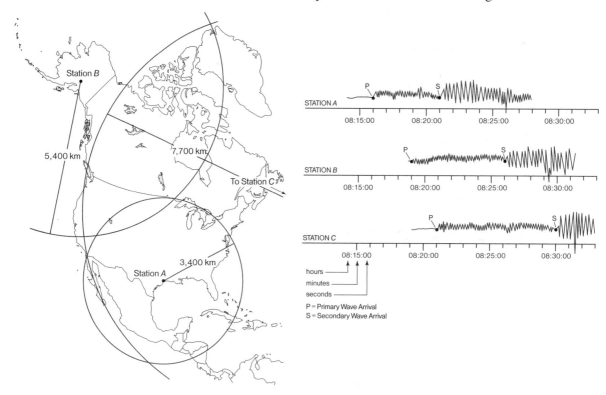

FIGURE 7.19 Three seismograms are required to "triangulate" and locate the epicenter of an earthquake. An analysis of each seismogram will provide the distance of each site from the epicenter. This distance is determined by measuring the time lapse between the arrival of the *P*-waves and the *S*-waves. A circle is then drawn around each station, using the map scale. The epicenter is located at the intersection of the three circles.

indicates that the point where the three circles intersect is the epicenter of the earthquake. A minimum of three seismograms is required to locate the epicenter.

MAGNITUDE AND INTENSITY

Two scales exist to describe the magnitude and the intensity of an earthquake. The *Richter magnitude scale* has been in use since the 1930s. The Richter scale describes an earthquake's magnitude (strength or amount of energy released), which is determined by studying the amplitudes (heights) of the waves recorded on seismographs located at three different stations. Since it is unlikely that a seismograph will be located precisely at the epicenter of an earthquake, a series of seismograms from seismographs at various distances from the epicenter is utilized to assess the magnitude of the earthquake at its epicenter.

Table 7.1 shows the relative frequency of earthquakes of each magnitude on the Richter scale and their impacts near the epicenter. The largest earthquakes have magnitudes above 9.0. In the past 100 years in the United States, earthquakes exceeding a magnitude of 8.0 include the great Alaskan earthquake of 1964 and the 1906 San Francisco earthquake. Earthquakes of this magnitude are capable of causing massive destruction.

One of the world's most destructive earthquakes occurred on December 26, 2004. Originating near the island of Sumatra, an island in western Indonesia, this earthquake registered 9.3 on the Richter scale. Studies in the months that followed revealed this to be the most powerful earthquake on modern record. Over a period of nearly 600 seconds, a 1400-kilometer gash was ripped in the Earth's crust, releasing the equivalent energy of a 100-gigaton hydrogen bomb! As virtually the entire planet vibrated, a tsunami formed and rushed across the Indian and Pacific oceans, traveling in excess of 800 km/hour, killing more than 225,000 people in 11 countries. In recorded history, only the 1960 Chilean earthquake exceeded this event, registering 9.5 on the Richter scale. Similarly, that event triggered a tsunami. However, the death toll was much lower than that experienced in the Sumatra event.

TABLE 7.1
RICHTER MAGNITUDE SCALE

Magnitude	Impact near Epicenter	Estimated Number Each Year (Worldwide)
<2.0	Cannot be detected by humans	Over 500,000
2.0–2.9	Can be detected, no damage	Over 250,000
3.0–3.9	Detected by many	50,000
4.0–4.9	Detected by all, shaking noticeable	6,000
5.0–5.9	Can cause damage	800
6.0–6.9	Destructive, damage to buildings and structures	270
7.0–7.9	Major event. Significant, widespread damage	20
≥8.0	Great earthquake. Total destruction near epicenter	<2

It is worth noting that, although the Far West and Alaska are generally associated with the largest earthquakes because of their proximity to plate boundaries, the greatest earthquake to hit the United States in recorded history actually occurred near New Madrid, Missouri. The New Madrid earthquake was actually a series of five earthquakes of magnitude 8.0 or greater, with epicenters in southeastern Missouri and northeastern Arkansas (see Figure 7.20), occurring between December 16, 1811, and February 7, 1812. Over 600,000 square kilometers were damaged as a result of this series of earthquakes. The New Madrid earthquakes actually changed the course of the Mississippi River. Fortunately, because of the sparse population of the region at that time, few lives were lost. If a similar earthquake were to occur today in that area, the results would be devastating. The New Madrid earthquake serves as a reminder that faults exist deep within Earth that we may not yet know about and regions that seem to be earthquake-proof are not necessarily so. The most likely region for the next big earthquake, though, is southern California.

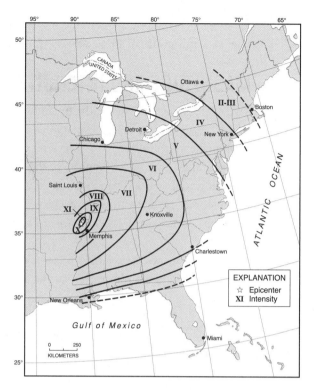

FIGURE 7.20 Map showing the epicenter of the first of the New Madrid earthquakes and the intensity of the earthquake as felt at a great distance from the epicenter. Source: Abridged from *Seismicity of the United States, 1568–1989* (Revised), by Carl W. Stover and Jerry L. Coffman, U.S. Geological Survey Professional Paper 1527, United States Government Printing Office, Washington, D.C.: 1993.

The *Mercalli intensity scale* (Table 7.2) offers a measure of earthquakes as gauged by the damage inflicted by the event. Since intensity is dependent upon several factors, including the magnitude of the earthquake, the distance from the epicenter, the type of rock underlying a region, and the design and construction of buildings, the intensity will vary from event to event and place to place even though the Richter scale magnitude is the same.

For example, a 7.0-magnitude earthquake in Turkey may cause massive destruction and great loss of life, whereas an earthquake of the same magnitude in San Francisco will cause much less damage and fever deaths. The key difference is the construction techniques used in the two areas and the structures' abilities to withstand a quake of such magnitude. In the 1989 San Francisco quake, despite a Richter scale magnitude of 7.1, the subway system survived and few lives were lost throughout the entire

TABLE 7.2
MERCALLI INTENSITY SCALE

Intensity	Description
I	Generally not felt
II	Felt by only a few, particularly in high-rise buildings
III	Felt by many indoors
IV	Creates the sensation of a truck hitting a building
V	Felt by all, awakens sleeping individuals
VI	Slight damage, including furniture moved and falling plaster
VII	Structural damage, particularly to poorly constructed buildings
VIII	Partial collapse of buildings not constructed to be "quake proof"
IX	"Quake-proof" buildings damaged, cracks noted in the ground
X	Most structures destroyed, large cracks in ground
XI	Few structures left standing, bridges destroyed
XII	Total destruction

region surrounding and including the city itself. In contrast, a slightly weaker quake a year earlier in Armenia resulted in the loss of 25,000 lives.

Special Topics

EARTHQUAKES—THE KEY TO UNDERSTANDING EARTH'S INTERIOR

Much of what we know about our planet's interior comes from the study of body (*P* and *S*) waves produced during an earthquake. Humans have never directly sensed (studied) anything deeper than the crust, or outermost layer, of Earth. Attempts to drill into the planet's interior are limited by the extreme pressures encountered as depth increases. Consequently, no drilling operation has reached the Mohorovic discontinuity, or simply Moho. The *Moho* is the boundary between the crust and the mantle of Earth (see Figure 7.21). The only direct experience humans have with the mantle occurs when magma reaches the surface through volcanoes that have very deep magma sources. Even then, the magma is probably chemically changed from its original state in the mantle.

This leaves us, in our endeavor to understand Earth's interior, with the

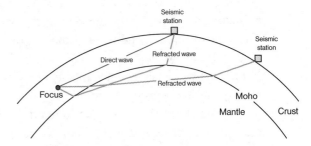

FIGURE 7.21 The Moho boundary. Source: *Let's Review: Earth Science—The Physical Setting,* 2nd Ed., Edward J. Denecke, Jr., Barron's Educational Series, Inc., 2002.

study of *P*-waves and *S*-waves as they travel through the planet after an earthquake. Although *P*-waves are able to pass through the entire planet, they are slowed and refracted as they reach the boundaries between crust and mantle, mantle and outer core, and outer core and inner core. By analyzing the behavior of these waves as they pass through the planet's interior, geologists have constructed a model of Earth's internal layers in the accuracy of which they have high confidence. Since the speed and behavior of body waves are affected by the density and rigidity of the material they pass through, geologists view the planet's interior as consisting of layers of increasing density from crust to core. At the boundary between each two layers, there is a sharp increase in density, indicated by a significant change in the behavior of the body waves. The outer core has been identified as a liquid since *S*-waves cannot pass through it; inability to penetrate liquids is a known property of these waves.

PREDICTING EARTHQUAKES

Scientists have identified the faults or plate boundaries that are under the greatest stress, and therefore are most likely to be sites of major earthquakes in the future. In the western United States, the San Andreas Fault and the Hayward Fault, which is parallel to the San Andreas just east of San Francisco, are prime candidates for trouble. The problem is, can we accurately predict an earthquake with enough advance warning to enable safe evacuation and save lives?

Researchers from several nations are at work on this problem. If short-range forecasts are to be accepted, they must be reliable and accurate, and predict the event to occur within a small window of time and a limited geographic region. For example, if a forecast states that sometime within the next month, somewhere in southern California, a significant earthquake is predicted, few people will respond.

Long-range forecasts, which are in actuality statistical probabilities, state the likelihood that an earthquake equal to or greater than a given magnitude will occur within a period of years. For example, the forecast might state that an earthquake equal to or exceeding magnitude 7.0 will occur in southern California within the next 100 years. This kind of forecast, although of interest to scientists and other researchers, is of little practical value to the population at large.

Police, fire, and emergency medical service officials regularly conduct drills and make plans in the event of a major earthquake in southern California. Plans are in place to quickly evacuate people from affected areas, to shut down natural gas lines to prevent widespread fire, and to redirect traffic to unaffected highways to facilitate the movement of emergency personnel.

TSUNAMIS

A *tsunami* is a water wave or series of water waves triggered when a large volume of water is rapidly displaced. Typical triggers include earthquakes, volcanic eruptions, and landslides (including underwater landslides). Tsunamis, from the Japanese "Harbor Wave," can be devastating when they make landfall. Japan in the western Pacific is no stranger to these impressive and deadly events. Most often formed along subduction zones, where convergent plate boundaries exist, their low amplitude and long wavelength often pass unnoticed when at sea. When they approach the shallow water along a coastline, tsunamis will grow to a height of 50 meters or more in extreme cases. After an earthquake, warnings may provide only a few minutes to run for safety. Careful observers may note a dramatic drawback in the water level just before the arrival of a tsunami. In 2004 in Phuket, Thailand, ten-year-old Tilly Smith of Surrey, England, noticed just this and warned her family and anyone who would listen that a tsunami was imminent. She is credited with saving dozens of lives as a result of her attentiveness to her science lessons that semester. Regions with a high risk of tsunamis generally participate in a tsunami warning system designed to warn the general populace before a tsunami makes landfall.

GEOLOGY OF OUR NATIONAL PARKS

The geographic sites for most, if not all, of our national parks were selected on the basis of their unique geology and natural beauty. Several national parks exhibit features that exemplify the processes discussed in this chapter. Specific examples of these parks include Mount Rainier, Crater Lake, Hawaii Volcanoes, Yellowstone, Shenandoah, and Great Smoky Mountains. Each of these parks provides excellent opportunities to study geology as it pertains to plate tectonics.

Mount Rainier and Crater Lake national parks illustrate volcanic features that form as a result of the subduction of an oceanic plate under a continental plate. Hawaii Volcanoes and Yellowstone national parks offer excellent examples of recent and ancient volcanic activity caused by hot spots under the crust. Shenandoah and Great Smoky Mountains national parks, both located in the Appalachian Mountains, exemplify features formed when the North American Plate converged with the African Plate over 250 million years ago.

Internet Resources to Study Current Earthquakes and Related Information

http://neic.usgs.gov/

The U.S. Geological Survey "National Earthquake Information Center" site provides access to real-time earthquake data in a variety of formats. Users can search earthquakes by date, magnitude, and other parameters. Map views are also available.

http://www.iris.edu/dms/seismon.htm

The Seismic Monitor provides access to near real-time earthquake information and maps along with information about plate tectonics and other educational links.

http://volcano.oregonstate.edu/

"Volcano World," self-advertised as the premier site for volcano information, is provided and maintained by Oregon State University.

http://pubs.usgs.gov/publications/text/dynamic.html

"This Dynamic Earth, The Story of Plate Tectonics" is a site created by the U.S. Geological Survey.

www.sciencecourseware.com/eec/Earthquake/

"Virtual Earthquake" is a great site for learning about earthquakes, including how epicenters are located and magnitudes are determined.

REVIEW EXERCISES FOR CHAPTER 7

WORD-STUDY CONNECTION

aftershock	fossils	plate
amplitude	glacial	plate tectonics
asthenosphere	hot spots	*P*-waves
continental drift	lithosphere	Richter scale
continental drift hypothesis	Love waves (*L*-waves)	rift
convection cell	magma	seismogram
convergent	mantle	seismograph
crust	Mercalli scale	Sonar
divergent	*Mesosaurus*	subduction
earthquake	Mid-Atlantic Ridge	*S*-waves
epicenter	Mohorovic discontinuity	transform fault boundary
focus	Pangaea	trench
foreshock		

SELF-TEST CONNECTION

PART A. Completion. *Write in the word or words that correctly complete the statement.*

1. _____ cells within the asthenosphere drive the lithospheric plates.

2. New material forms along _____ plate boundaries.

3. The San Andreas Fault is a classic example of a _____ plate boundary.

4. Major earthquakes are often followed by _____.

5. _____ is the only part of the Mid-Atlantic Ridge that is above sea level.

6. Magma frequently works its way to the surface and extrudes as lava at _____ plate boundaries.

7. The _____ is the actual site of an earthquake.

8. The _____ is the site on the surface directly above an earthquake.

9. The _____ measures the magnitude of an earthquake.

10. _____ are shock waves that can travel through all layers of Earth.

11. A _____ is a region where one plate is sliding beneath another.

12. A _____ is an instrument designed to record shock waves created by earthquakes.

13. Fossil correlations across continents provide evidence that supports the _____ theory.

14. A boundary where lithospheric plates slide past each other is called a _____.

15. The invention of _____ in the 1950s has enabled humans to study the seafloor.

PART B. Multiple Choice. *Circle the letter of the item that correctly completes the statement.*

1. The region within the mantle that exhibits movement, ultimately causing the plates, which ride on top of this layer, to slide across Earth's surface, is known as the
    (a) crust
    (b) lithosphere
    (c) asthenosphere
    (d) outer core
    (e) inner core

2. Convergent boundaries are sites where
    (a) new material is produced
    (b) material is destroyed
    (c) material is neither produced nor destroyed
    (d) material can be both produced and destroyed
    (e) none of the above

3. The waves generated by an earthquake that are often the most destructive are
   (a) Love waves
   (b) *S*-waves
   (c) *P*-waves
   (d) Both a and b
   (e) Both b and c

4. The minimum number of _____ stations with seismographs required to determine the epicenter of an earthquake is
   (a) 1
   (b) 2
   (c) 3
   (d) 4
   (e) 5

5. New material is often produced at _____ plate boundaries.
   (a) divergent
   (b) convergent
   (c) transform
   (d) subducting
   (e) stable

6. Wegener proposed the continental drift hypothesis, which later evolved into
   (a) the nebular hypothesis
   (b) the big bang theory
   (c) the planetary accretion theory
   (d) the plate tectonics theory
   (e) none of the above

7. Earthquakes that occur after a major quake are called
   (a) foreshocks
   (b) little earthquakes
   (c) Mercalli damage
   (d) Richter scale busters
   (e) aftershocks

8. A famous transform boundary is the
   (a) Mid-Atlantic Ridge
   (b) San Andreas Fault
   (c) Cascade Mountains
   (d) Appalachian Mountains
   (e) island called Iceland

9. A convergent boundary in the ocean is likely to exhibit a
    (a) ridge
    (b) trench
    (c) volcanic mountain
    (d) both a and b
    (e) both a and c

10. Evidence to support the plate tectonics theory has been found in
    (a) glacial deposits
    (b) fossils
    (c) ancient mountain ranges
    (d) seafloor features
    (e) all of the above

PART C. Modified True/False. *If a statement is true, write "true" for your answer. If a statement is incorrect, change the <u>underlined</u> expression to one that will make the statement true.*

1. Shock waves that sometimes precede a major earthquake are called <u>aftershocks</u>.

2. <u>Transform fault boundaries</u> are sites where new crustal material is neither produced nor destroyed.

3. The lithosphere is a rigid layer that rides on top of the <u>less</u> dense asthenosphere.

4. Wegener first proposed the <u>continental drift hypothesis</u>.

5. When continental crust and <u>oceanic crust</u> converge, the <u>oceanic crust</u> subducts under the continental crust.

6. <u>Pangaea</u>, the name given to the ancient landmass, means "all lands."

7. The <u>epicenter</u> is the actual location of an earthquake.

8. <u>Trenches</u> can be found in the vicinity of convergent plate boundaries.

9. <u>Radar</u> is a tool used to study the ocean floor.

10. Convection cells tend to be found in the <u>lithosphere</u>.

CONNECTING TO CONCEPTS

1. If new crustal material is created at divergent plate boundaries, why must old crustal material be destroyed at convergent plate boundaries?

2. Describe the difference between the lithosphere and the asthenosphere.

3. Describe the difference between the Richter and Mercalli scales. Which scale is more useful to the average person? Why?

4. Summarize the evidence that supports the plate tectonics theory.

5. Explain the dynamics that have produced the Himalayas.

6. Clarify the difference between earthquake magnitude and earthquake intensity.

7. Base your answers to the following questions on the information and map below and your knowledge of earth science.

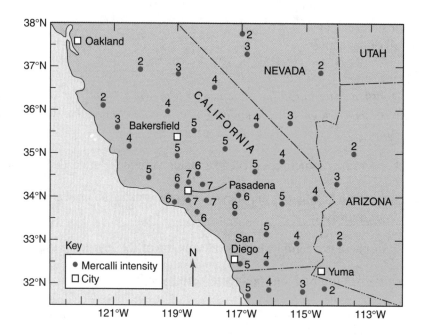

a. State the name of the city that is closest to the earthquake epicenter.
b. Identify the most likely cause of earthquakes that occur in the area shown on the map.

CONNECTING TO LIFE/JOB SKILLS

The study of earthquakes by geologists, seismologists, and other, related professionals is geared toward improving the general condition of human society. Efforts to predict the occurrence of earthquakes will eventually yield results. Once scientists have the ability to accurately predict earthquakes, humans will be able to prepare and lives will be saved. The science of seismology is still in its youth, and much remains to be learned.

ANSWERS
SELF-TEST CONNECTION

Part A

1. Convection
2. divergent
3. transform
4. aftershocks
5. Iceland
6. divergent
7. focus
8. epicenter
9. Richter scale
10. *P*-waves
11. subduction zone
12. seismograph
13. plate tectonics
14. transform fault boundary
15. Sonar

Part B

1. **(c)**
2. **(b)**
3. **(a)**
4. **(c)**
5. **(a)**
6. **(d)**
7. **(e)**
8. **(b)**
9. **(b)**
10. **(e)**

Part C

1. False; foreshocks
2. True
3. False; more
4. True
5. True
6. True
7. False; focus
8. True
9. False; Sonar
10. False; asthenosphere

CONNECTING TO CONCEPTS

1. The plates are part of a conveyor-belt system; thus material formed at convergent boundaries must be later destroyed at divergent boundaries.

2. The lithosphere is a rigid layer that rides on top of a denser, more pliable region of Earth's mantle called the asthenosphere.

3. The Richter scale reports the actual magnitude of an earthquake. It does not indicate the degree of damage at any one place. The Mercalli scale reports damage for each location affected by an earthquake. Therefore, the Mercalli scale is probably more useful to the average person.

4. Evidence supporting the plate tectonics theory includes fossil correlations between continents, glacial evidence, mountain ranges and sediments that match across continents now separated by oceans, the action occurring along mid-ocean ridges, and the action occurring along subduction zones.

5. The Himalayas formed and continue to grow as a result of a continent–continent convergence zone where the subcontinent of India is slowly and gradually sliding under Asia. As it does so, it is pushing up the Himalayas to ever-greater heights.

6. Earthquake magnitude is a measure of the energy released by an earthquake; earthquake intensity is an indication of how the earthquake is experienced on the surface.

7. a. Pasadena is surrounded by the highest values on the Mercalli scale.
   b. The San Andreas Fault runs right through the area shown.

8

Volcanoes

WHAT YOU WILL LEARN

This chapter focuses on volcanoes. In this chapter you will learn

- what the ring of fire is and how plate tectonic theory explains it;
- what lava is composed of and the kinds of eruptions that can occur;
- about intrusive features and how scientists interpret their formation;
- about certain case studies, including Mount Saint Helens and Kilauea.

SECTIONS IN THIS CHAPTER

- Volcanoes and Plate Tectonics—The Ring of Fire
- Volcanic Eruptions
- Types of Volcanoes
- Intrusive Features
- Case Studies of Volcanoes
- Internet Resources for Current Volcanic Activity
- Review Exercises for Chapter 8

Volcanoes and Plate Tectonics—The Ring of Fire

Volcanoes and volcanic eruptions have long fascinated humans. Volcanic eruptions are both visually spectacular and potentially deadly phenomena. Scientists have only recently begun to understand the source of energy behind these events and the movements within the lithospheric plate system that triggers eruptions.

Figure 8.1 shows all of the volcanoes classified as active. The pattern that emerges is not one of randomness, but instead outlines the boundaries that separate the lithospheric plates (see Figure 7.10). The region of volcanic activity that reaches northward along the entire western coastline of the Americas, across the Aleutian Islands, and then southward along the eastern coastline of Asia all the way to New Zealand is known collectively as the *Ring of Fire*, and is an easily recognized formation along the perimeter of the giant Pacific Plate. It is important to note the high degree of correlation between the volcanic activity illustrated in Figure 8.1 and the plate boundaries shown in Figure 7.10, particularly along the Pacific Rim, or Ring of Fire.

As discussed earlier, the theory of plate tectonics explains why earthquakes occur along plate boundaries and why volcanic activity is found both along (mid-ocean ridges) and near (island arcs and volcanic mountain ranges) plate boundaries.

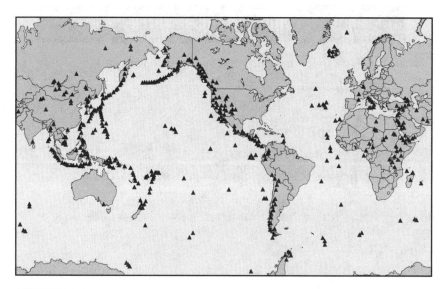

FIGURE 8.1 Earth's active volcanoes. Source: Adapted from the image at *http://nmnhwww.si.edu/gvp/volcano/index.htm*, Web site of the Global Volcanism Program, National Museum of Natural History, E-121, Smithsonian Institution, Washington, D.C.

Volcanic Eruptions

ORIGIN OF MAGMA

In order for a volcanic eruption to occur, molten material, or *magma*, must exist deep within Earth's crust or mantle. As much of the crust is composed of solid rock, a reasonable question to ask is, "What causes that rock to melt and form magma?" Magma seems to form when the friction created by one plate subducting beneath another produces heat. Another mechanism by which solid rock can melt is evident when hot "plastic" rock, or solid rock that is under great pressure in the mantle, flows upward into a region of lower pressure. The rock is initially heated by the process of radioactive decay occurring deep within Earth, but remains solid because of the great pressures in the mantle. As this hot rock flows very slowly upward toward the crust, however, pressures and consequently melting points decrease. Eventually the rock may reach a depth where it melts, forming magma.

GASES AND VISCOSITY IN MAGMA

Magma is a complex mixture of molten rock and gases. It is these gases, when present in sufficient quantity, that force magma to the surface and under the right conditions can result in explosive eruptions (see Figure 8.2). As the gases approach the surface, they attempt to expand, creating pressure on the molten rock with which they are mixed.

The magma's *viscosity*, or resistance to flow, determines whether or not a violent eruption occurs. In the Hawaiian Islands, low-viscosity magma allows smooth, gentle flows down the mountainside toward the ocean for a considerable distance before the lava cools and hardens. Pu`u ka Pele, on the southeast flank of Mauna Kea, is a classic example of a shield volcano (see page 197).

In contrast, Mount Saint Helens, located in the state of Washington, erupted in 1980 with an explosive force that blew part of the mountain off along with the highly viscous ejecta (ejected material). The magma inside Mount Saint Helens contained a much higher concentration of silica than does the magma located under Hawaii. Magma with a high concentration of silica tends to be highly viscous, thus resisting flow, and when combined with the presence of gases produces explosive eruptions.

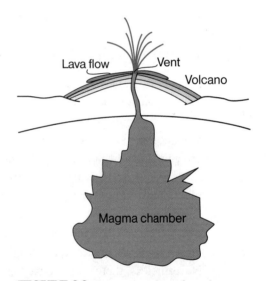

FIGURE 8.2 A cross section of a volcano. Magma rises from chambers in the upper mantle until it exits the crust through an opening called the *vent*. When magma reaches the surface it is called lava, and accumulates in a mound called a *volcano*. Source: *Let's Review: Earth Science—The Physical Setting*, 2nd Ed., Edward J. Denecke, Jr., Barron's Educational Series, Inc., 2002.

An excellent discussion exists at the following Web site: *http://mac.usgs.gov/mac/ isb/pubs/teachers-packets/volcanoes/poster/poster.html*. The Web site offers several excellent diagrams illustrating various types of volcanoes. There is also a sequence of four photographs, each a few seconds apart, illustrating the explosive eruption of Mount Saint Helens in 1980. Photographs follow that illustrate the extent of the damage caused by the eruption.

LAVA FLOWS

Lava flows, such as those seen in Hawaii, are composed of basaltic materials and are quite low in silica. The lava flows at a rate that is largely determined by the slope of the land down which it is moving. At an extreme, the lava can reach speeds of 30 kilometers per hour (20 mi/hr); however, speeds of 100 meters per hour (6.2 mi/hr) are more typical. Temperatures in the flowing lava typically exceed 1000°C. As the lava cools, it may form a smooth skin while the subsurface is still molten. This surface is called *pahoehoe*, first named for its smooth appearance by the Hawaiian natives. If the lava is more viscous, it flows more slowly; and as it cools, gases escape from its surface, making the surface rough. The Hawaiians named this type of flow *aa* for the sounds that a person may make while walking over it!

PYROCLASTIC MATERIALS

Pyroclastic materials, or *pyroclasts*, are pulverized rock, lava, and glass fragments that are explosively ejected from volcanoes that contain highly viscous lava. Pyroclasts may be as small as grains of sand or as large as boulders. There have been cases where vulcanologists, who sometimes venture dangerously close to the site of an erupting volcano, have been killed by falling debris from pyroclasts during an active eruption.

Pyroclastic ejecta has been known to travel at 200 kilometers per hour (125 mi/hr), making it difficult for a person caught too close to an erupting volcano to get out of harm's way. In 1980, when Mount Saint Helens erupted, most people had already been moved to a safe distance, as small tremors, signaling the movement of magma underground, enabled vulcanologists to issue warnings of an imminent eruption. Nevertheless, there were 50 casualties.

Residents living near Mount Saint Helens learned firsthand the problems presented by a volcano ejecting pyroclasts. Upward of 0.5 meter (18 in.) of ash collected on everything in nearby cities. Unlike snow, this mess did not melt away over time. The ash was also quite acidic, courtesy of the gases embedded in the lava, and destroyed the paint on many homes and cars.

Types of Volcanoes

The type of volcano that forms is largely determined by the kind of lava that flows through it. Four typical volcanic structures are illustrated in Figure 8.3.

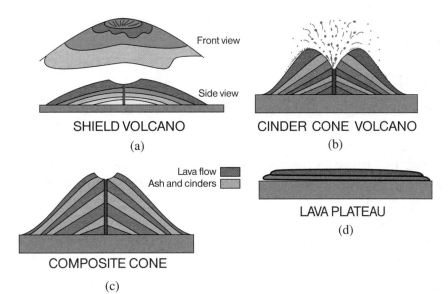

FIGURE 8.3 Typical volcanic structures. (a) Broad, gently sloping shield volcanoes form from successive lava flows. (b) Steep-sided cinder cones form from the buildup of ash and cinders. (c) Composite cones of intermediate slope form when periods of lava flows are interspersed with eruptions of ash and cinder. (d) Very fluid lava that erupts along a fissure spreads out to form a lava plateau. Source: *Let's Review: Earth Science— The Physical Setting*, 2nd Ed., Edward J. Denecke, Jr., Barron's Educational Series, Inc., 2002.

SHIELD VOLCANOES

Shield volcanoes (see Figure 8.3a) form when fluid, basaltic lava flows through the main vent, exits, and produces a broad, slightly domed structure. These volcanoes are so named because they resemble a warrior's shield.

Mauna Kea on the Big Island of Hawaii is a classic example of a shield volcano. It is also quite large, as it rises nearly 10,000 meters (30,000 ft) from the seafloor to a height of over 4100 meters (13,000 ft) above sea level.

CINDER CONES

Cinder cone volcanoes (see Figure 8.3b) form from ejected lava fragments. They are largely composed of loose pyroclastic material and, in contrast to the gently sloping shield volcanoes, can have slopes of upward of 40°.

In 1943, residents of Paricutin, Mexico, had the opportunity to observe firsthand the construction of a cinder cone volcano. The volcano began to build in a farmer's cornfield when tremors were followed by the emission of sulfurous gases from a small hole in the field. Hot rock fragments began to eject from the hole, and a cone was built. The cone reached a maximum height of 400 meters (1300 ft) over a period of 2 years, but grew as much as 40 meters (130 ft) in 1 day! Large amounts of ash were also reported to cover the entire area.

COMPOSITE CONES

Most of the *composite cone volcanoes* (see Figure 8.3c) line the Ring of Fire. Examples of these majestic volcanoes include Mount Saint Helens, Mount Rainier, and Mount Shasta. These volcanoes are also known as *stratovolcanoes*, as they are large, symmetrical structures composed of layers of lava flows and pyroclastic materials. As the layers reveal, composite cone volcanoes alternately emit viscous lava flows and pyroclastic materials.

Vesuvius is one such stratovolcano that can erupt violently, with little warning. In 79 A.D., the entire city of Pompeii was buried in 3 days by ash and larger pyroclastic materials. Before the eruption, the volcano had been dormant for centuries and was even covered with thick vegetation.

LAVA PLATEAUS

Lava plateaus (see Figure 8.3d) form where fissures allow fluid lava to spread, thus forming a plateau. They commonly occur where shield volcanoes are located. One of the most impressive lava plateaus is the Columbia Plateau. Now known as the Columbia Basin, this plateau is a large, slightly depressed lava plain, located in the Pacific Northwest.

Intrusive Features

BATHOLITHS

Magma does not always reach the surface. If gas pressures are not great enough to force the magma to the surface, or if overlying rocks cannot be penetrated by it, the magma may slowly cool and form a large pool of granite. Granite is easily recognizable by its large crystal structure and mix of minerals, including light-colored quartzes and darker feldspars and micas. A large region of intrusive igneous rock, located deep within the crust, is called a *batholith*. Batholiths can form the core of a mountain range. Figure 8.4 shows a batholith as well as other intrusive structures, all of which are discussed in upcoming sections of this chapter.

Weathering and erosion, along with regional uplift, can result in a batholith's becoming exposed at the surface. One of the most famous

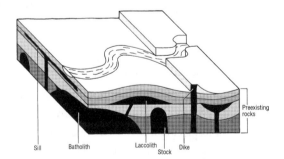

FIGURE 8.4 Some types of plutons. Source: *Let's Review: Earth Science—The Physical Setting*, 2nd Ed., Edward J. Denecke, Jr., Barron's Educational Series, Inc., 2002.

outcrops of granite from an ancient batholith is Granite Peak in the Grand Tetons. This batholith, once formed deep within Earth's crust, is now exposed at an altitude exceeding 3300 meters (10,000 ft) above sea level. The Black Hills of South Dakota are another classic example of this kind of geologic formation.

LACCOLITHS

A *laccolith* is similar to a batholith, but can cause the overlying rock layers to bulge as the magma forming the laccolith intrudes. A dome-shaped surface may reveal the presence of a laccolith beneath. Laccoliths and other igneous features can be observed in Big Bend National Park in southern Texas along the Rio Grande.

DIKES AND SILLS

Magma within a batholith is rarely stable and, while in a molten state, will try to force its way through preexisting rock layers. As it does so, *dikes* (vertical intrusions) and *sills* (horizontal intrusions) may form. A dike that reaches to the surface effectively becomes a vent through a volcano. Figure 8.4 shows the interrelationships among batholiths, laccoliths, dikes, and sills.

As noted earlier, the Palisades (see Figure 8.5), located in New Jersey along the Hudson River, are a classic example of a sill. As North America separated from Africa during the Triassic and Jurassic periods, these cliffs formed as magma flowed horizontally through preexisting layers of sandstone. Today, the overlying layer of sandstone has weathered away, exposing the sill along the surface.

Case Studies of Volcanoes

MOUNT SAINT HELENS

In the morning of May 18, 1980, at 8:32 A.M. Pacific Time, an earthquake measuring 5.1 on the Richter scale was recorded. This quake shook Mount Saint Helens, caused a massive rockslide, and was followed by a major eruption of pyroclastic material consisting of hot pumice and ash. Four hundred meters (1300 ft) of the mountainside blew outward that morning. Ultimately, over 60 square kilometers of the neighboring valley was filled with debris, and much more land, including areas used for recreation, stands of timber, and privately owned land, was impacted by the blast. Mud flows carrying debris and downed trees filled the rivers and flooded the lands downstream from the mountain. Despite extensive advance warnings and evacuations, the explosive eruption of Mount Saint Helens and the ensuing events caused the deaths of over 50 people.

In the weeks leading up to May 18, there were many signs of an impending eruption. Included were the emission of steam from vents in the mountain, a continuing series of earthquakes signaling the movement of magma beneath the surface, and a measurable bulge in the area that eventually blew out during the eruption.

(a)

(b)

FIGURE 8.5 Palisades cliffs along the Hudson River. (a) Illustrates the vertical cliffs. The fine texture of the rock indicates small crystal size, consistent with how the sill formed. (b) Illustrates the Palisades as they extend to the north of the George Washington Bridge, which spans the Hudson River between northern New Jersey and New York City.

Mount Saint Helens behaved according to character; its eruption consisted mostly of pyroclasts, as the magma was highly viscous and loaded with explosive gases.

KILAUEA

Kilauea is perhaps the world's most active volcano. It is the southeasternmost of all the volcanoes in the entire Hawaiian Island chain and is geologically the youngest. Kilauea's plumbing system extends from the surface to a depth of more than 60 kilometers. It is from these depths that Kilauea taps its highly fluid basaltic lava.

Kilauea's frequent eruptions make this crater a great study site for vulcanologists. In Hawaiian tradition, Kilauea is also the home of Pele, the Hawaiian volcano goddess. The nearly continuous eruptive activity during the nineteenth century and the 34 eruptions since the middle of the twentieth century have contributed to Kilauea's reputation, both as a geologic site of wonder and as a spiritual site of significance to some residents of Hawaii.

As the Pacific Plate continues to move, Kilauea will move farther away from the hot spot under the plate. A volcano dubbed Loihi has already been building beneath the ocean surface and is destined to become the next Hawaiian Island at some time in the future. As expected, Loihi is located to the southeast of the current Hawaiian Islands.

Internet Resources for Current Volcanic Activity

See the Internet section at the end of Chapter 7 for relevant links.

REVIEW EXERCISES FOR CHAPTER 8

WORD-STUDY CONNECTION

basaltic	intrusive	radioactive decay
batholith	Kilauea	Ring of Fire
cinder cone volcano	laccolith	shield volcano
composite cone volcano	lava	silica
crust	magma	sills
dikes	mantle	stratovolcano
ejecta	pahoehoe	viscous
gases	pyroclastic	volcanoes
granite	pyroclasts	vulcanologist

SELF-TEST CONNECTION

PART A. Completion. *Write in the word or words that correctly complete the sentences.*

1. _____ lava tends to have low viscosity and consequently to produce shield volcanoes.

2. Pyroclastic materials tend to be ejected when lava is high in _____ content.

3. _____ can be found in the Hawaiian Islands.

4. _____ form from ejected lava fragments.

5. Heat is generated in Earth's mantle by the _____ of unstable isotopes.

6. _____ are horizontal intrusions of magma into preexisting rock layers.

7. The _____ is a zone where much of Earth's volcanic activity occurs.

8. When magma reaches the surface, it is called _____.

9. The more _____ the magma, the more explosive the volcanic eruption.

10. A stratovolcano is also known as a _____.

11. A _____ can cause overlying rock layers to bulge.

12. Smooth lava is otherwise known as _____.

13. The world's most active volcano is _____.

14. _____ ejecta has been known to travel at speeds of up to 200 kilometers per hour.

15. _____ contained in magma are responsible for forcing it to the surface.

PART B. Multiple Choice. Circle the letter of the item that correctly completes the statement.

1. An example of an intrusive igneous rock is
   (a) basalt
   (b) mica
   (c) granite
   (d) feldspar
   (e) none of the above

2. Shield volcanoes form with
   (a) low-viscosity basaltic lava
   (b) high-viscosity basaltic lava
   (c) low-viscosity granitic lava
   (d) high-viscosity granitic lava
   (e) none of the above

3. The Ring of Fire, a region that exhibits most of the world's volcanoes, circles the
   (a) Atlantic Plate
   (b) North American Plate
   (c) Nazca Plate
   (d) Eurasian Plate
   (e) Pacific Plate

4. A scientist who specializes in the study of volcanoes is
   (a) a geologist
   (b) a seismologist
   (c) a meteorologist
   (d) a vulcanologist
   (e) an astronomer

5. The name given by native Hawaiians to smooth lava cooling on the surface is
   (a) pyroclasts
   (b) pyroclastic
   (c) pahoehoe
   (d) aa
   (e) ejecta

6. Highly viscous magma tends to result in
   (a) highly explosive eruptions and the formation of cinder cone volcanoes
   (b) highly explosive eruptions and the formation of shield volcanoes
   (c) gently flowing eruptions and the formation of cinder cone volcanoes
   (d) gently flowing eruptions and the formation of shield volcanoes
   (e) gently flowing eruptions and the formation of stratovolcanoes

7. Dikes are
   (a) horizontal intrusions through preexisting rock
   (b) vertical intrusions through preexisting rock
   (c) cracks in bedrock created by earthquakes
   (d) associated with composite cone volcanoes only
   (e) associated with cinder cone volcanoes only

8. Batholiths are found
   (a) along the Mid-Atlantic Ridge
   (b) within the San Andreas Fault
   (c) on the surface
   (d) deep within the roots of mountains
   (e) along with extrusive igneous rock

9. The Hawaiian Islands are a classic example of
   (a) cinder cone volcanoes
   (b) composite cone volcanoes
   (c) shield volcanoes
   (d) stratovolcanoes
   (e) none of the above

10. Mount Saint Helens is a classic example of a
   (a) cinder cone volcano
   (b) composite cone volcano
   (c) shield volcano
   (d) stratovolcano
   (e) both b and d

PART C. Modified True/False. *If a statement is true, write "true" for your answer. If a statement is incorrect, change the* <u>underlined</u> *expression to one that will make the statement true.*

1. Batholiths are composed of <u>intrusive</u> igneous rocks.

2. <u>Silica</u> determines the viscosity of magma.

3. Pyroclastic materials are often ejected from <u>shield</u> volcanoes.

4. The radioactive decay of <u>stable</u> isotopes creates enough heat within Earth's mantle to form magma.

5. <u>Batholiths</u> are often found in the core of ancient mountains.

6. Magma that is <u>high</u> in silica content will produce shield volcanoes.

7. Dikes are <u>horizontal</u> intrusions into preexisting rock layers.

8. The Hawaiian Islands are classic examples of <u>shield volcanoes</u>.

9. Volcanoes can grow over 100 feet in 1 <u>year</u>.

10. <u>Mount Vesuvius</u> is an example of a stratovolcano that erupted violently with little warning.

CONNECTING TO CONCEPTS

1. Compare cinder cone, composite cone, and shield volcanoes. How are they similar? Different?

2. What type of volcano is most likely to form when magma is high in silica? Why?

3. Where is the Ring of Fire located? What causes it to exist?

4. How quickly can a volcano form?

5. What are pyroclastic materials composed of?

6. Base your answers to the following questions on the diagram that follows. This diagram represents a cross section of a portion of Earth's crust. Letters *A* through *M* identify rock layers, structures, and boundaries within the cross section.

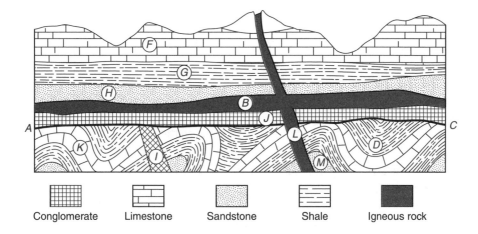

Conglomerate   Limestone   Sandstone   Shale   Igneous rock

a.   State the rock layer that best represents a dike.

b.   State the rock layer that best represents a sill.

c.   Place a circle around the rock structure that best represents a volcano.

d.   Explain how the igneous intrusion labeled *B* most likely affected the rock adjacent to it in rock layer *H*.

e.   State two likely sources of the heat originally contained in the igneous rock that formed the intrusions in the cross section.

CONNECTING TO LIFE/JOB SKILLS

Volcanoes have long fascinated humans. Their study by today's scientists may yield better forecasts as to when and where volcanic eruptions will occur. This knowledge will help to save lives and possibly property as well.

Those who choose to dedicate their lives to this study spend time both in the laboratory and in the field collecting data. Some vulcanologists have lost their lives in the pursuit of collecting pyroclastic samples and lava newly deposited on the surface. Although their vocation is dangerous and physically challenging, these scientists often indicate that there is nothing else they would rather be doing for a living!

ANSWERS
SELF-TEST CONNECTION

Part A

1. Basaltic
2. silica
3. Shield volcanoes
4. Cinder cone volcanoes
5. radioactive decay
6. Sills
7. Ring of Fire
8. lava
9. viscous
10. cinder cone volcano
11. laccolith
12. pahoehoe
13. Kilauea
14. Pyroclastic
15. Gases

Part B

1. (a)	3. (e)	5. (c)	7. (b)	9. (c)
2. (a)	4. (d)	6. (a)	8. (d)	10. (e)

Part C

1. True
2. True
3. False; cinder cone or composite cone
4. False; unstable
5. True
6. False; low
7. False; vertical
8. True
9. False; day
10. True

CONNECTING TO CONCEPTS

1. All volcanoes develop as a result of the existence of subsurface magma and tectonic activity. Magma is a complex mixture of liquids and gases that creates below the surface pressures that will result in an eruption when they build sufficiently. The type of volcano that develops is largely contingent upon the viscosity of the magma located beneath the surface. Shield volcanoes form when low-viscosity magma gently erupts and flows down the mountain side. Cinder cone volcanoes develop when viscous magma produces eruptions of pyroclastic materials. The magma is too viscous to flow, so eruptions are explosive and contain large amounts of hot ash and larger particles that are literally blown out of the volcano. Composite volcanoes result when both types of magma are present, so that some eruptions are explosive and others are more like those of a shield volcano.

2. High concentrations of silica are associated with high-viscosity magma; hence, cinder cone volcanoes are likely to form.

3. The Ring of Fire essentially outlines the entire Pacific Rim in both hemi- spheres—the Northern and the Southern. It is a zone of plate boundaries, along which many of Earth's earthquakes and volcanoes are known to originate.

4. According to the lesson learned in Paricutin, Mexico, in 1943, the best answer is on the order of weeks or months.

5. Pyroclastic materials are basically hot, partially molten rocks and ash that are ejected at high rates of speed from cinder cone volcanoes. They are composed of various minerals, but are likely to be high in silicates.

6. a. Rock layer *L* represents a dike.
   b. Rock layer *B* represents a sill. Note that it is actually an extension of layer *L* and is also igneous rock, as would be expected of a sill.
   c. The volcano can be recognized as a surface protrusion fed by a dike (layer *L*).
   d. As the sill formed, rock layer *H* was already present and may have been baked, thus experiencing contact metamorphism.
   e. Heat originates (1) as a by-product of radioactive decay and (2) from friction caused by plate collisions (resulting in subduction).

Earth's Surface

WHAT YOU WILL LEARN

This chapter focuses on Earth's surface and its dynamic nature. In this chapter you will learn

- what forces affect Earth's surface and how change has occurred over time;
- the difference between weathering and erosion;
- how mountains form and what types of mountains exist;
- how and in what ways land deforms.

SECTIONS IN THIS CHAPTER

- Our Dynamic Planet
- Forces Affecting Earth's Surface: Weathering and Erosion
- Types of Mountains
- Types of Deformation
- Review Exercises for Chapter 9

Our Dynamic Planet

When compared to other celestial bodies, Earth stands out in many ways. Perhaps one of the most extraordinary features is Earth's continual renewal of its surface. In comparing our planet to its closest neighbor, the Moon, scientists have discovered that, although the two formed around the same time, Earth's surface is much younger than that of the Moon. Much of the lunar surface is well over 2 billion years old. Much of Earth's surface, however, is only a few hundred million years old, and many parts are of lesser age.

This chapter explores the forces that continually renew our planet's surface. You will discover that many of these forces are not found elsewhere in our solar system and thus are unique to Earth.

Forces Affecting Earth's Surface: Weathering and Erosion

Weathering and erosion are the key processes responsible for reshaping Earth's surface. *Weathering* is the action by which a number of agents, most involving water in one form or another, break down preexisting rock into sediments. *Erosion* is the process of transporting the weathered material, eventually depositing it in a new location or environment.

RUNNING WATER

Earth is the only planet known to have flowing, liquid water. Mars, for example, has large quantities of the compound H_2O; but since temperatures are almost always below the freezing point of water on Mars, all the water known to exist on that planet is in the form of ice. In contrast, any water on Venus must occur as vapor because of the planet's extreme heat on the surface as well as in the atmosphere.

Moving water has the ability to weather and erode rock. This often occurs by a process known as *corrasion* or simply *abrasion* (see Figure 9.1). Corrasion occurs when sediment, including sand-sized or smaller particles, is carried by moving water. As these particles pass by preexisting rock, they wear it down over time. Rivers and streams can be very powerful weathering and erosional agents. Consider the Mississippi River, where a large delta has formed at the mouth of the river largely because the Mississippi has carried millions of tons of sediment and deposited them over time.

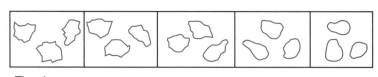

Time Increases ⟶

FIGURE 9.1 Rounding of particles caused by abrasion, or corrasion. Source: *Let's Review: Earth Science—The Physical Setting,* 2nd Ed., Edward J. Denecke, Jr., Barron's Educational Series, Inc., 2002.

WIND

Wind, although not as effective a weathering or erosional agent as water, can also move land. Sand storms and dust storms are classic examples of the kinds of action that wind can create. Wind is most effective when surface sediments, such as sands, silts, and clays, are loose or unconsolidated. Under these conditions, often found in desert or other arid regions, wind can move large quantities of surface material.

Extended drought conditions, such as those seen in the 1930s, when combined with steady winds, have produced dust storms that darken the skies and remove significant quantities of topsoil, particularly from the Great Plains. Dust can be transported over great distances, even across oceans. Larger particles, including sand, cannot be transported by wind as effectively as dust and silt. Wind can act to sculpt the land, however, forming sand dunes or hills of sand, and ripples in the surface of desert sands. Sand dunes form when sand piles up against stationary objects such as plant life or rocks.

Deposits of loose mineral particles, often yellow to red in color, are called *loess deposits*. Loess deposits in the Mississippi and Missouri river valleys are thought to have come from glacial outwash, or loose mineral sediments first deposited by glaciers. The wind may have carried these materials over some distance before depositing them again. Loess deposits can reach depths of up to 300 meters, as observed in parts of China.

Wind can also act to build waves, which can be erosional agents along shorelines. Persistent onshore winds can cause flooding at high tide, and the resulting high waters may remove valuable sediments (sands) along the shoreline. This has been an ongoing problem faced by coastal communities along the entire eastern seaboard of the United States. Coastal waves act to create a longshore current or drift. This phenomenon is discussed more fully in Chapter 11.

ANIMALS AND PLANTS

Both animals and plants play roles in the weathering of solid rock (see Figure 9.2). Earthworms, termites, and other animals burrow holes through soil and other loose

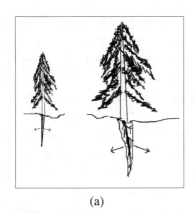

(a)

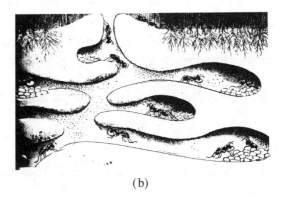

(b)

FIGURE 9.2 Plant (a) and animal (b) action on rocks. Source: *Let's Review: Earth Science— The Physical Setting,* 2nd Ed., Edward J. Denecke, Jr., Barron's Educational Series, Inc., 2002.

surface debris, thus allowing water to penetrate deeper and enhance the rate of weathering. Considering the number of organisms involved, this animal activity is a significant contribution toward the weathering process.

Plants weather rock a bit more directly. When a plant grows into soil along the surface, its roots can physically break rock near the surface down into smaller and smaller particles.

MASS WASTING

Mass wasting is the action of gravity upon rock and soil, causing downslope movement. A landslide is a spectacular example of mass wasting, which can result in the weathering and erosion of large areas of land in a very short time. Mass wasting is primarily an erosional process, as material already loosened by weathering agents such as rain and vegetation is carried to streams, which then transport the material to new sites of deposition.

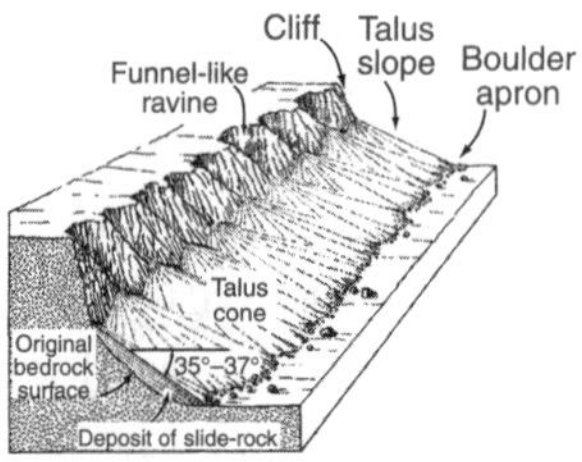

Mass wasting is facilitated by changes in temperature. In much of the United States, winter temperatures alternate between above freezing and below freezing. This frost-thaw cycle produces extensive "frost wedging" as a result of water's expansion upon freezing. Frost wedging expands cracks in rock, thus furthering the weathering process.

Talus slopes, such as those found along the Hudson River's spectacular Palisades, are examples of mass wasting. When the slope is too great, material breaks loose with the help of frost wedging, and the structure illustrated in Figure 9.3 is produced.

FIGURE 9.3 Talus slopes and deposits formed by mass movements. Source: *The Earth Sciences,* Arthur N. Strahler, Harper & Row, 1971.

GLACIERS

Earth's present climate supports extensive glaciation in polar latitudes and at high elevations. A glacier is essentially a large chunk of dirty ice. Glaciers are relatively permanent mixtures of ice and sediment. They form from hundreds or thousands of years of snow accumulation, which is then compacted and recrystallized. Glaciers can extend to a thickness of several thousand feet. Although a glacier appears motionless, it is actually sliding, albeit quite slowly, under its own weight on a thin layer of water formed from pressure and friction with the surface below (see Figures 9.4 and 9.5).

As the glacier slides or flows, it reshapes the surface beneath it (see Figure 9.6). The motion carries much material downstream from the source of the glacier. Much of the exposed rock in the interior Northeast and Midwest shows striations caused by glacial flow over the rock during the last ice age. In some areas, there are boulder fields composed of rock that has been carried several miles or tens of miles from its source. The only explanation for boulder fields is glacial action. Eastern Pennsylvania has several excellent examples of this kind of action, as well as parts of Yellowstone National Park in Wyoming.

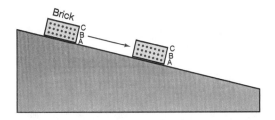

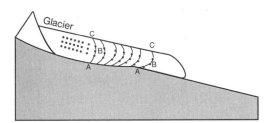

FIGURE 9.4 A brick and a glacier in motion. A glacier both slides along its base and flows like a fluid. The relative positions of particles inside the body of the ice change as the glacier flows downhill. Source: *Let's Review: Earth Science—The Physical Setting,* 2nd Ed., Edward J. Denecke, Jr., Barron's Educational Series, Inc., 2002.

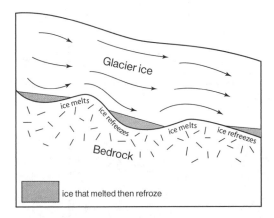

FIGURE 9.5 Meltwater at the base of glaciers is created by high pressure as the glacier gradually slides across the bedrock beneath. Source: *Let's Review: Earth Science—The Physical Setting,* 2nd Ed., Edward J. Denecke, Jr., Barron's Educational Series, Inc., 2002; adapted from *Glaciers and Landscape: A Geomorphological Approach,* David E. Sugden and Brian S. John, John Wiley, 1976.

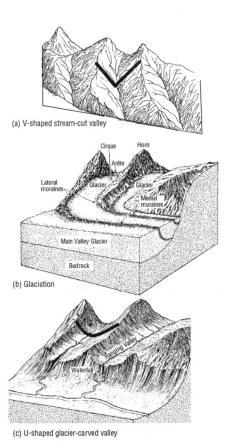

(a) V-shaped stream-cut valley

(b) Glaciation

(c) U-shaped glacier-carved valley

FIGURE 9.6 Landscape before, during, and after glaciation. (a) Valley before glaciation. (b) Glacial erosion forms cirques in hollows between slopes. Erosion in adjacent cirques forms sharp ridges called *arêtes* and converts rounded peaks into sharp horns. Eroded material carried along the ice horns. Eroded material carried along the ice margins forms moraines. (c) Hanging valleys form where tributary glaciers flowed into the main glacier. Source: *Earth,* 4th Ed., Frank Press and Raymond Seiver, W.H. Freeman, 1986.

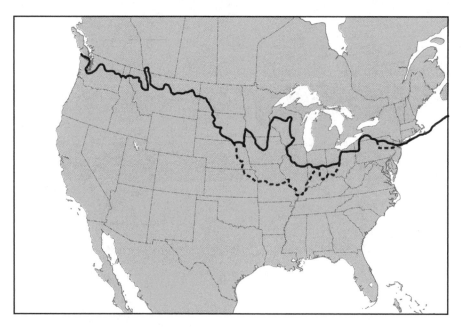

FIGURE 9.7 Maximum extent of glacial episodes across the United States. Lines show the extent of two glacial advances.

At the peak of the most recent ice age, about 18,000 years ago, glaciers covered much of North America. Figure 9.7 shows the extent of glacial coverage at the height of that ice age.

Glaciers can also leave behind recognizable deposits called *moraines*, which are composed of glacial till, or unconsolidated deposits of material transported by glaciers over time. The kinds of deposits that glaciers may leave include ground moraines, recessional moraines, terminal moraines, erratics, drumlins, and eskers (see Figure 9.8).

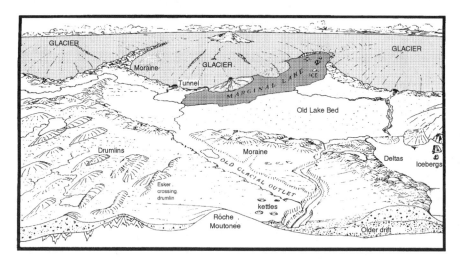

FIGURE 9.8 Landforms resulting from glacial deposits. Source: Educational Leaflet #28, The New York State Museum, Albany, New York.

In general a *moraine* is essentially rock material left behind by a glacier as it retreats or melts. A ground moraine is a large, fairly even blanket of sediment or till. Recessional moraines form as the glacier retreats. If it stays in one place for a time, a larger terminal, or end, moraine will form. A terminal moraine is the southernmost deposition of a glacier. Long Island, New York, and Cape Cod, Massachusetts, are examples of terminal moraines. Figure 9.9 illustrates the formation of an end or terminal moraine.

Erratics are large boulders carried from their native sites to remote sites by the glacier. The large boulder fields in eastern Pennsylvania and the northeastern section of Yellowstone National Park contain erratics. Erratics are commonly found wherever glaciers melted in the northeastern United States. In fact, they were a problem for the early farmers in this region.

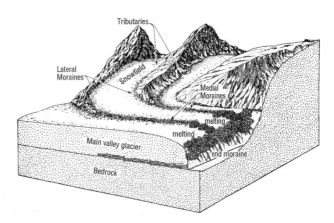

FIGURE 9.9 Formation of moraines. Lateral moraines consist of materials carried along the sides of a glacier. When two glaciers merge, their lateral moraines are welded together between the glaciers, forming a medial moraine. End, or terminal, moraines form where debris melts out of the ice at the end of the glacier. Source: *Earth*, 4th Ed., Frank Press and Raymond Seiver, W.H. Freeman, 1986.

Drumlins are groups of canoe-shaped hills that indicate the direction of glacial movement. Theories suggest that drumlins form when an advancing glacier overruns an earlier moraine. Eastern Wisconsin and portions of western and central New York State are locales where drumlins can be found.

As a glacier melts, the waters often run through crevasses or cracks in the ice. Streams form under the glacier and eventually carry sediment beyond the ice front. These winding channels of water partially fill with sands and gravel, and the water flow causes the materials to become stratified or layered. As the glacier retreats, the deposits slump down the sides of the channel and form long, winding ridges called *eskers*.

> **REMEMBER**
> Many factors on Earth contribute to shaping and changing its surface. Earth's abundance of major systems discussed in Chapter 1 contributes to the ongoing change discussed within this chapter.

Types of Mountains

FAULT-BLOCK MOUNTAINS

Fault-block mountains (see Figure 9.10) are the first of the four main types of mountains discussed in this section. Fault-block mountains, such as those seen in the Grand Teton Range of Wyoming, form as tensional stresses cause elongation and fracturing of the crust into large blocks. Movement along the faults then tilts the blocks, producing parallel mountain ranges. A large regional uplift results in the production of these tensional stresses and subsequent faulting.

FIGURE 9.10 Fault-block mountains. Source: *Macmillan Earth Science,* Eric Danielson and Edward J. Denecke, Jr., Macmillan, 1989.

The fault-block mountains found throughout the Basin and Range provinces, located across much of the interior far western states, formed after plate movements caused tensional forces on the order of 50 million years ago. Geologists suggest that a regional uplift began at that time as an oceanic plate subducted under this area and sank into the mantle, producing an upwelling of hot magma. The buoyancy of this magma forced the overlying crust to upwarp, resulting subsequently in the formation of fault-block mountains.

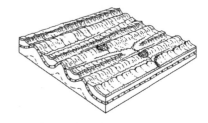

FOLDED MOUNTAINS

Folded mountains (see Figure 9.11), also known as complex mountain systems, are formed by compressional forces. The Alps, Urals, Appalachians, and Himalayas all formed as a result of convergence at plate boundaries. In the Appalachians, the folds in the crust are clearly visible from an airplane as one flies over these ancient mountains. Formed at the same time as Pangaea, about 250 million years ago, the Appalachians are quite weathered and eroded from their original majestic heights.

FIGURE 9.11 Folded mountains. Source: *The Earth Sciences,* Arthur N. Strahler, Harper & Row, 1971.

DOMED OR UPWARPED MOUNTAINS

Domed or upwarped Mountains are formed when broad arching occurs in the crust, or there is great vertical displacement along a fault. The Black Hills of South Dakota are an example of this kind of mountain building. These mountains consist of ancient igneous and metamorphic bedrock that has been eroded to form a *peneplain*, that is, a flat region, once mountainous, shaped by erosion. As new sediment covers the region, upwarping occurs, and erosion removes the sedimentary materials, leaving behind the igneous and metamorphic core that rises above the surrounding terrain.

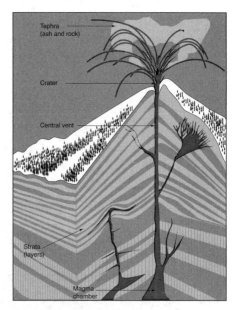

FIGURE 9.12 Sequential volcanic eruptions result in the formation of layers that act to build up the volcano over time. Source: National Mapping Division, U.S. Geological Survey.

VOLCANIC MOUNTAINS

Volcanic mountains, as described in Chapter 8, result when magma forms and works its way to the surface to erupt as lava through a series of dikes, or vertical faults. The magma formation can result from the nearby subduction of an oceanic plate under a continental plate or from a hot spot in the mantle.

When subduction occurs along a broad zone, as along the western United States, a volcanic mountain range such as the Cascades results (see Figure 9.12).

Types of Deformation

FOLDS

When bedrock is subjected to external forces created by plate movements and like phenomena, the rock will first deform or change elastically. Elastic changes are reversible. Under enough heat and pressure, solid rock will act like a rubber band and begin to flex and bend. There is a limit, though, to this kind of stress; it is known as the *elastic limit*. When the elastic limit is exceeded, rock will either deform plastically or fracture. Plastic deformation produces permanent change, often in rock that is buried and under considerable pressure. Surface rock is much more likely to fracture when subjected to external stress than it is to deform plastically.

Folds are quite observable, wavelike undulations in sedimentary rock. These are classic examples of the result of plastic deformation. This process can be replicated by taking a block of clay composed of multiple layers of different colors and compressing it gradually. The result will be folds much like those seen in various parts of the Appalachian Mountains in the eastern United States.

Five types of folds are illustrated in Figure 9.13. Roadway construction has revealed spectacular examples of anticlines and synclines. *Anticlines* are recognized by the upwarping of rock layers; *synclines* are troughs that exemplify downfolding.

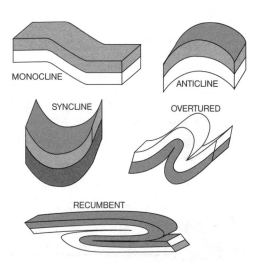

FIGURE 9.13 Various types of folds. Source: *Let's Review: Earth Science—The Physical Setting,* 2nd Ed., Edward J. Denecke, Jr., Barron's Educational Series, Inc., 2002.

FAULTS

Faults are zones where fractures have occurred in the crust. Displacement is often measurable along faults and can range from a few centimeters to much larger distances measured in kilometers. One of the most famous faults in the United States is the San

Andreas Fault. Oriented northwest to southeast through much of California, the region is actually a large fault zone, or area of interconnecting faults. Along the main fault, the San Andreas, displacement has been occurring for many millions of years.

Five types of faults are illustrated in Figure 9.14. Strike-slip faults tend to produce horizontal movement, whereas tensional faults and compressional faults produce vertical movement.

Regional faults, such as the San Andreas (see Figure 7.15), are found along plate boundaries and are often the sites of

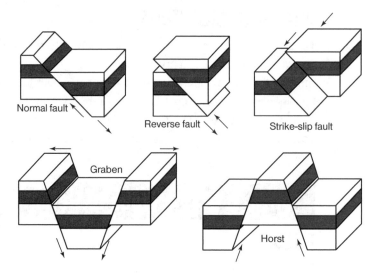

FIGURE 9.14 Various types of faults. Source: *Let's Review: Earth Science—The Physical Setting,* 2nd Ed., Edward J. Denecke, Jr., Barron's Educational Series, Inc., 2002.

earthquake activity. Chapter 8 provides more details on this topic.

JOINTS

Joints are fractures in rock where no appreciable movement has occurred. Joints can form when igneous rock cools and compresses. This type of jointing is known as *columnar jointing.* Jointing can also occur when deep layers of rock expand as they are unearthed by surface weathering and erosion. Most jointing occurs, however, in conjunction with stresses created by crustal movement and orogenic, or mountain-building, forces.

REVIEW EXERCISES FOR CHAPTER 9

WORD-STUDY CONNECTION

celestial	erosion	fracture
compressional	fault	glaciation
corrasion	fault-block	glaciers
dike	folded	hot spot
elastic change	folds	ice age

igneous
joint
Mars
mass wasting
metamorphic
moon
moraine
orogenic
peneplains

pyroclastic
recrystallized
rivers
sediment
streams
strike-slip
subduction
syncline
tensional stress

terminal moraine
uplift
upwarped
upwelling
Venus
vertical
weathering
wind

SELF-TEST CONNECTION

PART A. Completion. *Write in the word or words that correctly complete the sentence.*

1. Mountains weather and erode over periods of millions of years to form flat regions called _____.

2. The Cascade Mountains are a result of _____ occurring under the Pacific Ocean approximately 400 kilometers to the west.

3. _____ mountains form as tensional stresses cause elongation and fracturing of the crust into large blocks.

4. Gravity acting upon rocks and soil to cause landslides is called _____.

5. The elastic limit is the point where rock will either deform or _____.

6. Tensional faults and compressional faults produce _____ movement.

7. _____ occurs when moving water carries sediment, often sand-size or smaller particles.

8. Sand storms and dust storms are examples of _____ acting as an erosional agent.

9. _____ are major erosional forces that have impacted much of the northern United States and Canada.

10. Compressional forces tend to produce _____ mountains.

11. The Black Hills of South Dakota are an example of _____ mountains.

12. A _____ is identified at a site where rock fractures but no appreciable movements occur.

13. Mountain-building forces are also known as _____ forces.

14. _____ ejecta has been known to travel at speeds of up to 200 kilometers per hour.

15. _____ in magma are responsible for forcing it to the surface.

PART B. Multiple Choice. Circle the letter of the item that correctly completes the statement.

1. Earth is the only planet known to have _____.
   (a) an atmosphere
   (b) a surface with abundant ice
   (c) a surface with abundant running water
   (d) a solid surface
   (e) all of the above

2. Much of the work of weathering and erosion is done by
   (a) oceans
   (b) rivers and streams
   (c) wind
   (d) both a and b
   (e) none of the above

3. A landslide is an example of erosion by
   (a) mass wasting
   (b) rivers and streams
   (c) glaciers
   (d) wind
   (e) time

4. Convergence along plate boundaries can produce _____ mountains.
   (a) folded
   (b) fault-block
   (c) domed
   (d) upwarped
   (e) volcanic

5. Glaciers are found in
   (a) polar latitudes
   (b) high altitudes
   (c) all continents
   (d) both a and b
   (e) both a and c

6. The San Andreas Fault is an example of a _____ fault.
    (a) strike-slip
    (b) tensional
    (c) compressional
    (d) converging
    (e) diverging

7. Boulder fields indicate weathering and erosion by
    (a) wind
    (b) glaciers
    (c) rivers and streams
    (d) mass wasting
    (e) all of the above

8. Subduction zones can produce
    (a) fault-block mountains
    (b) folded mountains
    (c) upwarped mountains
    (d) volcanic mountains
    (e) flat plains

9. Vertical movement occurs along
    (a) strike-slip faults
    (b) compressional faults
    (c) tensional faults
    (d) both a and b
    (e) both b and c

10. The Cascade Mountains are an example of
    (a) fault-block mountains
    (b) folded mountains
    (c) upwarped mountains
    (d) volcanic mountains
    (e) flat plains

PART C. Modified True/False. *If a statement is true, write "true" for your answer. If a statement is incorrect, change the <u>underlined</u> expression to one that will make the statement true.*

1. <u>Erosion</u> is the action of breaking down preexisting rock.

2. <u>Upwarped mountains</u> are also known as complex mountain systems.

3. Faults are zones where fractures have occurred in the <u>crust</u>.

4. Tensional faults and compressional faults produce <u>horizontal</u> movement.

5. Earth's present climate supports extensive glaciation in <u>polar latitudes and at high elevations</u>.

6. Sand storms and dust storms are classic examples of the kind of action that <u>streams</u> can create.

7. <u>Landslides</u> are an example of mass wasting.

8. Earth is the only planet known to have <u>liquid water</u>.

9. Glaciers can extend to a thickness of <u>several feet</u>.

10. The <u>Grand Tetons</u> are an example of a fault-block mountain range.

CONNECTING TO CONCEPTS

1. Describe the major weathering and erosional forces on Earth. Which are most able to break down and move material?

2. Describe the four major types of mountains, and explain how each type forms.

3. What are folds, and how do they differ from faults?

4. Compare glaciers to mass wasting. How are their erosional actions similar? Different?

5. What conditions on Earth make it a unique planet capable of reshaping its surface on a regular basis?

6. Explain what is meant by the phrase "agent of erosion."

7. Discuss the following statement: Over time, all mountains are eventually reduced to rubble.

CONNECTING TO LIFE/JOB SKILLS

Weathering and erosion are major issues faced in certain regions. Portions of southern California periodically experience landslides. If a house or other structure is built in the wrong spot, it may slide with all the other material.

It is important, then, to research the geologic and other natural hazards that exist in an area before building or purchasing a home. No location can be certified as completely safe, but a little research can prove some sites to be safer than others.

ANSWERS
SELF-TEST CONNECTION

Part A

1. peneplains
2. subduction
3. Fault-block
4. mass wasting
5. fracture
6. vertical
7. Corrasion
8. wind
9. Glaciers
10. folded
11. upwarped
12. joint
13. orogenic
14. pyroclastic
15. Gases

Part B

1. **(a)**
2. **(a)**
3. **(e)**
4. **(a)**
5. **(c)**
6. **(a)**
7. **(b)**
8. **(d)**
9. (c)
10. (e)

Part C

1. False; Weathering
2. False; Folded mountains
3. True
4. False; vertical
5. True
6. False; wind
7. True
8. True
9. False; a thousand feet
10. True

CONNECTING TO CONCEPTS

1. Major weathering and erosional forces on Earth include water, ice, wind, and mass wasting (e.g., gravity-related mudslides). Water and ice are probably the two most effective forces. Together, they break down tons of rock and transport it across great expanses over time.

2. Fault-block mountains form as tensional stresses cause elongation and fracturing of Earth's crust into large blocks. Movement along the faults tilts the blocks, producing parallel mountain ranges.

   Folded mountains form as a result of compressional forces. For example, convergence at plate boundaries can cause compressional forces.

   Upwarped mountains form when broad arching occurs in Earth's crust, or great vertical displacement occurs along a fault.

   Volcanic mountains form when magma in Earth's interior works its way to the surface and erupts as lava through a series of dikes, or vertical faults.

3. Folds are ripples in the crust that form as a result of compressional forces. Faults are breaks in the crust along which earthquakes can occur.

4. Glaciers and mass wasting can cause profound changes in surface features. Both can scrape and scour underlying bedrock. However, glacial erosion occurs slowly over a long period of time, whereas mass wasting often occurs as a rapid, catastrophic event such as a landslide or a mudslide.

5. Earth is a dynamic planet with an atmosphere, a hydrosphere, a biosphere, and a cryosphere; all of these are capable of reshaping its surface on a variety of time scales.

6. An "agent of erosion" is a process or material that can cause erosion to occur. For example, all rivers are agents of erosion.

7. Our planet is continually changing. Landmasses that are thrust toward the skies by converging plate boundaries are eventually weathered and eroded to sea level by the natural erosional processes that occur continuously on Earth. Hence, no mountain is a permanent feature on the surface of this planet.

10

Water, Water Everywhere

WHAT YOU WILL LEARN

This chapter focuses on the distribution and importance of water on Earth. In this chapter you will learn

- why water is unique and its important properties;
- the role of water in modifying climate;
- the role of ocean currents in transporting massive amounts of heat energy;
- why groundwater exists and how we access this valuable resource;
- about streams and their role in shaping Earth's surface.

SECTIONS IN THIS CHAPTER

- Distribution of Water
- Important Properties of Water
- Groundwater
- Streams and Rivers
- Review Exercises for Chapter 10

Distribution of Water

In recent years, water and its availability as a natural resource have become increasingly important topics of discussion across the entire nation. Concerns about shortages of water as a result of the polluting of existing water supplies, regional drought, and population growth beyond the natural limits of a region have captured headlines across the United States from time to time. As our population increases, and consequently our need for freshwater grows, these issues are likely to become more serious in coming years.

Of all the water on our planet, only a small percentage is potable (drinkable.) About 97 percent of the liquid water on Earth is not freshwater. Most is seawater and is unacceptable for human intake without extensive and expensive processing. To complicate matters further, of the remaining 3 percent of freshwater, only about 15 percent is easily accessible. Much of the planet's freshwater, about 85 percent, is found in ice sheets and glaciers, neither of which are readily accessible.

Even though Earth is referred to as the water planet, and over 70 percent of its surface is covered with water, the supply of potable water is truly finite and limited. Without clean freshwater, no life-form can survive for long.

Important Properties of Water

Water has several interesting properties that occur as a result of its molecular structure. For example, water molecules have a natural affinity toward each other. This affinity, called *hydrogen bonding* (see Figure 10.1), results in properties such as surface tension, a relatively high boiling point, expansion upon freezing, and the ability to act as a universal solvent.

The shape of the water molecule helps to determine its unique set of properties. A water molecule is shaped like a V (see Figure 10.2). Chemists describe a molecule of this shape as a bent or angular molecule. The hydrogen atoms are at the "ends" of

FIGURE 10.1 The structure of water molecules causes hydrogen bonding to occur between water molecules. Water is a polar molecule; that is, one end of the molecule is weakly positively charged and the other is weakly negatively charged. Because opposites attract, the positive end of one water molecule is attracted to the negative end of another. This *intermolecular* attraction is known as a hydrogen bond. Many of the physical properties of water can be attributed to the hydrogen bonding that occurs. Physical properties affected include boiling point, surface tension, and viscosity. Source: *Marine Biology*, 3rd Ed., Peter Castro and Michael Huber, McGraw-Hill, 2000.

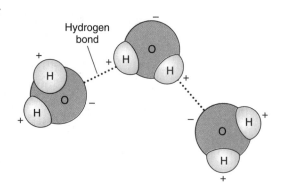

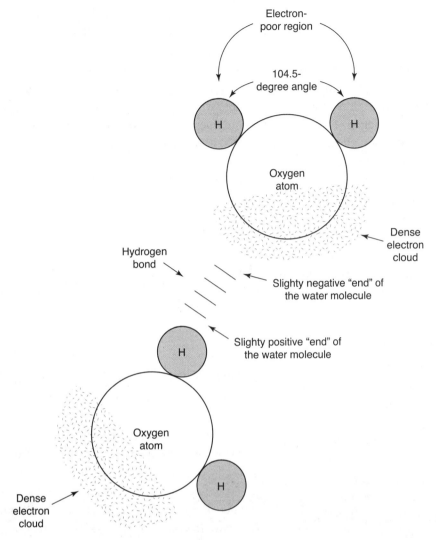

FIGURE 10.2 A water molecule has a bent or angular shape. Since the oxygen atom is a good competitor for the available electrons, the electron cloud becomes most dense toward the oxygen atom. This causes the water molecule to have a slightly negative charge near the oxygen atom. Water molecules are attracted to each other by an electrostatic force of attraction that exists as a result of the slightly positive hydrogen atom of one molecule "hydrogen bonding" to the slightly negative oxygen atom of another.

the molecule, and the oxygen atom is at the center. Oxygen is a very good competitor for electrons, giving it a somewhat negative charge within the molecule; the hydrogen atoms are left with a somewhat positive charge. This situation produces what chemists call a *polar* molecule, where one side of the molecule is more negatively charged than the other. The result is an attraction between the oxygen side of one water molecule and the hydrogen side of another. This attraction helps to hold water molecules together, thus creating the hydrogen bonding just discussed.

Any diver who has failed to execute a dive properly and has ended up doing a "belly flop" has learned firsthand about the *surface tension* of water! It is this same property that enables insects to seemingly walk on water.

Hydrogen bonding is responsible also for water's unusually *high boiling point* in relation to its size. Most substances that are composed of small molecules exist as gases at room temperature, but water does not. Water remains a liquid until a temperature of 100°C is reached. That air temperature does not naturally occur anywhere on this planet. Therefore, since the average global temperature is about 15°C, much of the H_2O on Earth exists in liquid form.

Water has a relatively narrow temperature range through which it remains a liquid; it freezes at 0°C (32°F) and boils at 100°C (212°F). In fact, if Earth were just a little warmer or colder, water would likely exist as water vapor or as ice, and life on this planet may never have taken hold (see Figure 10.3).

Water is one of the few substances that expands upon freezing. This property, also, results from its molecular structure. In Figure 10.3 the molecular structure of water is shown next to a graph of temperature versus time. Take note that water molecules in liquid form can reduce their intermolecular distance, thus changing a fundamental physical property known as *density*.

When a given quantity (mass) of water expands (increases in volume), it necessarily becomes less dense. The operational mathematical equation here is

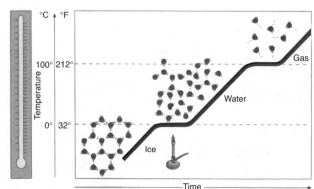

FIGURE 10.3 The melting and boiling behavior of water. The melting and freezing points of water can be identified easily because the temperature remains constant at 0°C during the phase change. The same can be said of water's behavior while changing phase from a liquid to a gas. As ice melts, the energy supplied to the ice is used to break down the crystalline structure present in ice. With the further addition of heat, water molecules overcome hydrogen bonds (attractive forces) as the water boils, forming water vapor. Source: *Marine Biology*, 3rd Ed., Peter Castro and Michael Huber, McGraw-Hill, 2000.

$$D = \frac{M}{V},$$

where D is density, M is mass, and V is volume.

Since ice (frozen water) has a lower density than water, ice floats on water. Consider how different things would be if ice sank to the bottom of a pond or lake rather than floating on its surface. During a severe winter in northern climates, lakes might freeze solid, a situation that never occurs today. When ice forms on a lake, it actually acts as a protective barrier between the cold air above and the organisms living in the lake below. Few if any life-forms could survive if a lake froze solid. An ecosystem would be severely disrupted if all life-forms living in a lake or pond failed to survive each winter.

Water's molecular structure also enables it to act as a *solvent* for many substances. A great number of ionic compounds and some molecular compounds dissolve in water; as a result, flowing water, such as that found in rivers and streams, can both dissolve and transport minerals over great distances. The transportation of dissolved minerals results in the formation of new rocks and minerals and the redistribution of the planet's natural resources. The salts and other ionic compounds found in the oceans are there as a result of water's ability to dissolve minerals.

The polarity of water also ensures its ability to dissolve like substances, that is, those that are polar or ionic. Sugar is a polar substance, and table salt (sodium chloride) is ionic; hence, both dissolve in water.

The ability of water to dissolve many substances explains the abundance on Earth of water that is too saline for human consumption. Even "freshwater" contains dissolved minerals; however, the concentration of these minerals is within tolerable limits for humans and other life-forms. In fact, the dissolved minerals found in water in minute quantity are vital for life. Distilled water, or water with the minerals removed, essentially "pure" water, is not good for humans as it disrupts cellular function.

Summarizing the discussion to this point, water is a small but, as noted, vital molecule. All life-forms require it to exist. Water in the liquid form, as noted frequently in this text, is unique to Earth, and so is life.

Another interesting property of water is its *high specific heat*. Specific heat is a measure of the amount of energy required to warm a given amount of a substance from one temperature to another. It is usually measured in calories or joules per gram-Celsius degree. The specific heat of water (conveniently) is 1.00 calorie per gram-Celsius degree. This means, in effect, that water requires 1 calorie of heat (energy) to raise the temperature of 1 gram of water 1 Celsius degree. Conversely, water releases to its surroundings, essentially the atmosphere, 1 calorie of heat for every 1 gram of water that cools 1-Celsius degree in temperature. When compared to other substances, water's specific heat is relatively high.

In practical terms, this means that large quantities of heat are required to warm a body of water, such as a lake or a portion of an ocean. Consequently, warm water must contain large amounts of heat. As a result, water is an important medium in maintaining and controlling global climate. Warm equatorial waters carrying large quantities of heat travel as currents in the ocean toward polar latitudes. As the warm waters encounter cooler air in higher latitudes, these waters release large quantities of heat to the atmosphere, thus modifying the climate of any region in the vicinity of this energy exchange.

Perhaps the most dynamic example of this process can be observed as the North Atlantic Ocean's Gulf Stream carries warm water from the Gulf of Mexico northward into the North Atlantic Ocean and eventually toward western Europe. England's climate, for example, is greatly influenced and modified by this warm water current as it warms the atmosphere of this nation, which otherwise would be much colder by virtue of its high latitude.

Portions of eastern Canada, situated at latitudes similar to those of England, experience much colder climatic conditions because the warm Gulf Stream waters do not affect them. Figure 10.4 compares two cities at nearly identical latitudes: Halifax in Canada, with minimal influence from the Gulf Stream, and Bordeaux, France, in western Europe, which is heavily influenced by warm waters off the Atlantic. Note the pronounced climatic differences, particularly during the winter season.

Recent research at the Woods Hole Oceanographic Institute in Massachusetts has raised concerns that the current warming trend in the Arctic and the subsequent melting of polar ice are changing the salinity of seawater at an alarming pace. Changes in salinity may act to alter or disrupt the Gulf Stream (Figure 10.5) and other major ocean currents. Should the Gulf Stream effectively cease to flow through the Atlantic, much of western Europe, and even portions of eastern North America, might cool as much as 10 Fahrenheit degrees in a very short period of time, perhaps in one decade. Conditions not seen since the "little ice age," which ended in the late 1700s, could return and devastate regions where hundreds of millions of people now live. If this scenario were to unfold, it would truly be an example of the unforeseen and unintended consequences of global warming, due in part, at least, to human activity. In this scenario, global warming could lead to another "mini ice age." Scientists at this time are simply uncertain that these events are likely to unfold as described here.

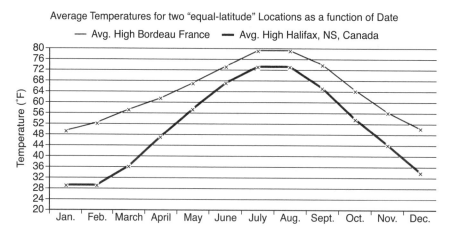

FIGURE 10.4 Comparison of the climates of Halifax, Nova Scotia, and Bordeaux, France, both located near 45° N. Note the significant difference, particularly during the winter months.

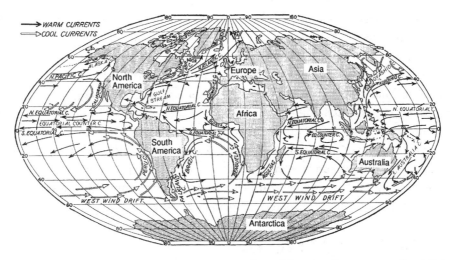

FIGURE 10.5 Surface ocean currents. Source: The State Education Department, *Earth Science Reference Tables*, 2001 Ed. (Albany, New York: The University of the State of New York).

Groundwater

At first glance, you might suspect that the earth beneath your feet is solid. This view is incorrect, however, as there are many breaks or joints in the bedrock; space even exists between sand grains. Water finds its way into all these tiny openings and moves through them. As the water travels, it is filtered so that surface pollutants are often removed before the water is later extracted for human consumption.

Groundwater is extremely important; it represents the greatest source of freshwater on the planet available to humans. By volume, glaciers and ice sheets are six times more plentiful, but their inaccessibility makes them poor alternatives to groundwater. Groundwater, found in underground aquifers, is the source of drinking water for more than half of our nation's population (Figure 10.6).

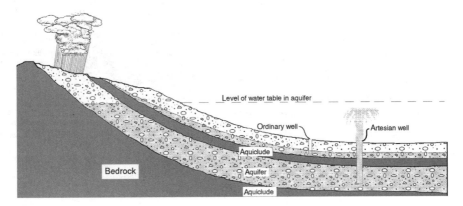

FIGURE 10.6 An Artesian formation. The water rises in an artesian well because of hydrostatic pressure created by the weight of the water in the aquifer. Source: *Let's Review: Earth Science—The Physical Setting*, 2nd Ed., Edward J. Denecke, Jr., Barron's Educational Series, Inc., 2002.

An ongoing concern in our society is the overuse of groundwater, which can lead and has led to contamination of natural freshwater sources as pollutants leach in or to depletion of the resource. This problem is particularly acute in the American Southwest. This region characteristically receives little rainfall, an unfortunate truth facing the millions who have migrated there because of its comfortable climate. Natural water sources, including lakes and aquifers, have been stressed to their limits, and natural pollutants, such as arsenic, have created health concerns.

Groundwater, one of these natural sources, is also an important erosional agent. Moving groundwater can create cavities that ultimately produce sinkholes on the surface. Caverns and caves may also result from the erosion of subsurface structures by groundwater. These natural formations are discussed in a later section of this chapter.

HOT SPRINGS AND GEYSERS

Water can be found in numerous places on this planet. Hot springs and geysers are certainly among the more interesting and pleasant of these places! Anyone who has visited a hot spring can testify to its soothing warmth, particularly when soaking in the middle of winter! Hot springs, common to many locales in the western United States, are sites where the groundwater is at least 6 Celsius (10 Fahrenheit) degrees warmer than the mean annual temperature for that site.

Hot springs form when groundwater at great depth is heated and is then circulated to the surface. The source of heat for the hot spring is slowly cooling igneous rock from "recent" volcanic activity. Such activity can have occurred many thousands of years ago, as in Yellowstone National Park, famous for its hot springs and geysers.

One of the most famous *geysers* is dubbed Old Faithful (see Figure 10.7) as a result of its consistent schedule of eruptions. Here, as with all geysers, the eruptions of water and steam are caused by the buildup of gases and hot water underground, which continues until the pressure becomes sufficient to cause the water to erupt violently from the opening at the surface. After an eruption, the continual supply of subsurface energy guarantees that steam and other gases will shortly build up sufficient pressure to produce another eruption—hence the name Old Faithful. A geyser may be thought of as an intermittent hot spring or as a fountain.

FIGURE 10.7 Old Faithful, Yellowstone National Park.

WELLS

Groundwater is ultimately extracted by means of wells. *Wells* are holes drilled into the zone of saturation (see Figure 10.8). This is a layer where all open spaces in sediment and rock are saturated with water. The well driller seeks to find an aquifer, that is, a layer where permeable rock (rock that can transmit water through pore spaces) or sediment allows groundwater to travel freely. A good aquifer can serve an entire city. The well creates a small reservoir of water that can be pumped to the surface. If the well is overpumped, however, the water table (level) drops and a new well has to be bored.

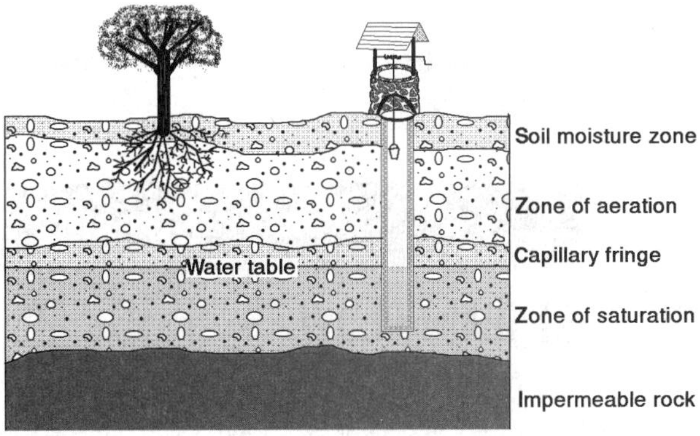

Artesian wells are wells that tap groundwater under pressure. If enough pressure exists, it will cause the groundwater to rise to the surface without the aid of external pumping. Artesian wells are usually found in an aquifer or in a layer of highly porous material that is filled with groundwater. Frequently, the porous rock lies between two layers of clay or other material that is not porous and hence does not let the water escape. If the aquifer is tilted, water will enter at the upper end and flow down the aquifer. This slowly flowing water creates the pressure that allows the artesian well to function as described.

FIGURE 10.8 Zones of soil water and groundwater. The water table is the boundary between the zone in which pores are filled with air and the zone in which pores are filled with water. Source: *Let's Review: Earth Science—The Physical Setting*, 2nd Ed., Edward J. Denecke, Jr., Barron's Educational Series, Inc., 2002.

CAVERNS AND CAVES

Caverns (see Figure 10.9) and *caves*, which are spectacular by-products of the flow of underground water, number in the thousands across the United States. Acids in groundwater can dissolve minerals, particularly from limestone. Limestone is rich in calcium carbonate, which readily dissolves if the water is acidic enough.

Caves and caverns tend to form at or below the water table within the zone of saturation.

Portions of the Appalachian Mountains are composed of limestone, so it should be no surprise to find caves, some of

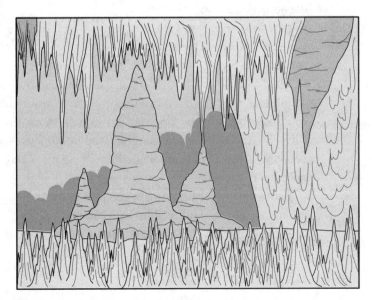

FIGURE 10.9 "Temple of the Sun" in the Big Room in Carlsbad Caverns National Park in New Mexico.

which have cavities that are quite large. Many larger caves and caverns, located in New York, Pennsylvania, Virginia, West Virginia, and Kentucky, are open to the public as commercial enterprises.

Streams and Rivers

STREAM FLOW

Today, streams and rivers are the major erosional agents on Earth's surface. Streams and rivers act to break down or erode bedrock into particles or dissolved minerals. These particles, through the process of corrasion, further scour the streambed as water flows downstream from a river's source to its mouth.

Rivers—and streams— often exhibit different stages from source to mouth (see Figure 10.10). Near the source, often in mountains, the river is *youthful*. The narrow channel and fast flow carve out a V-shaped valley. Farther downstream, the rate of flow slackens, and the river is then considered to be *mature*. The river's ability to carry larger rocks and boulders diminishes, and the river widens. The result is a wider, V-shaped valley and

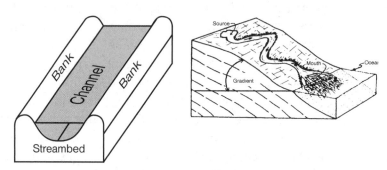

FIGURE 10.10 A stream from its source to its mouth. A stream begins at its *source,* flows along a path called a *channel,* and ends at its *mouth*. The channel consists of sides, called *banks*, and a bottom, called the *streambed*, whose slope is the stream's gradient. Source: *Let's Review: Earth Science—The Physical Setting*, 2nd Ed., Edward J. Denecke, Jr., Barron's Educational Series, Inc., 2002.

eventually a broad, wide, relatively flat floodplain (see Figure 10.11).

The Mississippi River offers an excellent example of this kind of action. In Minnesota, near the river's source, the water flows rapidly, and the floodplain is narrow and is bounded by steep valley walls. Farther downstream, in Iowa, Illinois, and Missouri, there is an increasingly broad floodplain that on more than one occasion has been the site of major flooding when spring rains and winter snowmelt combined to raise the river level.

Farther downstream, the Mississippi, like most major rivers, has a wide, flat delta as it empties into the Gulf of Mexico. The delta is composed of the finest particles and dissolved minerals that the river carries in its gentle flow toward its mouth.

DRAINAGE BASINS

A *drainage basin*, otherwise known as a *watershed* (see Figure 10.12), is a region that contributes to a stream or a river; all streams and rivers have drainage basins. A

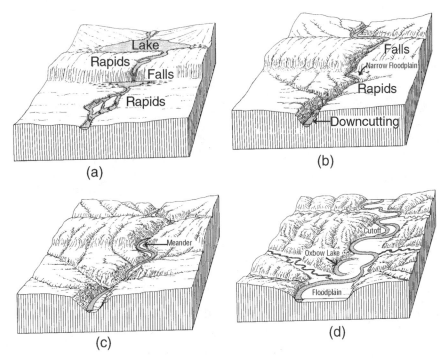

FIGURE 10.11 The three stages in the life cycle of a stream. (a) A youthful stream. The relief of the land is steep, and the stream has lakes, rapids, and falls. (b) Continued downcutting of the streambed eliminates falls, and cutting back of the valley walls allows a narrow floodplain to form. (c) Further downcutting and erosion of the valley walls enlarges the floodplain and allows the stream to meander widely between the valley walls. (d) A mature stream with a minimal gradient. The stream is characterized by shifting meanders that result in cutoffs and oxbow lakes. Source: *The Earth Sciences*, Arthur N. Strahler, Harper & Row, 1971.

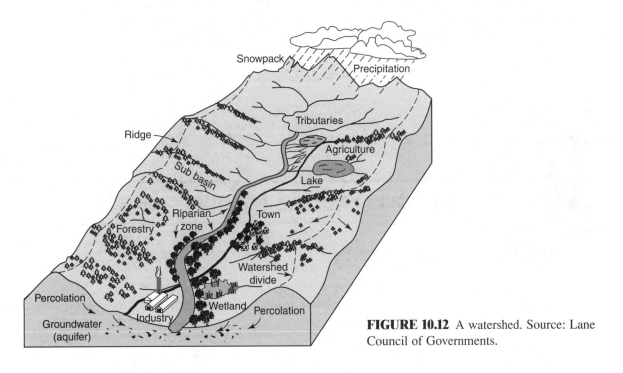

FIGURE 10.12 A watershed. Source: Lane Council of Governments.

FIGURE 10.13 The Continental Divide, situated along the spine of the Rocky Mountains. Note the flow of rivers from this natural separation of two major watersheds. West of the Continental Divide, all rivers eventually flow to the Pacific Ocean; east of the divide, all rivers eventually flow toward the Atlantic Ocean.

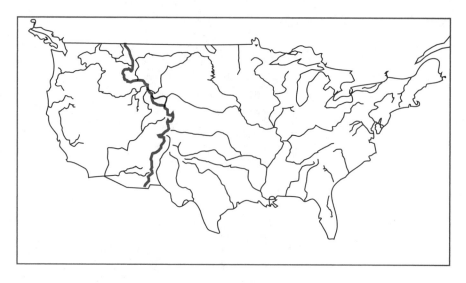

simplistic model that will help you to visualize a drainage basin is, of all things, a swimming pool cover! Refer to the following Web site: *http://ga.water.usgs.gov/edu/ pictureshtml/swpool.html* for the full story. In brief, the entire pool cover represents a watershed; that is, precipitation collects in this "basin" and flows toward the lowest part of it (in this example, that is usually the center of the pool cover). The water collects to form a lake; and if there is a natural route to lower ground, a river will flow from the lake.

Divides separate drainage basins. On a small scale, a *divide* can be a ridge along which water flows in two different directions. The Continental Divide, illustrated in Figure 10.13, is a divide on the grandest of scales. The Continental Divide, which runs along the spine of the Rocky Mountains, separates all rivers that eventually flow to the Pacific Ocean from all rivers that eventually flow to the Atlantic Ocean. A smaller divide in the heart of the Appalachian Mountains separates rivers that flow directly to the Atlantic Ocean from those that flow to the Gulf of Mexico, which empties into the Atlantic Ocean.

REVIEW EXERCISES FOR CHAPTER 10

WORD-STUDY CONNECTION

angular	boiling point	corrasion
aquifer	calcium carbonate	density
bedrock	cavern	divide

drainage basin	ionic compounds	reservoir
ecosystem	joints	rivers
erosion	limestone	saline
freezing point	mass	salt
gases	mature	sink hole
geyser	minerals	streams
glaciers	molecular shape	solvent
groundwater	molecular structure	water
hot spring	polar molecule	well
hydrogen bonding	pollutant	youthful stage
ice	potable	

SELF-TEST CONNECTION

PART A. Completion. *Write in the word or words that correctly complete the sentence.*

1. A _____ is an intermittently "erupting" stream of hot water and steam.

2. A _____ separates drainage basins.

3. _____ is probably the greatest source of drinking water for our nation.

4. A substance's _____ is the temperature at which a phase change occurs from the liquid to the gaseous state.

5. Since water is a highly polar molecule, it exhibits _____ between the molecules.

6. Water's molecular shape is best described as bent or _____.

7. Most small molecules have relatively low boiling points and therefore occur as _____ at room temperatures.

8. When water is pure enough to drink, it is said to be _____.

9. Distilled water is not good for human consumption as it lacks the _____ necessary for certain cellular functions.

10. Most of the world's water is too _____ for humans to consume.

11. The Continental Divide separates _____ that flow to the Pacific Ocean from those that flow to the Atlantic Ocean.

12. When groundwater erodes limestone, a _____ can form on the surface.

13. Rivers are usually in their _____ near their source in the mountains.

14. Breaks in bedrock are also known as _____ .

15. When water freezes, its _____ becomes less.

PART B. Multiple Choice. Circle the letter of the item that correctly completes the statement.

1. The temperature at which a liquid changes phase and becomes a solid is known as its
   (a) boiling point
   (b) condensation point
   (c) evaporation point
   (d) freezing point
   (e) melting point

2. A solvent is capable of _____ other substances.
   (a) breaking down
   (b) dissolving
   (c) mixing with
   (d) reacting with
   (e) both b and c

3. A large aquifer can supply an entire
   (a) apartment building
   (b) city block
   (c) city
   (d) state
   (e) nation

4. When a river begins to widen and the average slope declines, the river is said to be entering a _____ stage.
   (a) mature
   (b) youthful
   (c) frozen
   (d) slow-moving
   (e) immature

5. Glaciers are a potential supply of
   (a) saline water
   (b) salty water
   (c) freshwater
   (d) rocks and minerals
   (e) ice for lemonade

6. In regard to our water supply, the most significant issue(s) we face as a society include(s) _____.
   - (a) overpopulation of certain regions
   - (b) pollution of existing supplies of freshwater
   - (c) periodic and regional droughts
   - (d) Both a and b
   - (e) all of the above

7. Hot springs and geysers owe their warm waters to energy released from
   - (a) radioactive decay
   - (b) slowly cooling sedimentary rock
   - (c) slowly cooling igneous rock
   - (d) giant heat vents beneath the surface
   - (e) all of the above

8. Of all the water on Earth, approximately _____ percent is freshwater.
   - (a) 97
   - (b) 90
   - (c) 85
   - (d) 15
   - (e) 3

9. Water has high melting and boiling points in relation to its molecular size. This property can be attributed to water's
   - (a) surface tension
   - (b) polarity
   - (c) hydrogen bonding
   - (d) both a and b
   - (e) both b and c

10. Aquifers are sites where
    - (a) hot springs are often found
    - (b) much industrial activity is likely to take place
    - (c) freshwater can be obtained
    - (d) highly saline water is found
    - (e) polluted water is found

PART C. Modified True/False. If a statement is true, write "true" for your answer. If a statement is incorrect, change the <u>underlined</u> expression to one that will make the statement true.

1. <u>Aquifers</u> are layers within the Earth where permeable rock allows groundwater to flow freely.

2. When water freezes, it becomes <u>less</u> dense.

3. A <u>hot spring</u> is a surface feature that periodically erupts as a stream of hot water and steam.

4. A <u>divide</u> separates drainage basins.

5. Caverns form when moving water dissolves <u>igneous rock</u>.

6. A river's <u>mature</u> stage is characterized by many rapids and fast-flowing waters.

7. Most rivers have a <u>narrow</u> floodplain in their mature stages.

8. Surface tension is a property of water that results from <u>hydrogen bonding</u>.

9. Water can dissolve <u>polar</u> substances because "like dissolves like."

10. All life on Earth requires <u>oxygen</u> to survive.

CONNECTING TO CONCEPTS

1. Create a pie chart illustrating the relative distribution of major water sources as described in the introductory section of this chapter.

2. Describe each of the properties of water that result from the fact that it is an angular, polar molecule.

3. If water were a linear, nonpolar molecule, how would the world be different?

4. Research what types of ionic compounds are most soluble in water and which are least soluble.

5. When hot springs and geysers are found on the surface, what does this tell you about what is occurring underground?

6. What is a divide?

CONNECTING TO LIFE/JOB SKILLS

The subject of water—accessing fresh sources, conserving our current supplies, and avoiding polluting these supplies—is central to the survival of our species and many others. Most if not all forms of life that live on Earth's surface require fresh, clean water. As the human population continues to grow and as climatic patterns seemingly are becoming increasingly variable, our society faces a significant challenge to keep humanity supplied with as much fresh, unpolluted water as is needed.

ANSWERS
SELF-TEST CONNECTION

Part A

1. geyser
2. divide
3. Groundwater
4. boiling point
5. hydrogen bonding

6. angular
7. gases
8. potable
9. minerals
10. saline

11. rivers
12. sink hole
13. youthful stage
14. joints
15. density

Part B

1. **(d)**
2. **(b)**

3. **(c)**
4. **(a)**

5. **(c)**
6. **(e)**

7. **(c)**
8. **(e)**

9. **(e)**
10. **(c)**

Part C

1. True
2. True
3. False; geyser
4. True

5. False; limestone
6. False; youthful
7. False; wide

8. True
9. True
10. False; water

CONNECTING TO CONCEPTS

1. The pie chart should emphasize the relative abundance of water in the oceans as compared to other sources.

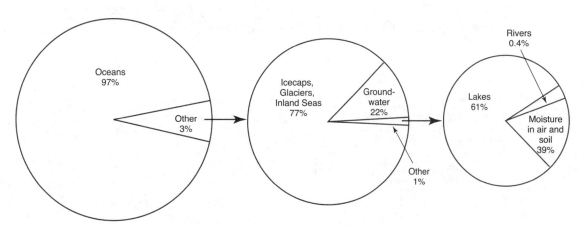

2. Water's molecular shape and polarity are critical to its macroscopic, that is, observable to the naked eye, behavior. As a result of their angular shape, water molecules, upon changing phase from the liquid to the frozen state, assume a less efficient arrangement. This phenomenon assures that ice is less dense than water

and therefore floats on water. As ice forms on water bodies, the water and life-forms within it are protected from the extreme cold above.

The polarity of water assures its ability to dissolve a number of minerals or salts. This property explains the presence of dissolved minerals or salts in water and also the abundance of salts in ocean waters. It also enables water to act as a solvent for many substances.

3. Water would not behave as a "universal solvent" and therefore would not contain dissolved minerals. Water would be less dense than ice. Upon freezing, ice would sink. Shallow-water bodies would freeze solid during prolonged periods of cold weather, and most life-forms within these water bodies would be destroyed as a result.

4. Generally, alkali metal salts (e.g., sodium chloride—all sodium salts for that matter) are soluble. Additionally, all nitrates are soluble. A classic example is potassium nitrate, known as "potash" when processed commercially. Other ions are soluble under certain conditions. Least soluble substances include many of the "heavy metal" salts.

5. Hot springs and geysers are indications of large amounts of subsurface heat. This could be caused by tectonic activity (plate movement along faults) or to a thin spot in the crust where hot mantle material reaches toward the surface and heats sub-surface waters to create hot springs and geysers on the surface.

6. A divide is a physical feature, usually a mountain range that separates watersheds and the flow of rivers. The most significant divide in the United States is the Continental Divide. Located along the spine of the Rocky Mountains, the Continental Divide separates all rivers that ultimately flow to the Pacific Ocean from those that ultimately flow to the Atlantic Ocean.

11

The World's Oceans

WHAT YOU WILL LEARN

This chapter focuses on the world's oceans. In this chapter you will learn

- what seawater is composed of;
- how the ocean is layered;
- what El Niño is and how it impacts the climate globally;
- about seafloor features and how the ocean floor is mapped.

SECTIONS IN THIS CHAPTER

- Seawater
- Mapping the Oceans
- Features of the Seafloor
- Review Exercises for Chapter 11

Seawater

COMPOSITION

From space, Earth appears to be a blue planet. Oceans that extend to an average depth of several kilometers cover over 70 percent of our planet's surface. Ocean water is composed of more than pure water; the oceans contain a complex mixture of salts. The salinity, or average concentration of these salts, is approximately 35 parts per thousand, or 35 grams of dissolved minerals per 1000 grams of seawater. Because of the immense amount of water in the oceans, this concentration, which may seem small, would yield a huge amount of salt if the oceans were ever to evaporate.

The dissolved salts are a mixture of cations and anions that remain in solution together. These include, in order of abundance, chloride (19,000 ppm), sodium (10,500 ppm), magnesium (1350 ppm), sulfate (885 ppm), calcium (400 ppm), potassium (380 ppm), and many others in smaller quantities. Although only seven elements occur in abundance, over seventy naturally occurring elements have been found in seawater. Figure 11.1 shows the composition of seawater in detail.

It is interesting that, while the above concentrations of ions do not change, the overall salinity of water can vary from 33 to 37 parts per thousand. Lower salinities are found in regions where precipitation is great and evaporation is minimal; high salinities are typical of dry climates with high rates of evaporation. Figure 11.2 indicates regions of high and low ocean-surface salinity.

Two Web sites provide additional information about ocean salinity. Project ARGO is discussed at

www-argo.ucsd.edu/.

Project ARGO, which is supported by the National Oceanic and Atmospheric Administration, will eventually deploy 3000 free-drifting floats throughout the world's oceans. These floats are designed to measure temperature and salinity in the upper 2000 meters of the ocean. A database, called "Global Temperature-Salinity Profile Program (GTSPP) Database," has been created and can be found at

www.nodc.noaa.gov/GTSPP/gtspp-home.html.

SOURCES OF MINERALS

Where do all the dissolved minerals or salts found in seawater come from? Rivers, streams, and underground water help to chemically weather rocks on the continents. Chemical weathering dissolves minerals, thus releasing ions into the water and turning it from freshwater to saline water. The landmasses, however, are not the only source of the minerals found in oceans; another major source is Earth's interior. Outgassing, the process of releasing gases from the interior, occurs from volcanic activity along the seafloor, as well as from other hydrothermal events.

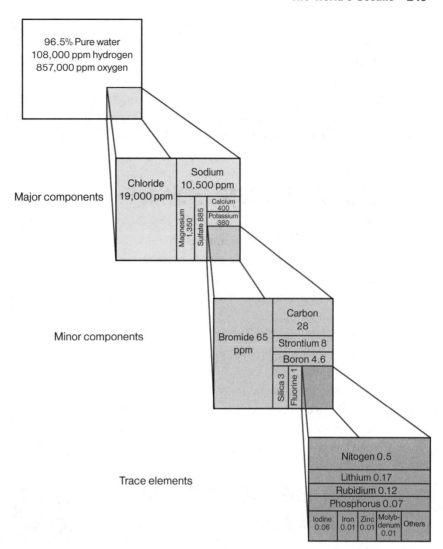

FIGURE 11.1 Materials dissolved in seawater. Source: *Introduction to Oceanography*, 3rd Ed., David A. Ross, Prentice-Hall, 1982. Used by permission.

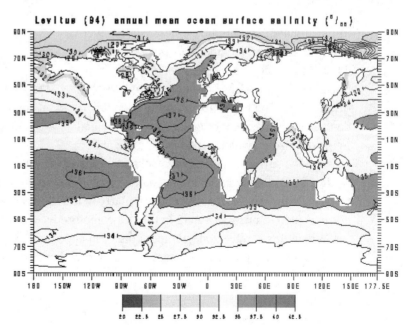

FIGURE 11.2 Ocean salinity as it varies by latitude. Average ocean salinity is 35 ppt (parts per thousand), with values ranging from about 30 to 38 ppt on the map. High values are highly saline waters, low are low salinity waters. A value of 35 ppt means that there are 35 grams of salt per 1000 grams of seawater. Source: International Research Institute Data Library.

As new minerals are added to the ocean, plants and animals remove other minerals just as quickly. Salts enter the oceans from various sources, including underwater volcanoes, the erosion of mineral-rich rocks, and the shells of some marine organisms when they die. Salts are removed from the oceans when plants and animals, acting as natural sinks, incorporate them into their structures. Precipitation of minerals also removes salts from seawater. The net result is a balance between input by sources and outflow by natural sinks that has produced stable salinity levels for the past couple of hundred million years.

RESOURCES FROM SEAWATER

The oceans supply many commercial products as a result of their vast stores of minerals. Among the most common are ordinary table salt and other sea salts. Salts season foods and also preserve them. Magnesium metal and bromine are also harvested from the sea. Magnesium is an important trace mineral for human consumption. Bromine is used as an additive to gasoline.

Freshwater can be gleaned from seawater through *desalination*. This process removes salts and other chemicals from seawater, making it potable and also useful to agriculture and industry. From a cost perspective, however, desalination remains prohibitive as a means of making seawater a common source of potable water.

THE LAYERED STRUCTURE OF THE OCEANS

The oceans can be divided into three distinct and somewhat separated layers, on zones, particularly in the midlatitudes and the tropics. Where layers are observed, they include a *surface mixed* zone; a transition zone, or *thermocline*; and a *deep*, benthic zone. Figure 11.3 shows the relative location of each layer and the temperature profile with depth. Over 80 percent of ocean water is located in the deep zone, which varies by latitude and

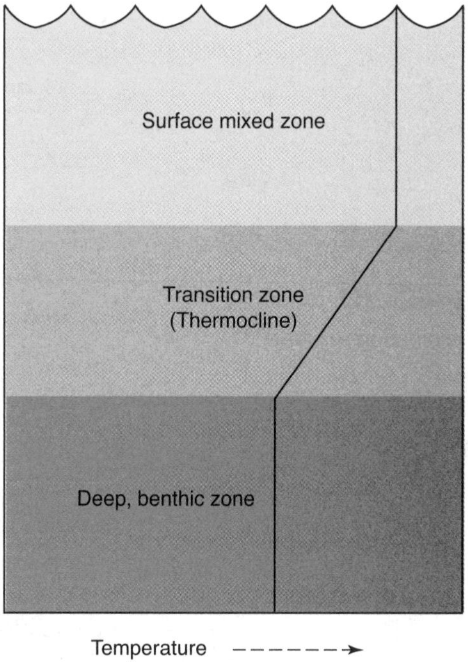

FIGURE 11.3 Oceanic layers. Note: Not shown to scale here, the surface mixed zone is razor thin when compared to the benthic zone.

season. The surface mixed zone reaches its greatest depths in the tropical latitudes.

Relatively warm temperatures characterize the surface mixed zone as the Sun's energy warms the water surface, particularly in the summer season. Waves and surface currents help to mix the waters in this zone and produce strong vertical energy transfer. Within this surface zone, temperatures can approach 26°C and may be even higher in some isolated spots with strong sunlight and without cool surface currents.

Below the surface mixed zone lies the thermocline, which is characterized by rapid temperature change with depth. The thermocline generally begins at depths of 150 meters in the tropics and up to 300 meters in midlatitude or subtropical waters, and it leads to the cold, dark waters of the deep zone.

At depths greater than 1500 meters in the midlatitudes and lower latitudes, water temperatures remain at just below 4°C. In the high latitudes, these temperatures are observed from the surface to great depths because of the cold surface conditions. In contrast to the midlatitudes and tropics, where surface waters warm for at least part of the year, polar surface waters remain cold year-round.

The deep zone does not exhibit any significant mixing or vertical transport of energy with the exception of an important global ocean current that passes through these deep, cold waters. This current, known as the *thermohaline circulation*, essentially creates a global ocean conveyor belt (see Figure 11.4), or deep ocean current system that ultimately transports cold, dense waters toward the Equator. The

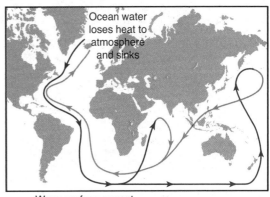

Warm surface current
Cold and salty deep ocean current

FIGURE 11.4 Deep ocean currents.

current originates in the surface or near-surface waters of the polar regions. Two factors, low temperatures and high salinity, combine to increase the density of polar waters. It is clear why the low temperatures exist; the high salinities develop as ocean water freezes. The ice does not absorb the salts in ocean water as it freezes. Rather, these salts stay in the remaining water, raising its salinity and consequently its density. Dense polar waters sink slowly, thus initiating the thermohaline circulation. This current travels along the ocean floor toward the Equator, where upwelling (see Figure 11.5) along the western coastlines of the Americas and Africa brings these cold, dense waters to the surface.

Upwelling acts to regulate Earth's climatic zones. Furthermore, upwelling impacts the biosphere, as deep, cold waters are rich in nutrients. Where upwelling transports nutrients to the surface, marine life flourishes. Chapter 12 contains a discussion of upwelling and its effect on coastal climates.

The phenomenon known as El Niño or, more formally, as the El Niño Southern Oscillation (ENSO), effectively shuts down upwelling along the west coast of South America for periods of up to 2 years per event. Fishermen have known for years when an El Niño is beginning as their catch diminishes significantly when the upwelling currents cease. Widespread climatic changes occur during an El Niño as sea surface temperatures diverge from their normal patterns throughout the Pacific Ocean. The El Niño/La Niña phenomenon is discussed in greater detail in Chapter 12. Figure 11.6 illustrates El Niño's impact on sea surface temperatures.

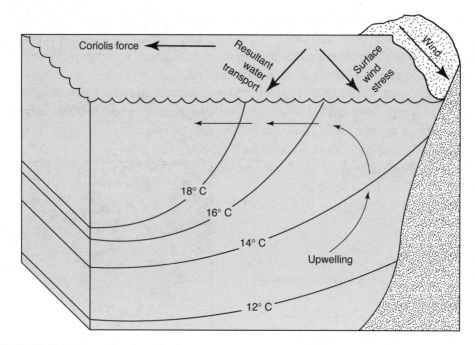

FIGURE 11.5 Upwelling is an important process in the oceans that acts to supply nutrients to surface waters as well as modify climate on a global scale. Wind-driven currents create upwelling, bringing cool, nutrient-rich waters at depth to the surface.

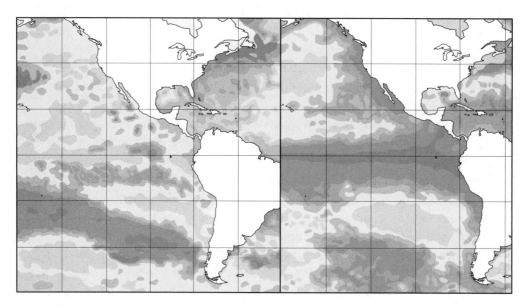

FIGURE 11.6 Study the changes in sea-surface temperature patterns that occur in the eastern Pacific Ocean during an El Niño (right) as compared to La Niña or cold-phase conditions (left). Note that sea-surface heights correlate directly to sea-surface temperatures: the higher the sea-surface height, the higher the sea-surface temperature. Dark indicates higher temperatures, light lower temperatures.

Recently, another oscillation, the North Atlantic Oscillation (NAO), has received increased attention. Various scientific teams, including those at the National Oceanic and Atmospheric Administration, are currently studying the NAO to identify its natural rhythms and its impact on populated regions of North America and Europe. Since the Atlantic Ocean is significantly smaller than the Pacific, the NAO is not as great an influence as the ENSO on global climate patterns. This explains why El Niño gets all the headlines!

> **REMEMBER**
> The world's oceans exert great control over climate, as they transport great quantities of heat via convection currents.

MARINE-LIFE ZONES

The study of the oceans would not be complete without considering the life-forms found within them. Such factors as seawater temperature, salinity, availability of sunlight, and the presence of other seafloor organisms determine the kinds of plants and animals that are likely to thrive in the various marine zones (see Figure 11.7).

One method for categorizing the ocean zones is based upon the availability of light. Light penetrates only the uppermost reaches of the oceans. This region, the *photic zone*, extends to a depth of about 200 meters (600 ft). The depth of the photic zone is limited by the degree of clarity of the seawater. The presence of air bubbles, decaying organic matter, and microorganisms can limit the depth of the photic zone even further.

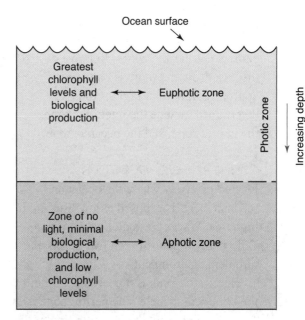

FIGURE 11.7 Diagram of euphotic, photic, and aphotic zones.

The *euphotic zone* lies within the uppermost portion of the photic zone. This is a region where sunlight is bright enough to support photosynthesis. Phytoplankton, tiny organisms of the alga family, are present in great quantities in this zone. Phytoplankton are truly the bottom of the marine food chain, supplying food for zooplankton, which serve, in turn, as food for more complex organisms. Phytoplankton are also important because they produce large quantities of oxygen, much more than the tropical rain forests, which are often touted as an important source of oxygen for our planet.

The photic zone has an abundance of life as compared to the *aphotic zone*, where the only light is generated by bioluminescent bacteria and other life-forms. Even today, large portions of this dark, mysterious region remain unexplored. Recent discoveries have revealed surprising new and previously unknown forms of life. These life-forms have been found near hydrothermal vents along the seafloor in a region known also as the *benthic zone*. A *hydrothermal vent* is an opening in mid-ocean ridges that releases superheated water, gases, and minerals from Earth's interior. These materials warm the surrounding waters and provide important nutrients for the simple life found in these regions. Among the more exotic species recently discovered are giant tubeworms that grow at astonishingly rapid rates, as much as 1 meter per year. This activity occurs at depths of several kilometers, in the absence of light and at incredible pressures.

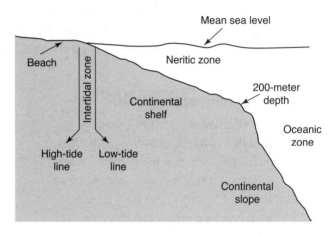

FIGURE 11.8 Relative locations of intertidal, neritic, and oceanic zones.

Other methods are also used to categorize marine-life zones. Relative distance from the shore is one approach. Important regions thus determined include the intertidal zone, the neritic zone, and the oceanic zone (see Figure 11.8).

The *intertidal zone*, which is located precisely where the ocean and the land meet, is exposed at low tide and is underwater at high tide. Waves crash along this region much of the time. Only hardy organisms can live in this extremely variable and turbulent environment.

The *neritic zone* is a near-shore, shallow region along the continental shelf. The neritic zone is always underwater as it stretches from the low-tide line to the edge of the continental shelf, where ocean depths are about 200 meters (650 ft). This region has great biomass and is rich in biodiversity. Most of the world's commercial fisheries operate within this region. The neritic zone usually has a sandbar located just offshore. The sandbar forms from sand carried from the beach and acts as a reservoir of sand for the beach. As wave activity varies, sand may be transported either to or from the sandbar. Beaches with a pronounced longshore current (see Figure 11.9), such as those along the eastern seaboard of the United States, tend to experience a net loss of sand as the longshore current carries sand away from the beach.

The sand transported by the longshore current ends up filling in bays or coves or forming new features, such as

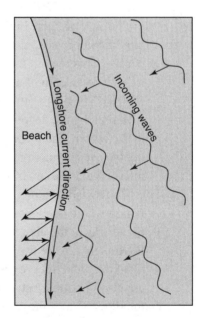

FIGURE 11.9 The longshore current. Source: USGS, *http://pubs.usgs.gov/circ/c1075/longshore.html.*

spits. A *spit* is a sandbar that has reached above sea level and is connected to the headlands or nearby coastal beaches. When the spit begins to curve toward its end, the result is called a *hook*. A classic example, called Sandy Hook (see Figure 11.10), is seen along the northern New Jersey shoreline. Sandy Hook is part of the Gateway National Recreation Area and is one of only two national urban seashore parks. The other is located along the California coastline in and near San Francisco.

The *oceanic zone* lies beyond the continental shelf. This region is best described as a biological desert as it is sparsely populated. Most of the nutrients carried from the land are deposited in the neritic zone, and depths are too great in the oceanic zone to supply surface waters with the nutrients required for extensive populations.

FIGURE 11.10 Sandy Hook, N.J. Note the northwestward-oriented hook at Gateway National Recreation Area, otherwise known as Sandy Hook. This is the northernmost tip of the New Jersey shoreline.

Mapping the Oceans

EARLY DISCOVERIES

The first organized global attempt at mapping the seafloor was made by H.M.S. *Challenger* between the years 1872 and 1876. The journey covered over 100,000 kilometers and carried the vessel to every ocean other than the Arctic. Ocean depths were determined by means of a weighted line. This method provided the first glimpse of seafloor topography. It revealed a seafloor, previously thought to be flat and unimpressive, that is quite varied and features mountains and valleys.

Additional details resulted from the invention in the 1920s of a sounding instrument called the *echo sounder*. This instrument, also known as sonar, which is a precursor of radar, employs a transmitter to emit sound waves aimed at the ocean floor. A finely tuned receiver detects the signals reflected from the seafloor. The time between emission and reception is used to assess the depth of the seafloor at that point.

Although seafloor mapping began in the nineteenth century (see Figures 11.11–11.13), it was not until the International Geophysical Year in 1957–58 that a concentrated effort resulted in the extensive and detailed picture that we have today.

The 1950s also saw a study, conducted in Europe, of a phenomenon known as *paleomagnetism* that revealed convincing evidence of continental drift. Scientists determined that the age of lava flows on the seafloor is related to the distance of those flows from the mid-ocean ridge. The youngest flows are located near the

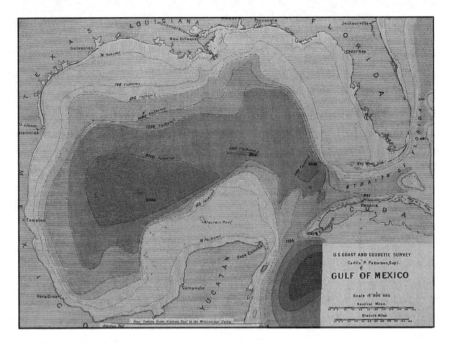

FIGURE 11.11 An early bathymetric map of the Gulf of Mexico. This map was created by a wireline sounding machine (a machine that sends sound waves down to the seafloor in an effort to determine depth). This machine was developed in the 1870s by British physicist William Thomson, later known as Baron Kelvin. Source: NOAA

FIGURE 11.12 Wireline sounding machine installation, circa 1942.

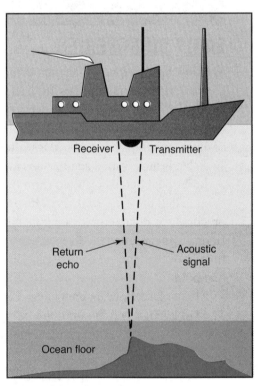

FIGURE 11.13 Operation of a seafloor sounder. Sonar sends sound waves through ocean water and times their return to assess distance.

ridge, and the age of the samples increases with increasing distance from the mid-ocean ridge.

Earth's magnetic field causes lava, when it solidifies, to assume a magnetic alignment, with the minerals oriented toward magnetic north. Scientists identified an apparent drift in the magnetic north pole since the magnetic orientation seemed to shift or vary in the older samples.

When the apparent position of magnetic north over the past 500 million years was plotted, scientists were able to identify the apparent movement of the continents. Their conclusion was based on the assumption that magnetic north had not moved but the continents had.

MORE RECENT PROJECTS

Recent technological breakthroughs have made possible more detailed study of the oceans than ever before. A few of the satellites launched or planned for launch are discussed below.

Established in the early 1990s and first launched in August 1993, the Sea-viewing Wide-Field-of-view Sensor (SeaWiFS) project is designed to conduct a quantitative study of global ocean bio-optical properties. Subtle changes in ocean color reveal information as to the various types and quantities of marine phytoplankton (microscopic marine plants) present in surface waters.

Launched in November 1997, the Tropical Rainfall Measuring Mission (TRMM) is a joint mission of NASA and the National Space Development Agency of Japan. Its goal is to monitor and study tropical rainfall and the release of energy in the tropics that helps to drive the entire global atmospheric circulation system.

Launched in December 2001, the Jason-1 project is designed to monitor global ocean circulation, ocean eddies, and El Niño events. Armed with a radar altimeter, it can study sea surface temperatures in real time. France is sharing the responsibility with the United States for the Jason-1 project. TOPEX-Poseidon, launched in 1992, has a nearly identical set of responsibilities and goals attached to its mission.

Aqua, launched May 4, 2002, is designed to measure the atmosphere and oceans to increase our understanding of Earth's climate and its fluctuations. It has sensors to measure clouds, precipitation, temperature (atmosphere and sea surface), snow cover, and extent of sea ice. The project is international to the extent that Brazil and Japan have each provided an instrument for the satellite.

The National Polar-Orbiting Operational Environmental Satellite System (NPOESS) Preparatory Project (NPP) satellite launched in February 2009. This satellite is monitoring ocean biological productivity, among other phenomena. More details about this mission and others can be found at the Earth Observing System Web site, located at

http://eospso.gsfc.nasa.gov/eos_homepage/mission_profiles/index.php.

The Global Precipitation Measurement (GPM) project is slated for launch in June 2013. The primary mission of the GPM is precipitation measurement with an eye toward increasing our understanding of ocean-atmosphere interactions, including latent heating (energy absorbed as water evaporates and released as water vapor condenses) and global climate.

Additional missions can be explored and studied at the NASA Earth Systems Enterprise (Mission to Planet Earth) Web site, located at

http://nasascience.nasa.gov/missions.

Current projects are available for study at NOAA's Ocean Explorer Web site. As of this writing, a mission to explore deep-sea corals in March 2009 is a highlighted exploration. See *http://oceanexplorer.noaa.gov.*

Features of the Seafloor

The seafloor is marked by a number of types of features, each of which will be detailed in the sections that follow.

DEEP OCEAN TRENCHES

Deep ocean trenches are the deepest sites in the oceans. Depths range to 11,000 meters below sea level at a site called the Challenger Deep (see Figure 11.14) in the Mariana Trench, one of several trenches in the Pacific Ocean. Deep ocean trenches are narrow, linear features generally associated with converging tectonic plates and subduction

Major Ocean Trenches (depths in feet)

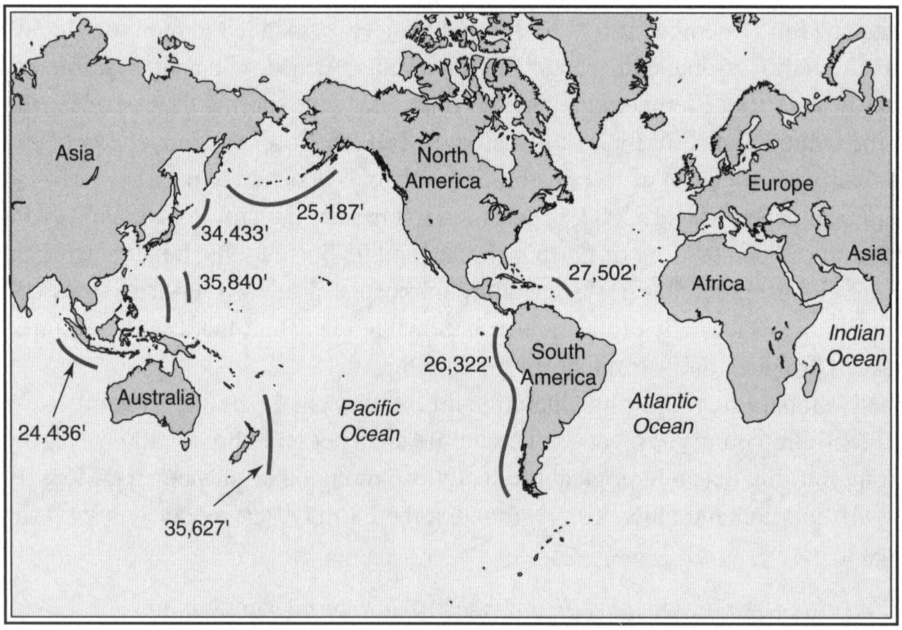

FIGURE 11.14 Global distribution of ocean trenches.

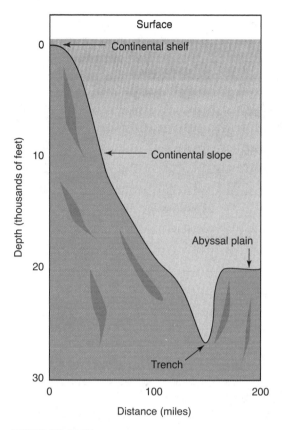

FIGURE 11.15 Cutaway profile of an ocean trench.

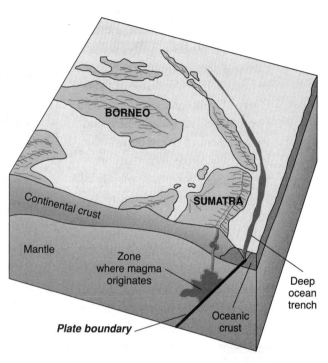

FIGURE 11.16 Island arc environments. Sumatra formed as an island arc as a result of plate subduction just to the east.

zones (see Figure 11.15). The trench is the site where crustal material is being destroyed as the plate is subducted into the mantle.

Trenches also tend to parallel volcanic island arcs. This relationship is not due to chance. As the plate subducts, it does so on an angle. When it reaches great depth, it begins to melt because of great pressure. The magma and gases can force their way back to the surface, forming a volcanic island arc (see Figure 11.16).

A classic example of this kind of action is Japan, which formed as the Pacific Plate subducted under the Philippine Plate. As this process continues, Japan will remain a site where earthquakes and volcanic activity are relatively common events.

ABYSSAL PLAINS

The *abyssal plains* are what most persons would call the ocean floor. Their depth varies from 2200 to 5500 meters (7200–18,000 ft), and they lie between the continental rise and the mid-ocean ridges. Abyssal plains (see Figure 11.17) make up most of the seafloor in terms of square kilometers. Other features, such as trenches, mountains, and seamounts, comprise a relatively small percentage of underwater topography when compared to the abyssal plains.

The abyssal plains are extremely flat. This is due to the extensive accumulation of sediment, which has buried a seafloor that would otherwise offer some variation in its topography (see Figure 11.18). The result of these layers of sediment is a region with virtually no variation in elevation for distances of over 1000 kilometers. Many of the sediments are fine-grained, consisting mainly of clays and silt. Turbidity currents carry these materials in channels called *submarine canyons* from the continental margins to the deeper waters. Additional sediment collects from dust blown out to sea and the remains of small marine plants and animals. If sediments are trapped in the submarine trenches, as occurs in parts of the Pacific, abyssal plains fail to form.

Features of the Underwater World

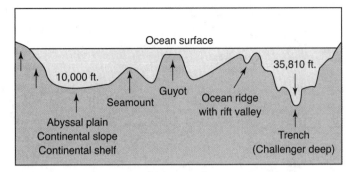

(Not to scale)

FIGURE 11.17 Abyssal plains, the vast ocean floor.

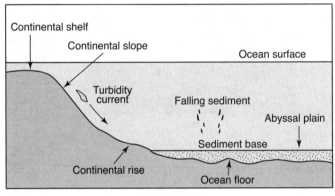

(Not to scale)

FIGURE 11.18 Abyssal plains with overlying sediments.

SEAMOUNTS

Seamounts are isolated volcanic peaks that emerge from the abyssal plain, but remain below sea level (see Figure 11.19). Seamounts are most common in the Pacific Ocean and can rise several hundred meters above the surrounding ocean floor. Some seamounts occur in chains. The Hawaiian Islands were once a chain of seamounts but have now risen above sea level. A hot spot in the mantle

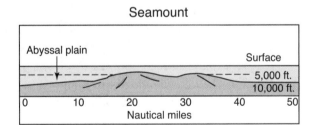

FIGURE 11.19 Seamounts.

has caused volcanic activity over a long but narrow region from the Aleutian Trench to the Hawaiian Islands. As the Pacific Plate slid over this hot spot (see Figure 11.20), a chain of volcanoes formed. Most remain below sea level and therefore are referred to as seamounts.

To the northwest of the Hawaiian Islands lies a chain of underwater seamounts, known collectively as the Emperor seamounts, that are no longer tectonically active.

They appear to have formed as the Pacific Plate passed over the same hot spot that was responsible for the Hawaiian Islands.

A *guyot* is a special type of seamount that grew to a height above sea level and existed as an island for perhaps millions of years. As wave action and running water erode the island to sea level, the flat top characteristic of a guyot forms.

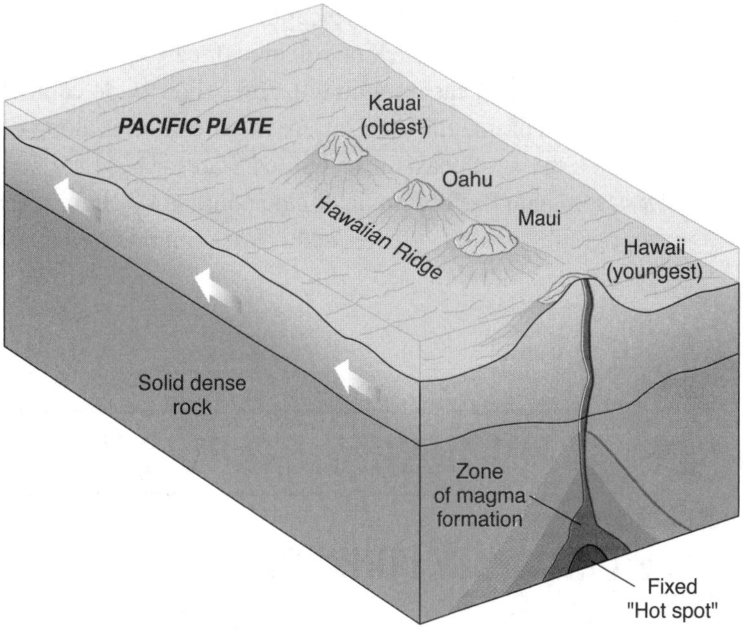

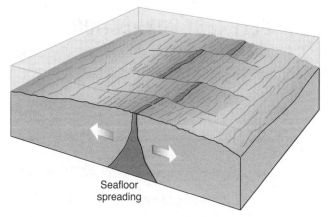

FIGURE 11.20 The Hawaiian Islands formed as a result of rising magma within the crust and the mantle. The Hawaiian Islands are not located along or near a major plate boundary.

MID-OCEAN RIDGES

Mid-ocean ridges might more appropriately be called the *global interconnected ocean ridges system*. This system connects over 70,000 kilometers of largely underwater mountains, many of which are not located in the middle of an ocean.

The Mid-Atlantic Ridge is a ridge system that does happen to fall in the middle of an ocean. Generally, all of these ridges are associated with plate divergence and the production of new crustal material. Specifically, basalt is produced along the ridges and is then transported away by plate movement. A more extensive discussion of this process is given in Chapter 7.

A ridge system is not continuous; rather, it is broken by numerous transform faults. These faults create what look like stair steps on a map of the ridge system (see Figure 11.21).

CONTINENTAL MARGINS

A comparison of the eastern and western seaboards of the United States reveals two different types of *continental margins*. Continental margins can be classified as passive or active.

FIGURE 11.21 Mid-ocean ridge systems. Seafloor spreading takes place or originates along the mid-ocean ridges.

The U.S. East Coast is typical of regions with passive margins. This type of margin is characterized by a lack of volcanic or earthquake activity. Sediments accumulate and yield wide beaches with few cliffs and a wide continental shelf.

Active continental margins are located near or along plate boundaries. This would include the west coast of the United States. These are newer coastlines, and therefore less sediment has accumulated, the continental shelf is demonstrably narrower, and the topography exhibits greater relief. It is common to see cliffs near or along an active continental margin.

REVIEW EXERCISES FOR CHAPTER 11

WORD-STUDY CONNECTION

abyssal plain	euphotic	passive
algae	evaporation	photic
active	guyot	phytoplankton
aphotic	H.M.S. *Challenger*	salinity
benthic zone	hydrothermal vents	seamount
bioluminescent	intertidal	solution
chemical weathering	ions	sonar
concentration	low tide	subduction zones
continental margins	mid-ocean ridges	surface mixed zone
deep zone	neritic	thermocline
desalination	ocean trench	transition zone
dissolved	oceanic	volcanic island arc
echo sounder	outgassing	zooplankton

SELF-TEST CONNECTION

PART A. Completion. *Write in the word or words that correctly completes the sentence.*

1. Hydrothermal vents are found in the _____ zone of the ocean.

2. The _____ zone is a region of the ocean where light penetrates.

3. _____, an additive to gasoline, is found within the oceans.

4. Freshwater can be obtained from seawater through the process of _____.

5. The _____ is best described as an oceanic desert.

6. Early attempts to map the ocean floor were conducted by _____.

7. The _____ zone receives rich deposits of minerals and nutrients from the land.

8. Trenches are generally associated with _____.

9. The scientific name for the ocean floor is the _____.

10. A _____ is a flat-topped seamount that grew to a height above sea level and existed as an island for perhaps millions of years.

11. _____ are sites where new oceanic crust is being produced.

12. Large portions of the _____ zone remain unexplored.

13. The intertidal zone is a region that is exposed at _____.

14. Wide beaches and flat topography characterize _____ continental margins.

15. Numerous _____ create breaks in the mid-ocean ridge system.

PART B. Multiple Choice. *Circle the letter of the item that correctly completes the statement.*

1. The most abundant ion in seawater is
   (a) sodium
   (b) chloride
   (c) oxygen
   (d) nitrogen
   (e) calcium

2. Most life within the oceans is found in the
   (a) surface mixed zone
   (b) thermocline
   (c) deep zone
   (d) aphotic zone
   (e) both c and d

3. The presence of sea life is often dependent upon
   (a) seawater temperature
   (b) salinity
   (c) availability of sunlight
   (d) presence of seafloor dwellers
   (e) all of the above

4. The main problem with desalination is
    (a) its limited effectiveness
    (b) the lack of public support for such projects
    (c) the expense involved in running the process
    (d) there is no need for it
    (e) it can't remove enough salt to make seawater potable

5. The neritic zone is found along
    (a) the continental shelf
    (b) the abyssal plain
    (c) seamounts
    (d) guyots
    (e) trenches

6. The continental margin along the West Coast of the United States is best described as
    (a) active
    (b) passive
    (c) younger than the continental margin along the East Coast
    (d) older than the continental margin along the East Coast
    (e) both a and c

7. _____ act to transport dissolved minerals into the ocean.
    (a) Rivers
    (b) Streams
    (c) Underground water
    (d) Hydrothermal vents
    (e) All of the above

8. The salinity of water tends to be greater in regions where
    (a) evaporation exceeds precipitation
    (b) rainfall is sparse
    (c) seawater is freezing into ocean ice
    (d) the climate is dry
    (e) all of the above

9. The only region of the oceans capable of supporting photosynthesis is the
    (a) euphotic zone
    (b) photic zone
    (c) aphotic zone
    (d) intertidal zone
    (e) surface mixed zone

10. The greatest salinities observed in ocean water are
    (a) about 37 percent
    (b) about 37 parts per thousand
    (c) about 37 parts per million
    (d) about 37 parts per billion
    (e) unknown values

PART C. Modified True/False. *If a statement is true, write "true" for your answer.*
If a statement is incorrect, change the <u>underlined</u> *expression to one that will make the*
statement true.

1. <u>Outgassing</u>, in addition to rivers and streams, supplies minerals to the world's
   oceans.

2. Plants and animals <u>add</u> minerals *to* the world's oceans.

3. <u>Sodium</u> is the most abundant ion in the oceans.

4. The <u>benthic zone</u> is a region in the oceans where temperature changes rapidly
   with depth.

5. The surface mixed zone is also known as the <u>aphotic</u> zone.

6. <u>Desalination</u> is the process of converting salt water to freshwater.

7. <u>Zooplankton</u> are the bottom of the food chain.

8. The <u>benthic zone</u> is located at the bottom of the ocean.

9. The <u>intertidal</u> zone has a very low biomass with few species and little
   biodiversity.

10. Trace minerals, such as <u>magnesium</u>, are harvested from the ocean.

CONNECTING TO CONCEPTS

1. The chapter details several ways of looking at the oceans. Compare the various
   methods by which ocean depth is classified.

2. Do some research to discover the most common ions found in rocks within Earth's
   crust. Do they match the ions found in seawater? If so, why? If not, why not?

3. Do some research on desalination facilities. Are any desalination plants currently
   operating in the world? If so, where? What is the cost to produce freshwater from
   salt water?

4. Create a concept map illustrating the importance of phytoplankton to the world's ecosystems.

5. How did studies of the ocean floor change our ideas about plate tectonics?

6. In which part(s) of the ocean is new landmass being created? Where is solid earth being destroyed?

7. Ocean salinity is known to vary from region to region. What condition(s) contribute(s) to high ocean salinity?

CONNECTING TO LIFE/JOB SKILLS

The oceans provide a fascinating study from many points of view. Biologists and oceanographers are interested in the potential of oceans for their biomass as oceans may represent a greater percentage of our food supply in the future. Geologists study the oceans to better understand plate tectonics. Environmentalists study the oceans for many reasons, which include tracking pollution and detecting its sources. Climatologists are interested in the oceans because they are integral to Earth's climate systems.

ANSWERS
SELF-TEST CONNECTION

Part A

1. benthic	6. H.M.S. *Challenger*	11. Mid-ocean ridges
2. photic	7. neritic	12. aphotic
3. Bromine	8. subduction zones	13. low tide
4. desalination	9. abyssal plain	14. passive
5. oceanic	10. guyot	15. transform faults

Part B

1. **(b)**	3. **(e)**	5. **(a)**	7. **(e)**	9. **(a)**
2. **(a)**	4. **(c)**	6. **(e)**	8. **(e)**	10. **(b)**

Part C

1. True	5. False; photic	8. True
2. False; remove ... from	6. True	9. False; oceanic
3. False; Chloride	7. False; Phytoplankton	10. True
4. False; thermocline		

CONNECTING TO CONCEPTS

1. One method is to classify depths by layers as follows: a surface mixed zone, characterized by relatively warm temperatures; a transition zone, or thermocline, where temperatures fall rapidly with increasing depth; and a deep, cold, benthic zone.

    Another scheme classifies depths by available light. The top 200 meters in most ocean waters is classified as the photic zone. In the uppermost portion of the photic zone, called the euphotic zone, sunlight is strong enough to support photosynthesis. The zone beneath the photic zone, where no light reaches, is classified as the aphotic zone. The deepest waters are known as the benthic zone.

    A third method for classifying ocean depths uses the ocean's relationship to the coastline. Coastal waters, where waves crash and land is exposed for a portion of the tidal cycle, are known as intertidal waters. The neritic zone stretches from the intertidal zone to the oceanic zone. Land in this zone is permanently submerged. Depths approach 200 meters at the edge of the continental shelf. Deeper waters, beyond the continental shelf, are classified as the oceanic zone.

2. The most common ions found within Earth's crust are silicon and oxygen. In contrast, the oceans have high concentrations of sodium and chloride. The presence of the ions found in seawater must be due at least in part to the solubility of these ions. Silicon is generally found in highly insoluble minerals. Oxygen, although somewhat soluble, is taken up by marine organisms.

3. A few Middle Eastern nations have implemented desalination plants to supply freshwater to their populations. The warm and dry conditions throughout much of this region limit the number of traditional sources for potable water. Costs will vary. According to recent research reports costs are now as low as $1.95 per thousand gallons. Improvements in desalination technology have reduced costs from $15 to $20 per thousand gallons in the 1940s and 1950s.

4. Maps will vary. In general, however, the map should emphasize the interdependence of life-forms upon each other, with phytoplankton at the base of the food chain.

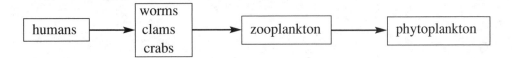

5. Studies of the seafloor have revealed extensive evidence of plate tectonic activity. This evidence includes the varying ages of seafloor sediment, from the youngest at mid-ocean ridges to the oldest along coastlines and near subduction zones; the volcanic activity along mid-ocean ridges; the occurrence of earthquakes with increasing depth along subduction zones; and the magnetic reversals of igneous material along the seafloor.

6. New landmass is created along mid-ocean ridges. Old crust (solid earth) is destroyed along subduction zones.

7. Ocean salinity is controlled by precipitation and air temperature. Hot, dry climates result in excessive evaporation and yield high salinities. Interestingly, arctic waters are also highly saline. When sea ice forms, dissolved salts remain in the liquid water, thus raising the salinity of the remaining water.

The Dynamic Oceans

12

WHAT YOU WILL LEARN

This chapter focuses on the dynamic nature of the oceans. In this chapter you will learn

- circulation patterns in the oceans;

- about deep ocean circulation and the transport of heat energy;

- how tides occur and tidal patterns;

- the importance of our beaches and human efforts to restore and protect them.

SECTIONS IN THIS CHAPTER

- Surface Currents
- Deep Ocean Circulation
- Tides
- Waves
- Human Efforts to Restore Our Beaches
- Related Internet Resources for Great Images and Information
- Review Exercises for Chapter 12

Surface Currents

OCEAN CIRCULATION PATTERNS

The oceans are not just large bodies of heterogeneously mixed water. The oceans also contain an intricate network of currents flowing across their surface, at depth, and from the surface to great depths (and back). These currents, produced by persistent wind patterns, are responsible for the transport of large amounts of energy from tropical latitudes to polar regions. They are also an ongoing testimony to the great degree of interaction that occurs between the hydrosphere and the atmosphere. Global winds help to create ocean currents, and ocean currents, through their transport of warm and cold waters, act to modify climates worldwide. Any significant change in ocean current patterns will likely bring about major climatic shifts.

The lower atmosphere is characterized by very consistent large-scale patterns of wind flow. These winds create a frictional drag along the surface of the oceans. The net result is a series of surface ocean currents that parallel the wind flows above them (see Figure 12.1).

The *Coriolis effect*, a force created by Earth's rotation, influences ocean currents and global winds. Currents are deflected to the right in the Northern Hemisphere and to the left in the Southern Hemisphere by the Coriolis effect (see Figure 12.2).

UPWELLING

Upwelling, discussed also in Chapter 11, illustrates the action of vertical ocean currents. When cold waters rise from depth within the oceans and reach the surface, upwelling has occurred. This phenomenon is most common along the western seaboards of continents as ocean currents return to the Equator. Wind flow in these regions causes surface waters to deflect away from the coastline, thus allowing colder and denser waters to rise, or upwell, to the surface (see Figure 12.3).

Clear evidence of upwelling, or at least of its effects, can be seen when ocean temperatures just off the coast of southern California are compared to those just off the coast of South Carolina. Although both locations are situated at similar latitudes, and experience similar sea surface temperatures during the winter months, there is a great disparity in summer and autumn sea surface temperatures (see Figure 12.4). Persistent upwelling along the U.S. West Coast, in conjunction with a cool water current flowing toward the Equator, serves to cap sea surface temperatures.

Upwelling also tends to supply fresh nutrients to the surface waters where it occurs. These nutrients supply marine life in the area, resulting in abundant fish in these regions.

As stated in Chapter 11, the phenomenon known as *El Niño*, one facet of which is the cessation of upwelling along the coast of Peru, was recognized hundreds of years ago by fishermen when their daily catch of fish began to diminish significantly. El Niño is discussed in more detail in the section headed "El Niño and La Niña."

Surface Ocean Currents

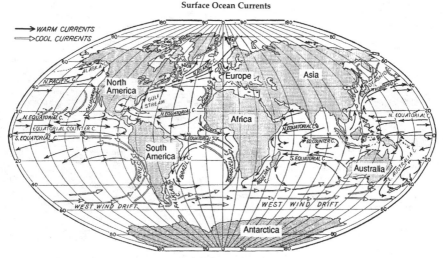

(a) Surface ocean currents.

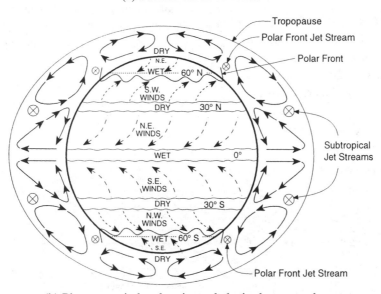

(b) Planetary wind and moisture belts in the troposphere.

FIGURE 12.1 Compare the ocean currents to the action of global winds at similar latitudes. Note, for example, the westward drift of the equatorial current (a) and the westward movement of air caused by the northeast trade winds. Farther north, in the midlatitudes, wind currents also parallel ocean currents (b). Source: The State Education Department, *Earth Science Reference Tables,* 2001 ed. (Albany, New York: The University of the State of New York).

SURFACE OCEAN CURRENTS AND CLIMATIC PATTERNS

Surface ocean currents can have a profound impact upon adjacent landmasses, particularly when prevailing winds carry air modified by an ocean current to the landmass. In the United States, this happens primarily along the West Coast. The cool waters located just offshore heavily influence summers in coastal California, Oregon, and Washington.

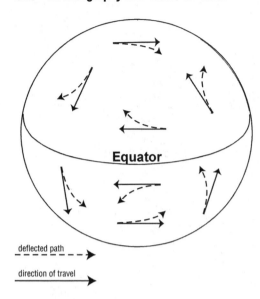

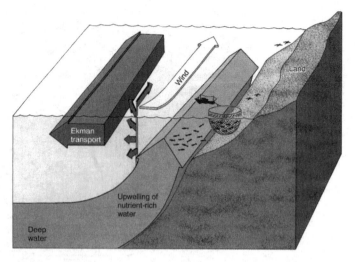

FIGURE 12.2 The Coriolis effect causes objects (flowing air and water) to be deflected to the right from their points of origin in the Northern Hemisphere and to the left from their points of origin in the Southern Hemisphere. Source: *Let's Review: Earth Science—The Physical Setting,* 2nd Ed., Edward J. Denecke, Jr., Barron's Educational Series, Inc., 2002.

FIGURE 12.3 Wind-driven currents create upwelling, bringing cool, nutrient-rich waters at depth to the surface. The arrow labeled "Ekman transport" illustrates the net direction of water transport. The Ekman transport is created by an interaction between surface winds, the drag of subsurface waters, and the Coriolis effect. Source: *Marine Biology,* 3rd Ed., Peter Castro and Michael Huber, McGraw-Hill, 2000.

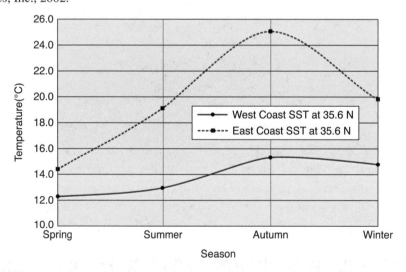

FIGURE 12.4 Sea surface temperature (SST) data for similar latitudes along the east and west coasts of the United States. Note the annual temperature range along the East Coast in comparison to that on the West Coast. Sea surface temperatures remain relatively cool year-round along the coast of California. Because of this fact, in conjunction with the persistent onshore flow of air, coastal air temperatures are relatively constant year-round. See Figure 12.5 for further clarification of this point.

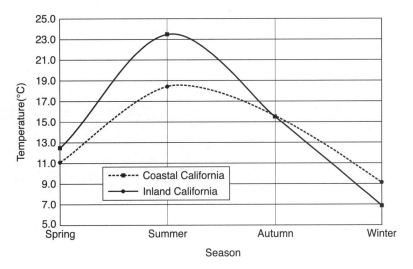

Figure 12.5 Coastal temperature ranges are severely limited by persistent onshore winds and sea surface temperatures that hover near 15°C year-round. At 360 kilometers to the east, summers are much warmer and winters are much colder.

While the landmass heats up throughout much of the contiguous United States, often causing temperatures to exceed 32°C (90°F), conditions are moderated along the immediate coastline.

Figure 12.5 shows temperature patterns at coastal and inland California at 40.5°N latitude. The two locations differ by approximately 360 kilometers (200 mi). The coastal California location reports data from 124.5°W longitude; the inland California location, data from 120.0°W longitude. The coastal range is partially responsible for separating the air masses that prevail along the Pacific coast from those farther inland.

Mark Twain once remarked, "The coldest winter I ever spent was summer in San Francisco." Summer temperatures in San Francisco rarely reach 21°C (70°F). Data and charts to support this statement are given in Chapter 16.

EL NIÑO AND LA NIÑA

Temporary changes in ocean and atmospheric circulations can have pronounced effects upon landmasses and populations across the world. One such phenomenon that has captured national headlines in recent years is El Niño and its "reflection," La Niña. Most people know that El Niño can cause significant changes in weather patterns, some of which cause millions of dollars' worth of damage to property. When pressed to explain exactly what El Niño is, however, they are at a loss for a clear, concise answer.

In the most succinct terms, El Niño is a warming of the surface waters off the coast of western South America, with the greatest warming near Peru. What causes this change in sea surface conditions and how it affects all of Earth's systems for periods of months to years are the key questions. There has been extensive research in an

effort to improve the prediction of this
phenomenon so that
regions and populations
can better prepare for
the changes, generally
in weather conditions, that
it brings.

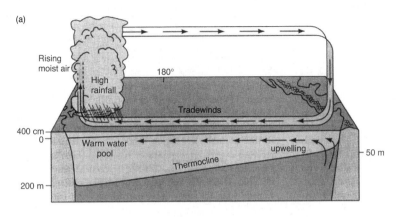

What starts an El Niño
is somewhat of a "What
comes first, the chicken
or the egg?" question.
The warming of surface
waters in the eastern
equatorial Pacific is
accompanied by a relax-
ation in the persistent
southeasterly trade winds,
and it is not well under-
stood which event causes
which. In either case, the
net result is a global
change in weather pat-
terns that affects diverse
regions (see Figure 12.6).
Most directly, with warm-
ing of the sea surface
near Peru, there is a cooling of the sea surface in
the western Pacific. This

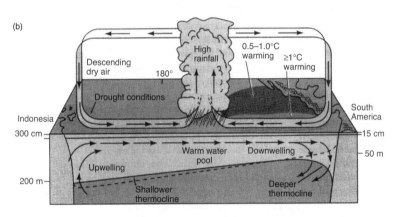

FIGURE 12.6 El Niño versus normal conditions. Notice the eastward
shift in warm surface waters and the accompanying eastward shift in
convective thunderstorms. Populated regions in the western equatorial
Pacific are left with drought conditions as the convective activity
migrates eastward. Source: *Environmental Geology,* Barbara W. Murck,
Brian J. Skinner, and Stephen C. Porter, John Wiley, 1996.

results in a drop in convective activity, or thunderstorms, over portions of Australia
and Indonesia. Since thunderstorms are responsible for much of the rainfall in these
regions, El Niño's impact, that is, the cessation of convective activity, causes
drought and human suffering.

El Niño's impacts extend well beyond the tropical Pacific. In the United States,
an active storm track develops. This storm track originates in the Pacific and brings
flooding rains to much of central and southern California. Newspaper headlines during
an El Niño often report extensive flooding in this region, along with destructive mudslides. Stormy conditions also prevail across the southeastern United States. The rainy
weather brings above-normal cloud cover, which results in colder temperatures for
many of the Gulf Coast states. In the northern United States, however, El Niño winters
are milder than normal as arctic air is frequently "locked up" north of this country's
borders for much of the winter.

La Niña, a cooling of the eastern equatorial Pacific Ocean waters to below normal levels, often follows El Niño. Scientists are actively engaged in the study of how La Niña impacts global weather patterns. Many areas that experience flooding rains during El Niño revert to drought during La Niña. In the United States, the extreme weather conditions that both of these phenomena are known to cause are responsible for millions of dollars of property damage.

Deep Ocean Circulation

Ocean circulation, as noted earlier, is not limited to surface currents. Circulation deep within the ocean is driven by density differences and by gravity rather than by winds, as is the case with surface currents.

Temperature and salinity combine to affect the density of water. Water near a temperature of 4°C is denser than water at any other temperature. More salts in the ocean water also raise the density.

A *thermohaline circulation* (see Figure 12.7) driven by sinking dense surface waters originates in both the Arctic and the Antarctic oceans. These waters are both cold and very saline. The high salinity can be attributed to the fact that, as water freezes into ice, the salts do not remain in the ice but instead add to the saline concentration in the remaining liquid. These cold, dense waters sink and travel toward the Equator at great depths. Several

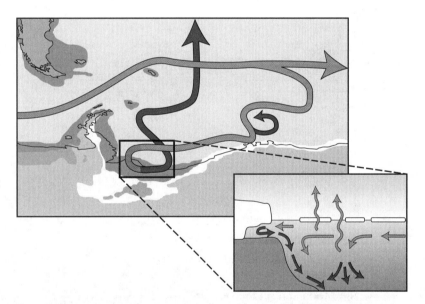

FIGURE 12.7 Deep ocean currents in the vicinity of Antarctica. Once these cold, dense waters sink, they travel toward the Equator, ultimately rising to the surface and gaining energy from the intense tropical Sun.

centuries later, these waters surface and continue the conveyor-belt-like movement, transporting large amounts of water in the process.

The deep ocean circulation has a pronounced impact upon global climate as warm water is displaced by cold in upwelling currents. The variation in surface temperatures across the Equator acts to drive the trade winds, which are an integral part of the global atmospheric circulation system.

Tides

CAUSES OF TIDES

Tides, that is, the rhythmic rise and fall of ocean waters along coastlines, are one of the most noticeable phenomena to anyone visiting a beach for any length of time. The difference in height between high tide and low tide is called the *tidal range*. The Bay of Fundy in Nova Scotia, Canada, offers an extreme in tidal range with a 20-meter difference between high and low tides. High and low tides occur on a regular and predictable basis. The section headed "Tidal Patterns" below discusses the various patterns known to occur across the globe.

Tides are caused by gravitational attraction by the Moon and to a lesser extent by the Sun. The Moon (see Figure 12.8), is, of course, much smaller than the Sun and therefore has much less gravity. The Moon, however, is about 400 times closer to Earth than the Sun is, making the Moon the primary celestial body that produces tides.

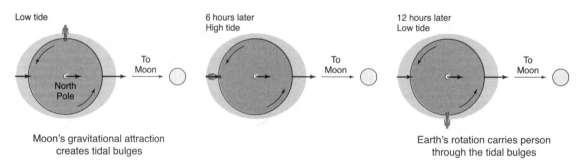

FIGURE 12.8 Tides and the Moon.

If Earth is thought of as a basketball, and one envisions the Moon pulling on that ball, the ball becomes somewhat oblong.

A tidal bulge is produced on two sides of Earth, on the side facing the Moon and on the side opposite the Moon (see Figure 12.9). The usual result is two high tides (and two low tides) each day as Earth rotates through these bulges (refer to Figure 12.8). The planet's crust is also affected by the Moon's gravitational pull, but movement is minimal because the crust is significantly denser than the water in the oceans.

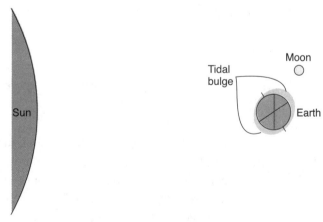

FIGURE 12.9 Tidal bulge caused by the Moon.

The two high tides observed at many locations each day are not necessarily of equal height. The Moon's position may cause one tidal bulge in the Northern Hemisphere and one in the Southern. Refer to Figure 12.9 for an illustration of this.

SPRING AND NEAP TIDES

Throughout each month tidal heights vary as the relative positions of Earth, the Moon, and the Sun change. The two extreme conditions are identified as spring tides and neap tides (see Figure 12.10).

Spring tides occur when the Earth, Moon, and Sun are all in alignment, that is, during the full Moon and new Moon phases. With this alignment, Earth is under the greatest possible gravitational pull. Tidal ranges are at a maximum, and high tides are at their highest. The high tide that occurs during the full Moon phase is often called an *astronomical high tide*; it is then that coastal flooding is most likely to occur. When onshore winds, sometimes caused by a storm system out at sea, coincide with the astronomical high tide, coastal flooding is likely even if a drop of rain does not fall. A persistent onshore wind has the ability to raise tides about 1 meter over their normal height.

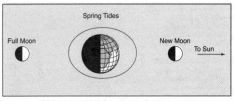

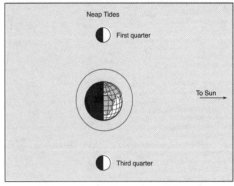

FIGURE 12.10 Spring and neap tides. Source: *Let's Review: Earth Science— The Physical Setting,* 2nd Ed., Edward J. Denecke, Jr., Barron's Educational Series, Inc., 2002.

When the Sun, Moon, and Earth align at right angles, *neap tides* occur. Since the gravitational stresses are not aligned, however, tidal range is at a minimum, and the lowest high tides of the month occur.

TIDAL PATTERNS

Three types of tidal patterns are known to occur: semidiurnal, diurnal, and mixed (see Figure 12.11). The shape of the coastline and the water body involved determines the type of tide observed.

Semidiurnal tides (Figure 12.11a), common to the eastern seaboard of the United States, exhibit two high tides and two low tides in a 24-hour period. The heights of the high tides are roughly equal, as are the heights of the low tides.

Diurnal tides (Figure 12.11b), common to smaller water bodies such as the Gulf of Mexico, exhibit one high tide and one low tide each day. The tidal range is often limited in these waters also.

Mixed tides (Figure 12.11c), common to the West Coast of the United States, exhibit two high tides and two low tides each day. In contrast to semidiurnal tides, however, the heights of the high tides vary significantly, as do the heights of the low tides.

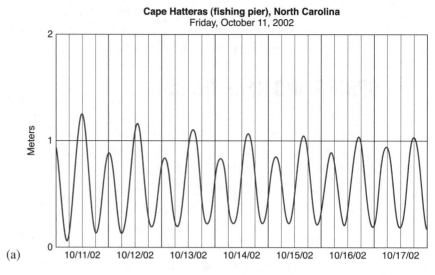

(a)

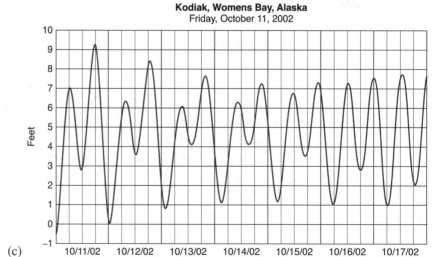

(b)

(c)

FIGURE 12.11 Tidal patterns. (a) Semi-diurnal tides. Note two high and two low tides daily. This typical pattern occurs along much of the U.S. eastern seaboard. (b) Diurnal tides. Note one high and one low tide daily. This pattern is typical along much of the Gulf Coast. (c) Mixed tides. These tides vary widely from one cycle to the next. Certain Pacific coast sites experience this pattern.

Waves

CAUSES OF WAVES

Waves, in contrast to tides, are not highly predictable. Waves are the result of wind action on the sea surface. Three distinct factors—the intensity, the duration, and the fetch of the wind—influence the height, frequency, and type of waves that are experienced on a coastline. *Intensity* is simply the velocity or the rate of speed at which the wind is blowing from a particular direction. *Duration* is the period of time for which the wind is blowing. *Fetch* is the distance over which the wind blows in a particular direction.

Residents of the U.S. eastern seaboard from Maine to Florida are periodically warned of rough coastal waters even though the weather conditions along the shore are perfectly calm. Hurricanes, sometimes hundreds of miles out to sea, can provide the intensity, duration, and fetch required to generate large swells that achieve great heights when they reach the shallow waters along the immediate shore.

CHARACTERISTICS OF WAVES

All waves can be described by certain characteristics. Study Figure 12.12, showing a typical transverse wave, as each characteristic is described. The *wave height* is the distance between the crest (peak) of the wave and the trough (minimum) of the wave. The *wave period* is the time interval between waves as observed, for example, by a person standing in shallow water near the shore. The *wavelength* is the distance between crests (or troughs). The wave height, wave period, and wavelength are all determined by the factors discussed earlier.

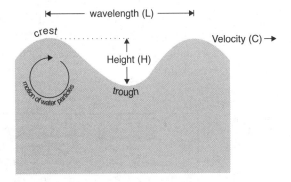

FIGURE 12.12 A typical transverse wave. Source: An Introduction to Oceanography, David A. Ross, Prentice-Hall, 1970.

As wind energy increases, so does wave energy. As wave energy increases, wave height increases. At some point, shallow water cannot support the wave, or the base of the wave slows because of ground friction while the crest continues. In the end, the waves crash onto the shore. When very high waves occur, *white caps*, or *breakers*, at some distance from the shore are observed. Under these conditions, wave heights are too high to support the waves, and they break in the deeper waters offshore.

Normal wave action is illustrated in Figure 12.13. As waves approach the coastline, ocean depths decrease, forcing the waves upward. The base of the wave begins to drag on the seafloor, while the crest continues to move toward land. Eventually, the wave topples over itself.

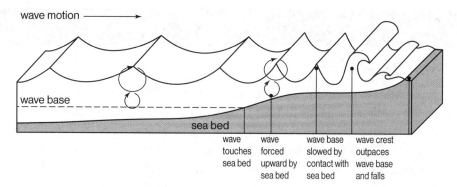

FIGURE 12.13 Mechanics of a breaking wave. Source: *Earth Science on File* by the Diagram Group. Copyright © 1988 by The Diagram Group. Reprinted by permission of Facts on File, Inc.

WAVE EROSION

Waves can act as erosional agents, particularly during storms and other times of high waves, in two distinct ways. First, high waves crashing on the shore loosen and break up materials. Along the U.S. eastern seaboard, where much of the coastline is composed of loose sand, erosion is a particularly serious problem. Waves often crash onto the shore at an angle other than the perpendicular. This tends to produce a net flow of water called a *longshore current* (see Figure 12.14) in one direction or the other along the coastline.

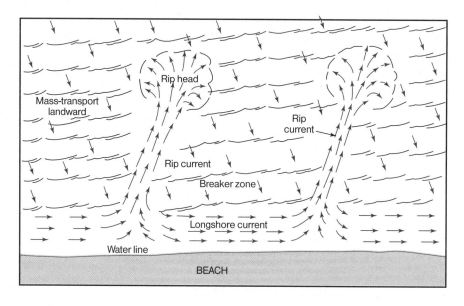

FIGURE 12.14 Formation of a longshore current. Source: *The Earth Sciences*, Arthur N. Strahler, Harper and Row, 1971.

Second, once the waves have loosened materials, they carry these sediments to a position just offshore where the longshore current can then transport them for miles. Aerial photographs of the northernmost New Jersey shore reveal a barrier island, connected to the mainland, that has been created by the longshore current. Aerial and space shuttle photographs reveal that this island, called a *spit*, extends into Raritan Bay. This region, known as Sandy Hook (see Figure 12.15), has grown by over a mile in just 200 years. Considering the pace at which geologic change usually occurs, this is an extremely rapid rate. The most obvious evidence is the location of the lighthouse, built in the early 1800s. Originally only a short distance inland, it is now more than a mile away from the tip of the hook. Figure 12.16 is a map of the region; note the location of the lighthouse and, using the distance scale on the map, verify that there is now about 1 mile of land around the lighthouse.

While Sandy Hook has grown, the coastal beach communities just to its south have lost most of their sand. As beaches are a major attraction to tourists, this natural process of erosion has had a profound impact upon the local economies. Nor is the problem of sand transport by the longshore current limited to New Jersey. Various other sites along the East Coast from Maine to Florida face the same problem. The next section discusses human efforts to ameliorate the situation.

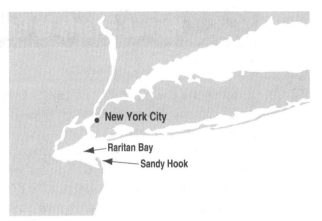

FIGURE 12.15 Sandy Hook and Raritan Bay, N.J.

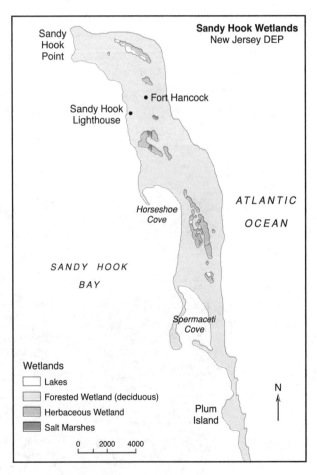

FIGURE 12.16 Gateway National Recreation Area, Sandy Hook Unit, N.J.

Human Efforts to Restore Our Beaches

GROINS

Obviously, for both aesthetic and financial reasons, coastal communities and states with extensive shorelines have a vested interest in maintaining wide, sandy beaches. Early efforts to manage the shore involved the construction of *groins*, sometimes referred to as *jetties* (see Figure 12.17).

The concept was simple: build a rock wall, approximately 1–2 meters above the high-tide line and extending on a perpendicular approximately 50 meters from the beach. The hope was to capture sand otherwise being transported by the longshore current. The net result, however, was an uneven shoreline where one side of the groin had an abundance of sand as the longshore current piled it up, and the other side a dearth of sand. Furthermore, the town just "downstream" from the groin was now sand starved, as sand was still being removed from its beaches, but no sand was being carried along the shore to replace it. Figure 12.18 illustrates

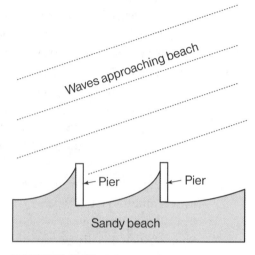

FIGURE 12.17 Efforts at beach management. Jetties or groins are built to trap sediment. Source: *Let's Review: Earth Science—The Physical Setting,* 2nd Ed., Edward J. Denecke, Jr., Barron's Educational Series, Inc., 2002.

FIGURE 12.18 Photograph of groins in use along a coastline.

the sawtooth pattern that forms when groins are in use. In northern New Jersey, to cite a specific example, where the longshore current travels northward along the coastline, each community was soon forced to build groins to preserve what sand it had. The entire effort proved to be a losing battle.

SEAWALLS

Building a *seawall* (see Figure 12.19) may be considered an action of last resort. It is the equivalent of using a Band-Aid on a broken limb. Erection of a seawall, which separates a beach from property and populations, virtually dooms the beach to destruction.

As described earlier, storm waves tend to crash on the beach at greater height and with more force than normal waves. During these times, waves will hit the seawall with their full force and thus have a greater capability to transport sand out to sea. On beaches whose gradual slope diminishes wave energy, however, sand transport is limited.

After a series of storms that produce high waves, a beach with a seawall is reduced to a fraction of its former size. In Sea Bright, New Jersey, the problem became so acute that at high tide there *was no beach*; the water simply crashed on the seawall.

FIGURE 12.19 A seawall in use. Seawalls are often an action of last resort because wave action in front of the seawall acts to remove the remaining beach. The seawall's basic function is to protect life and property behind it.

Over time this action caused significant damage to the wall itself, and storm waves regularly crashed over the wall, flooding the populated area behind it.

Clearly, new approaches for coastal shoreline management were called for. Two such approaches are described in the sections that follow.

DUNES

Sand dunes provide a natural and effective barrier to protect lives and property along the shoreline. Over time, sand dunes tend to develop naturally. In recent years, however, coastal communities have planted new dunes in the hopes that their beaches can be preserved.

Dunes act to preserve beaches in two significant ways. First, as waves crash onshore and roll up the beach, the dunes provide a sloped surface designed to dissipate wave energy, thus limiting the amount of erosion. Second, dune grass creates a network of roots that act to hold the sand together and prevent its loss by erosion.

Overall, dunes are a relatively inexpensive, effective, and aesthetically pleasing method for preserving a beach.

BEACH RESTORATION

In extreme cases, such as the situation described along the northern New Jersey shore-line, beach restoration procedures (see Figure 12.20) are implemented. This is a slow and expensive process as sand is dredged from near-shore locations, transported to the

FIGURE 12.20 Beach restoration.

beach to be restored, and pumped onto that beach. Although the results are often quite spectacular, they are also temporary. Restoration crews often return to the same site within 2 years to repeat much of the process. One procedure that has helped somewhat is to combine beach restoration with the planting of new dunes.

Related Internet Resources for Great Images and Information

http://rs.gso.uri.edu/avhrr-archive/archive.html
"Sea Surface Temperature Satellite Image Archive." The University of Rhode Island Graduate School of Oceanography has developed and maintains this Web site.

www.ssec.wisc.edu/data/sst

Beautiful imagery is available at the University of Wisconsin's "Space Science and Engineering Center" Web site.

http://rucool.marine.rutgers.edu/index.php/Cool-Data/Cool-Data.html

"Coastal Oceanography Observation Lab." Rutgers University developed this Web site to provide current and archived sea surface temperatures, along with satellite imagery.

www.pbs.org/wgbh/nova/elnino/anatomy/machine.html

"Tracking El Niño—Global Weather Machine" is a site dedicated to the study of El Niño. Designed by the Public Broadcasting System, it explains the phenomenon and contains links to much information.

www.elnino.noaa.gov/ani.html

El Nino animations and graphics. NOAA has prepared this directory to connect the reader with several outstanding animations and graphics.

Review Exercises for Chapter 12

WORD-STUDY CONNECTION

astronomical high tide	crest	Equator
axis	currents	erosion
convection	diurnal tide	full Moon
Coriolis effect	El Niño	groins

La Niña   seawall   trough
longshore current   semidiurnal tide   upwelling
mixed tide   spit   wave duration
neap tide   spring tide   wave fetch
new Moon   temperature   wave height
prevailing winds   thermohaline circulation wave intensity
rotation   tidal range   wave period
salinity   tides   wavelength
sand dune   trade winds   waves

SELF-TEST CONNECTION

PART A. Completion. *Write in the word or words that correctly complete the sentence.*

1. Erecting a _____ is equivalent to putting a Band-Aid on the remaining beach.

2. The _____ causes global winds to deflect to the right in the Northern Hemisphere.

3. The peak of a wave is called its _____.

4. _____ is the concentration of dissolved minerals or salts in water.

5. _____ are installed along beaches to impede the transport of sand caused by the longshore current.

6. _____ is a cooling of seawater temperatures in the eastern equatorial Pacific Ocean to below normal levels.

7. _____ are strongest during the full Moon.

8. Coastal locations with only one high and one low tide each day are experiencing _____.

9. The _____ is a deep ocean current driven by density differences.

10. The time interval between ocean waves passing a point is the _____.

11. A _____ is a barrier island created by the longshore current.

12. The distance over which wind blows to create waves is known as the _____.

13. _____ is the process whereby cold, dense, deep waters are driven to the surface.

14. The distance between crests is known as the _____.

15. The _____ is the difference in height between high tide and low tide.

PART B. Multiple Choice. *Circle the letter of the item that correctly completes the statement.*

1. El Niño is best described as a
   (a) cooling of the waters in the eastern equatorial Pacific Ocean
   (b) warming of the waters in the eastern equatorial Pacific Ocean
   (c) cooling of the waters in the western equatorial Pacific Ocean
   (d) warming of the waters in the western equatorial Pacific Ocean
   (e) cooling of the waters in the eastern equatorial Atlantic Ocean

2. Surface waters in the Arctic
   (a) are very cold
   (b) have low salinity
   (c) have high salinity
   (d) both a and b
   (e) both a and c

3. Tides are caused primarily by
   (a) the Sun
   (b) the Moon
   (c) Jupiter
   (d) winds
   (e) all of the above

4. Spring tides occur during the
   (a) full Moon
   (b) quarter Moon
   (c) new Moon
   (d) both a and b
   (e) both a and c

5. Semidiurnal tidal patterns
   (a) produce one high tide and one low tide each day
   (b) produce one high tide and two low tides each day
   (c) produce two high tides and one low tide each day
   (d) produce two high tides and two low tides each day
   (e) vary from season to season

6. Waves are created by
   (a) the intensity of the wind
   (b) the duration of the wind
   (c) the fetch of the wind
   (d) both a and b
   (e) a, b, and c

7. Once a _____ is constructed, the beach is likely to completely erode away.
   (a) groin
   (b) sand dune
   (c) seawall
   (d) sand dune
   (e) beach restoration

8. _____ provides a natural and effective barrier to preserve a beach and protect property behind the beach.
   (a) A groin
   (b) A sand dune
   (c) A seawall
   (d) A longshore current
   (e) Beach restoration

9. The Coriolis effect
   (a) causes global winds to deflect to the right in the Northern Hemisphere
   (b) causes global winds to deflect to the left in the Northern Hemisphere
   (c) causes global winds to deflect toward the Equator in the Northern Hemisphere
   (d) causes global winds to deflect toward the North Pole in the Northern Hemisphere
   (e) has no effect on global winds

10. The astronomical high tide can occur during
    (a) spring tides
    (b) neap tides
    (c) the full Moon
    (d) both a and b
    (e) both a and c

PART C. Modified True/False. *If a statement is true, write "true" for your answer. If a statement is incorrect, change the underlined expression to one that will make the statement true.*

1. <u>Sand dunes</u> represent a last-ditch effort to save property from the ocean.

2. Neap tides occur during the <u>quarter</u> phase of the moon.

3. <u>Fetch</u> is the time interval between waves.

4. Wave height is controlled by the fetch, <u>duration</u>, and intensity of the winds.

5. Regions with <u>semidiurnal</u> tidal patterns experience two high tides within a 24-hour period.

6. Upwelling is the rise of <u>warm waters</u> to the surface.

7. <u>Seawalls</u> are both a natural and human method of beach management.

8. Arctic waters, with <u>low</u> salinity, are effectively the point of origin of the thermohaline circulation.

9. Astronomical high tides can occur only during <u>neap</u> tides.

10. <u>Diurnal tides</u> are common to smaller water bodies such as the Gulf of Mexico.

CONNECTING TO CONCEPTS

1. Describe the various methods employed by humans to manage the shoreline. Which has the least negative impact on the natural environment? Why?

2. Research and name two factors that may be responsible for creating the extreme tidal ranges in the Bay of Fundy.

3. Do some research on the New Jersey shore. Is the money being expended to save the beaches and restore them worth the investment?

4. Do some research on El Niño. How does it affect the United States?

5. What portion of the continental United States typically experiences the most moderate temperatures year-round? Describe the roles of the ocean and prevailing winds in causing this situation.

6. How can a diurnal tidal pattern be recognized? How is this pattern different from that of semidiurnal tides?

7. What conditions produce large waves in the oceans?

CONNECTING TO LIFE/JOB SKILLS

This chapter deals with the dynamics of the oceans and the ways in which other systems are impacted when normal wave operation changes. El Niño is a phenomenon that has captured headlines in recent years as lives have been lost and millions of dollars of property damage have resulted from the changes it causes.

Teams of scientists across the world are working to better understand how the hydrosphere functions and to predict when normal functioning of the ocean is likely to change. Most important, scientists are striving to understand how these changes are likely to impact the atmosphere and biosphere. Specifically, the ability to predict how precipitation patterns change with changes in sea surface temperature patterns is a major objective of current research.

ANSWERS
SELF-TEST CONNECTION

Part A

1. seawall
2. Coriolis effect
3. crest
4. Salinity
5. Groins
6. La Niña
7. Spring tides
8. diurnal tides
9. themohaline circulation
10. wave period
11. spit
12. fetch
13. Upwelling
14. wavelength
15. tidal range

Part B

1. **(b)**
2. **(e)**
3. **(b)**
4. **(e)**
5. **(d)**
6. **(e)**
7. **(c)**
8. **(b)**
9. **(a)**
10. **(e)**

Part C

1. False; Seawalls
2. True
3. False; period
4. True
5. True
6. False; cold waters
7. False; Sand dunes
8. False; high
9. False; spring
10. True

CONNECTING TO CONCEPTS

1. Beach management is big business. Beaches are major tourist attractions, as well as a main reason for people to purchase homes near the ocean. Efforts to manage the natural resources include building groins and seawalls, planting dunes, and replenishing the beach. Of the four, dunes have the least negative impact on the natural environment. Dunes act to minimize wave energy without incurring great expense or creating other unwanted problems. Groins cause a sawtooth beach pattern to develop where sand is not evenly distributed along the coastline. Seawalls, while protecting the shoreline, ultimately result in the destruction of the beach in front of the wall. Beach restoration is an expensive and temporary solution.

2. The Bay of Fundy has an extreme tidal range as a result of the funnel shape of its boundaries and the slope of its subsurface. When the tide comes in, it has nowhere to go but up.

3. This is a highly debatable question. Tourism represents millions of dollars to the state and coastal communities each summer. However, solutions are often temporary, and the many residents of the state who do not visit coastal communities never see any direct benefit from the money and effort expended.

4. El Niño winters tend to bring stormy weather to the West Coast, especially central and southern California. The south-central and southeastern states also experience more rainfall than during a normal winter. The northern United States is generally drier and warmer than normal. Since each El Niño is different, however, no one can speak with certainty about how a particular El Niño will affect the United States.

5. The immediate coastline along the Pacific Ocean experiences a very moderate climate with mild winters and cool summers. Prevailing winds, which blow onshore from the Pacific Ocean located just to the west, act to keep temperatures quite stable year-round.

6. A diurnal tidal pattern produces a high tide approximately every 24 hours. In a semidiurnal pattern, high tides occur approximately every 12 hours.

7. Ocean waves are created by the wind. The higher the wind velocity, over a distance (fetch), for a period of time (duration), the larger the observed waves will be.

13

Earth's Atmosphere

WHAT YOU WILL LEARN

This chapter focuses on Earth's atmosphere and its interaction with incoming solar radiation. In this chapter you will learn

- what gases are found within our atmosphere;
- how our atmosphere is layered;
- what happens to the Sun's energy when it reaches our atmosphere;
- what causes seasons;
- how convection creates and transports storm systems.

SECTIONS IN THIS CHAPTER

- Composition of the Atmosphere
- Structure of the Atmosphere
- Important Earth-Sun relationships
- The Dynamic Earth
- The Earth Rotates on a Tilted Axis
- Circulation of the Atmosphere—Global Convection Cells
- Review Exercises for Chapter 13

Composition of the Atmosphere

MAJOR COMPONENTS

Our planet's atmosphere is an amazingly consistent, yet complex, mixture of several different gases (see Table 13.1). Nitrogen (78%) and oxygen (21%) constitute most of the atmosphere. Argon accounts for most of the remaining 1 percent. Our atmosphere also contains several trace gases. Although present in small (trace) concentrations, these gases, including carbon dioxide (actual concentration, 0.03%), can be quite important. Carbon dioxide has the ability to trap heat and therefore is largely responsible for creating a natural greenhouse effect.

NOTES:
An excellent primer on the greenhouse effect is posted by UCAR at *www.ucar.edu/learn/1_3_1.htm.*

VARIABLE COMPONENTS

Some components of our atmosphere vary significantly from time to time and place to place. The most important trace gas that varies in this way is water vapor, which can compose up to 4 percent of the atmosphere at times, particularly in certain tropical locations.

Water vapor is important because it is an intricate part of the water cycle. Under the right conditions, this trace gas can condense to form clouds and precipitation. Water vapor also acts as a greenhouse gas with an ability, similar to that of carbon dioxide, to trap heat. In addition, water vapor transports heat that it absorbs during the process of

TABLE 13.1
COMPOSITION OF AIR IN THE TROPOSPHERE

Gas	Percent by Volume in Dry Air
Nitrogen	78
Oxygen	21
Argon	Almost 1
Neon	Trace
Helium	Trace
Krypton	Trace
Xenon	Trace
Hydrogen	Trace
Ozone	Trace
Carbon dioxide	Trace
Nitrous oxide	Trace
Methane	Trace

evaporation and releases heat during condensation. Have you ever stepped out of a pool and felt cold until you dried off? This chill is caused by water absorbing energy from your body as it evaporates.

Ozone is another important trace gas. Ozone is an *allotrope* of oxygen; it is composed of the same element, oxygen, but has three atoms bonded together rather than two. This small molecular change makes a great difference in the properties of these two gases. Ozone has a long and complex story. In the following sections, we will clarify the difference between ground-level ozone, an unwanted pollutant prevalent in and near our cities, and stratospheric ozone, which is the same molecule but, when present in the upper atmosphere, plays a vital role for all life on the surface of our planet.

Tiny liquid and solid particles called *aerosols* also populate our atmosphere. Aerosols include airborne dust, pollen, sea salts, microorganisms, and human-introduced pollutants. Aerosols provide surfaces onto which water vapor can condense; thus aerosols play a role in the global water cycle. These tiny particles also act to scatter sunlight, sometimes causing spectacular sunsets in areas with excessive levels of aerosols.

THE EARLY ATMOSPHERE

Earth's atmosphere has not always been a life-supporting mixture of nitrogen (N_2) and oxygen (O_2). Today's atmosphere, considered an "oxidizing" atmosphere by atmospheric chemists because of the presence of oxygen, is believed to have started as a "reducing" atmosphere, that is, one devoid of free oxygen. In order for life to have evolved on the planet, the atmosphere probably contained oxygen in the form of carbon dioxide (CO_2), nitrogen in the form of free nitrogen (N_2) or of ammonia (NH_3), and water vapor (H_2O). Carbon dioxide, a strong greenhouse gas, would have raised temperatures to levels near or above today's despite a younger Sun that emitted perhaps as much as 25 percent less energy than the Earth receives today. Nitrogen is found in all forms of life; since it is the basis of all protein molecules, it was likely present at that time. Water vapor probably provided the source material for what became the oceans as the planet cooled during its first billion or so years of existence. Some water vapor very likely resulted from impacts with comets and meteorites, as these objects are known to contain ice. Early Earth was also quite tectonically active, so outgassing likely added large quantities of water vapor to the developing atmosphere.

The rock record suggests that free oxygen, O_2, first appeared around 2 billion years ago. This is the age of the first sulfate rocks. Older rocks that contain sulfur do not contain oxygen and as such are "sulfides." For example, there is no iron sulfate ($FeSO_4$) older than 2 billion years; before that, geologists find iron sulfide (FeS).

Probably the first free oxygen resulted from the photosynthetic activity of primitive life-forms such as phytoplankton and bacteria. Photosynthesis produces oxygen, and even today large amounts are produced by simple marine plant life. Other theories that attempt to explain the appearance of free oxygen in the atmosphere include the chemical breakdown of carbon dioxide and water vapor. None of these theories, however, is favored over the notion that early plant life was responsible.

Once the atmosphere contained free oxygen, some of the oxygen in the stratosphere decomposed into atomic (one-atom) oxygen and bonded with diatomic (free) oxygen to form ozone (O_3). Ozone in the stratosphere acts to protect all life on Earth's surface from dangerous high-energy solar radiation. Shortly after (in geologic time) the appearance of ozone in the stratosphere, there was a proliferation of life on the planet's surface.

Structure of the Atmosphere

Like the oceans and Earth's interior, the atmosphere is layered (see Figure 13.1).

TROPOSPHERE

The *troposphere* constitutes approximately the lowest 16 kilometers (10 mi) of the atmosphere. Its thickness varies, from a minimum near the poles to a maximum near the Equator. Everything that we are familiar with—all forms of life, mountains and oceans, clouds and storms—exists in the troposphere.

When you read about the composition of the atmosphere, as in the preceding section, you are really reading about the composition of the troposphere. Other layers do not have the same mixture of gases.

The troposphere contains most of the air found in our atmosphere. As noted earlier, air is a mixture of gases composed primarily of nitrogen and oxygen. The atmosphere, held to Earth by the planet's gravitational field, thins out quite quickly with altitude (see Figure 13.2). Average barometric pressures, caused by the motion of air molecules, as illustrated in Figure 13.3, are just over 1000 millibars (this unit of air pressure, often used by meteorologists, is equivalent to $14.7\,lb/in^2$). By the *tropopause*, a boundary separating the troposphere from the stratosphere (see Figure 13.4), air pressures are reduced to approximately 100 millibars, or one-tenth of the pressures observed at the surface.

Temperatures also gradually fall with height in the troposphere (Figure 13.4). The greatest heating occurs at Earth's surface as a result of the absorption of solar radiation by the uppermost portions of the surface and the subsequent emission of infrared radiation. Solar radiation is often referred to as shortwave radiation because it contains great amounts of visible radiation and some ultraviolet radiation, both of which are

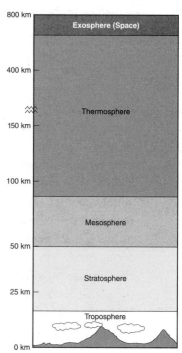

FIGURE 13.1 The structure of Earth's atmosphere. Source: *Astronomy Explained*, Gerald North, Springer-Verlag, 1997.

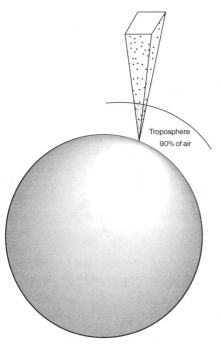

FIGURE 13.2 A sectional slice through Earth's atmosphere. Most of the gas molecules in the atmosphere are clustered near the surface within the troposphere. Only the most energetic molecules can overcome gravity and reach higher levels in the atmosphere. Source: *Let's Review: Earth Science—The Physical Setting*, 2nd Ed., Edward J. Denecke, Jr., Barron's Educational Series, Inc., 2002.

shorter wavelengths than the infrared radiation emitted by the uppermost crust. Additionally, thin air cannot hold as much heat as air near the surface; hence temperature falls with elevation until the tropopause is reached.

The *tropopause* is easily identified as a boundary; above it, in a layer known as the stratosphere, with a height from the tropopause of about 30 kilometers (18 mi), the atmosphere is nearly isothermal. In other words, this is a region with no temperature change with increasing altitude (Figure 13.4). In the upper stratosphere, temperature actually *increases* with height because ozone in this region absorbs incoming solar energy.

The tropopause acts as a barrier to keep most human-induced pollutants in the troposphere from entering the stratosphere. Ground-level ozone never interacts with stratospheric ozone, partially for this reason. However, one pollutant, chlorofluorocarbons, used in the refrigeration industry and elsewhere, have penetrated the tropopause to some extent and have wreaked havoc with stratospheric ozone. Chapter 16 discusses stratospheric ozone and its interaction with chlorofluorocarbons in greater detail.

STRATOSPHERE

Above the troposphere and tropopause lies the *stratosphere*, so named because of high-velocity winds that circumnavigate the globe and exhibit only horizontal movement, with very little vertical motion. This is in contrast to the vertical and horizontal wind motions observed in the troposphere.

The stratosphere is characterized as a region where temperatures generally rise with height after an initial region near the tropopause where temperatures remain constant with increasing altitude. Temperatures actu-

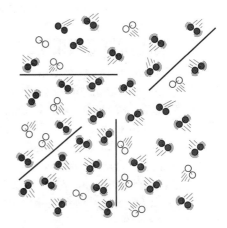

FIGURE 13.3 Gas molecules in a container. Regardless of the angle at which a surface is held, air molecules collide with it, exerting the same pressure in all directions. Source: *Let's Review: Earth Science—The Physical Setting*, 2nd Ed., Edward J. Denecke, Jr., Barron's Educational Series, Inc., 2002.

ally rise from approximately −60°C at the tropopause to nearly 20°C near the uppermost stratosphere, or stratopause, at a height of nearly 50 kilometers. As noted above, the main cause of this warming is the presence of ozone in the stratosphere. This ozone is a vital gas as it traps the high-energy radiation emitted by the Sun. This solar energy, if it reached the ground, would cause serious harm to any life-form exposed to it for any length of time.

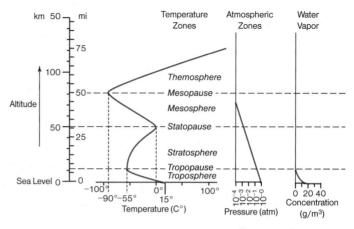

FIGURE 13.4 Selected properties of Earth's atmosphere. Source: The State Department, *Earth Science Reference Tables*, 2001 Ed. (Albany, New York: The University of the State of New York).

MESOSPHERE

In the *mesosphere*, thermal patterns are similar to those observed in the troposphere. Temperatures fall from their comfortable levels at the stratopause to nearly −100°C at the top of the mesosphere, approximately 80 kilometers above Earth's surface. Although air in the mesosphere is quite thin, it is substantial enough to cause enough friction to burn up most of the meteoroids that enter our atmosphere.

THERMOSPHERE

The *thermosphere*, composed of very rarified air, mostly atoms of nitrogen and oxygen charged atoms or ions, has no real upper limit. Molecular motion becomes excited because of high-energy solar radiation. Technically, temperatures are quite high because temperature is a measure of average kinetic energy or molecular motion. The air is so thin, however, that heat would never be felt as it is on Earth's surface. Thus, although temperatures approach 1000°C, the air in this layer or region would not feel hot to the touch.

> **REMEMBER**
> The composition of Earth's atmosphere has evolved over time.

Important Earth-Sun Relationships

RADIATION

Energy travels from the Sun to Earth by means of invisible waves (see Figure 13.5). These waves are commonly referred to as *radiation*, a mode of energy transfer that requires no medium between the sender and the receiver; hence radiation can travel through space. The electromagnetic radiation spectrum describes all the various forms of radiation, virtually all of them emitted by the Sun, that exist.

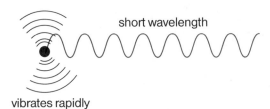

FIGURE 13.5 Electromagnetic waves are produced by particles vibrating at different rates. The higher the temperature, the faster the particle vibrates and the shorter the wavelength produced. Source: *Let's Review: Earth Science—The Physical Setting*, 2nd Ed., Edward J. Denecke, Jr., Barron's Educational Series, Inc., 2002.

As Figure 13.6 illustrates, the electromagnetic radiation spectrum can be broken down into several regions. These include gamma rays, X-rays, ultraviolet waves, visible waves, infrared waves, microwaves, and radio waves. The electromagnetic radiation spectrum is a continuum, and the energy emitted by the radiation gradually diminishes from gamma rays to radio waves. Visible light is just a small portion of the spectrum, and the only thing unique to this portion is our ability to directly sense these wavelengths. We see each wavelength in the visible portion of the spectrum as a different color. As one might suspect, violet borders the ultraviolet region, and red borders the infrared. Although we cannot see ultraviolet radiation, we witness its effects as it darkens skin that is exposed to the Sun. We can also sense infrared radiation as heat.

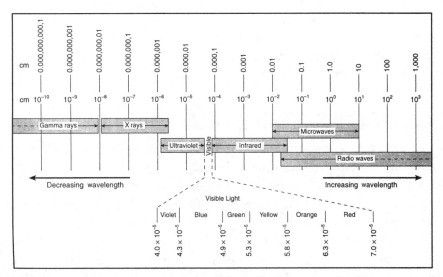

FIGURE 13.6 The electromagnetic radiation spectrum. Wavelength increases to the right. The electromagnetic radiation spectrum is broken into several regions from gramma rays (high-energy radiation) to radio waves (low-energy radiation.) Source: The State Department, *Earth Science Reference Tables*, 2001 Ed. (Albany, New York: The University of the State of New York).

If we were to conduct a detailed study of the Sun, we would discover that it is a strong emitter of ultraviolet and visible radiation, as shown in Figure 13.7. Although the Sun emits radiation ranging in wavelength from X-rays to infrared waves (and beyond), it emits most effectively around the wavelength that we see as yellow. This explains why the Sun appears to be yellow.

From an Earth systems science perspective, the most interesting question is what happens to the Sun's solar radiation when it reaches our planet. Fortunately for all life here on Earth's surface, much of the high-energy radiation is absorbed by either the Van Allen radiation belts or by ozone in the stratosphere.

The Van Allen radiation belts are two belts of high-energy, charged particles above the thermosphere that protect Earth from gamma-ray and X-ray radiation. Ozone is also an excellent absorber of ultraviolet radiation. If the ozone in the stratosphere, often referred to as the ozone layer, were destroyed, lethal amounts of ultraviolet radiation would reach the planet's surface.

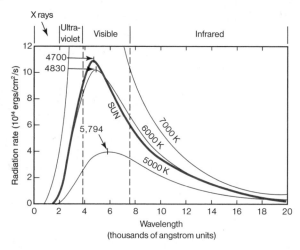

FIGURE 13.7 Electromagnetic radiation emissions from the Sun. The Sun emits electromagnetic radiation ranging in wavelength from infrared to X-rays. The majority of this radiation is in the visible portion of the electromagnetic radiation spectrum. The peak is at 4700 angstrom units, where 1 angstrom is equal to 1 billionth of a meter. This is very close to the theoretical output of a body at a temperature of 6000 kelvin, which is why the Sun's surface temperature is thought to be near 6000 K. Source: *Let's Review: Earth Science—The Physical Setting*, 2nd Ed., Edward J. Denecke, Jr., Barron's Educational Series, Inc., 2002.

PATHS TAKEN BY INCOMING SOLAR RADIATION

Incoming solar radiation, or *insolation*, is the visible and ultraviolet radiation that does reach the troposphere and surface of Earth. This radiation is reflected, scattered, or absorbed when it interacts with the planet's atmosphere and surface. On average, about 30 percent of the insolation, referred to as Earth's *albedo*, is reflected right back out to space. This energy does not heat the planet significantly, as the energy is not absorbed by the atmosphere, anything in it, or the surface of the planet.

The remaining 70 percent is either absorbed by clouds, absorbed by the surface, or scattered by the atmosphere and later absorbed by clouds and the surface. In Table 13.2, which shows the planet's "global radiation budget," note that the "net incoming energy" gained by the Earth-atmosphere system is eventually lost to space and as such equals the "net outgoing energy."

TABLE 13.2
GLOBAL RADIATION BUDGET*

	Percent	
Incoming Solar Radiation	Gain	Loss
Reflection from clouds to space		21
Diffuse reflection (scattering) to space		5
Direct reflection from Earth's surface		6
Net energy loss back to space		32
Absorbed by clouds	3	
Absorbed by molecules, dust, water vapor, and CO_2	15	
Absorbed by Earth's surface	50	
Net incoming energy gained by Earth-atmosphere system	**68%**	

	Percent	
Outgoing Terrestrial Radiation	Gain	Loss
Total infrared radiation emitted by Earth's surface	98	
Absorbed by atmosphere	90	
Lost to space		8
Total infrared radiation emitted by the atmosphere	137	
Absorbed by Earth's surface	77	
Lost to space		60
Net outgoing energy lost from entire Earth-atmosphere system		**68%**
Net energy leaving *Earth's surface* (98% emitted—77% reabsorbed)		21
Net energy leaving the *atmosphere* (137% emitted—90% reabsorbed)		47

*Source: *The Earth Sciences*, Arthur N. Strahler, Harper & Row, 1971.

The Dynamic Earth

ALBEDO AND RERADIATED ENERGY

Since Earth cannot retain the absorbed insolation forever, it reradiates or emits it. A key element of the process is that the planet emits the radiation at a longer, or lower energy, wavelength than the one at which it was absorbed. The absorbed energy is often referred to as *short-wave radiation*, and the emitted energy as *long-wave radiation*.

THE NATURAL GREENHOUSE EFFECT

The process gets even more interesting. The long-wave radiation emitted by the planet is absorbed by certain gases known as greenhouse gases. These gases have a resonance frequency that matches the frequency of the long-wave radiation emitted by Earth's surface. When the long-wave radiation reaches the greenhouse gases, they begin to resonate or vibrate, thus increasing their kinetic energy. The energy absorbed by the gases can then spread throughout the atmosphere, producing a net warming effect that is known as the *natural greenhouse effect* (see Figure 13.8).

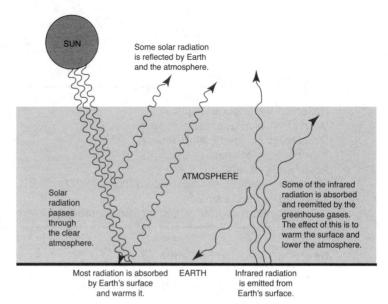

FIGURE 13.8 The greenhouse effect. This diagram illustrates incoming solar radiation (short-wave radiation) passing through the atmosphere, the absorption of some of that energy, and then the subsequent emission of long-wave radiation from the surface. Some of the long-wave radiation is trapped. This is known as the greenhouse effect. Source: The Intergovernmental Panel on Climate Change.

Were it not for the natural greenhouse effect, Earth would be approximately 33C° colder than it is. This would be a critical change since the average global temperature would be −18°C (0°F) instead of 15°C (60°F). Consider what our planet would be like if it were not for the natural greenhouse effect!

The greenhouse gases include carbon dioxide, ozone, water vapor, nitrogen dioxide, and methane. All of these gases occur naturally in our atmosphere. Humans, however, are adding more of each to the atmosphere, as well as a group that has no natural source, namely, chlorofluorocarbons (CFCs). The input of additional concentrations of the natural gases and the introduction of new ones have caused great concern worldwide among scientists as to the long-term effects and a possible enhanced global warming that may already be in its early stages.

CONVECTION

The warming of Earth's lower atmosphere by emitted long-wave radiation is dependent upon the type of surface that is absorbing the short-wave radiation and the angle of the Sun in the sky, which affects the intensity of the insolation.

As described earlier, the planet's albedo is 30 percent; this figure can vary greatly, however, for individual surfaces. Freshly fallen snow can have an albedo of 90 percent. This type of surface absorbs little short-wave radiation and consequently can emit little long-wave radiation. In contrast, blacktop roads can have an albedo as small as 5 percent! Since so much short-wave energy is absorbed by the blacktop, these surfaces emit great amounts of long-wave radiation. Many studies have documented climatic differences between cities, with much concrete and pavement, and the surrounding suburbs that have trees and grasses with higher albedos.

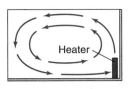

Convection

FIGURE 13.9 Convection is a form of energy transfer that requires the flow of a fluid (usually air or water) to transport energy (or heat) from a region that has an excess of heat to a region with a deficit. Source: *Physics*, 2nd Ed., Arthur Beiser, Benjamin Cummings, 1978.

The elevation of the Sun in the sky also affects how much energy is absorbed and later emitted. Tropical latitudes, where the Sun is typically nearly overhead during midday hours, receive great intensity of insolation. In contrast, in polar latitudes, where the Sun remains close to the horizon even at midday, the intensity of insolation is much lower.

As a result of the unequal global heating, a net surplus of energy builds up near the Equator and a net deficit occurs near the poles. The result is the creation of another form of energy transfer known as convection (see Figure 13.9). *Convection* is the transfer of energy through the flow of a fluid. The fluids in this case are the oceans and the atmosphere, both of which carry heat from the Equator to the poles and then back.

The currents in the oceans and atmosphere that result from this transfer of heat are called *convection currents* or *convection cells*. Figure 13.10 illustrates global convection cells as they transport energy through flowing air at equatorial and polar latitudes.

A simplified view of global convection cells is provided in Figure 13.11, where the action at all latitudes (polar, midlatitudes, and tropical) is depicted. This topic is treated in somewhat greater detail later in the section headed "Circulations of the Atmosphere—Global Convection Cells."

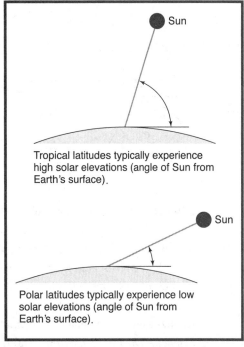

Tropical latitudes typically experience high solar elevations (angle of Sun from Earth's surface).

Polar latitudes typically experience low solar elevations (angle of Sun from Earth's surface).

REMEMBER
The three major forms of energy transportation are radiation, convection, and conduction.

The Earth Rotates on a Tilted Axis

SOLSTICES, EQUINOXES, AND SEASONS

Our planet exhibits two primary types of motion in space, rotation and revolution. *Rotation*, illustrated in Figure 13.12, is the spinning of Earth on its axis. Earth completes one rotation every 24 hours, which, by definition, is a day. The interesting

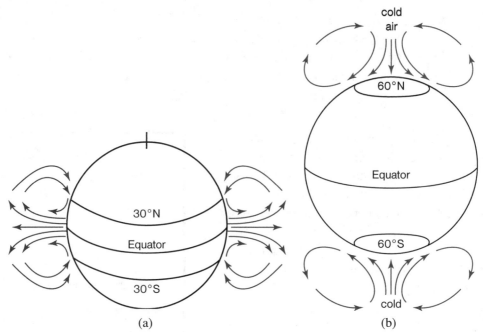

FIGURE 13.10 Convection cells near the Equator (a) and near the poles (b). Air near the Equator is heated by more intense insolation, becomes less dense, and floats upward in the surrounding denser air. As the air rises, it expands and cools, causing it to sink back toward the surface. The result is a circular movement of air, or convection cell. Air near the poles cools because of less intense insolation, becomes more dense, and sinks downward. The sinking air spreads outward at the surface, is warmed as it moves away from the poles, and rises. The result, again, is a circular movement of air, or convection cell. Source: *Let's Review: Earth Science—The Physical Setting*, 2nd Ed., Edward J. Denecke, Jr., Barron's Educational Series, Inc., 2002.

aspect of this rotational behavior is the fact that, as Earth spins, its axis is tilted $23\frac{1}{2}°$ from the vertical. The implications of this fact are enormous when Earth's *revolution* about the Sun is considered. As Earth revolves about the Sun, the orientation of each hemisphere, the Northern and the Southern, as it relates to the Sun, changes throughout the year (see Figure 13.13).

In essence, during summer, the hemisphere is tilted toward the Sun, and during winter, the hemisphere is tilted away from the Sun. Seasons result from the changing

FIGURE 13.11 Global convection cells—a simplified view. Source: *Let's Review: Earth Science—The Physical Setting*, 2nd Ed., Edward J. Denecke, Jr., Barron's Educational Series, Inc., 2002.

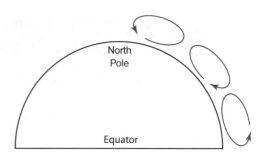

positions of the Sun in the sky. In summer, the Sun appears to be higher in the sky at midday, the incoming solar energy is more direct or intense, and days are longer. This effect is most apparent in the midlatitudes, where solar elevations, that is, the elevations above the horizon achieved by the Sun, vary significantly between the winter and summer seasons, as illustrated in Figure 13.14. Note the angle labeled "altitude of the Sun at noon." This altitude is quite low in the winter and high in the summer. Figure 13.15 shows the changes in solar elevation that occur between the summer and winter solstices for an observer in the Northern Hemisphere. The zenith represents the point in the sky that is overhead or, in geometrical terms, the point that is 90° above the horizon.

The summer *solstice* occurs on the day when the Northern Hemisphere is tilted most directly toward the Sun. On this day, on or

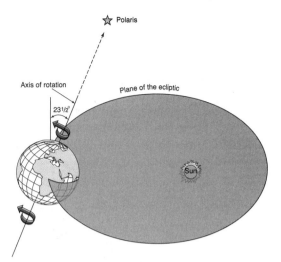

FIGURE 13.12 Rotation. Earth rotates from west to east once every 24 hours around an axis that runs through the poles. Source: *Let's Review: Earth Science—The Physical Setting*, 2nd Ed., Edward J. Denecke, Jr., Barron's Educational Series, Inc., 2002.

about June 21 each year, all locations north of the Equator experience the longest daylight of the year. The sun is observed most directly overhead at the Tropic of Cancer (23.5° N), and daylight occurs for a continuous 24 hours at all locations along and north of the Arctic Circle. The North Pole, in fact, receives 6 months of continuous daylight, beginning on the vernal equinox in March and ending on the autumnal equinox in September. North of the Arctic Circle, however, temperatures remain relatively cold, as the Sun never reaches a position in the sky very far above the horizon and the insolation remains rather indirect. Surface heating is minimal.

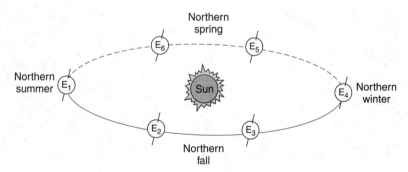

FIGURE 13.13 Earth revolves about the Sun. Note positions E_1, where the Northern Hemisphere is tilted toward the Sun, and E_4, where the Northern Hemisphere is tilted away from the Sun. Source: *Astronomy Explained*, Gerald North, Springer-Verlag, 1997. Used with permission.

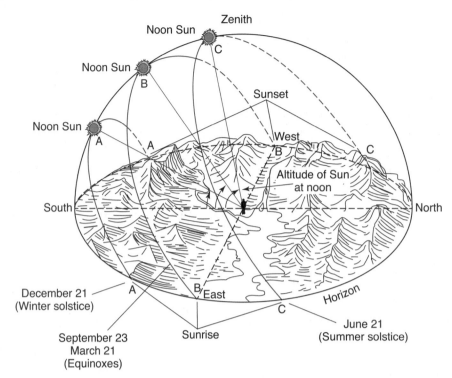

FIGURE 13.14 The Sun's apparent path as it varies throughout the year. An observer at most mid-latitude locations would see the Sun move along different paths across the sky as the seasons progress. Source: *Earth Science: A Study of a Changing Planet*, Daley, Higham, and Matthias, Prentice-Hall, 1986, and *The Story of Maps*, Lloyd Brown, 1977.

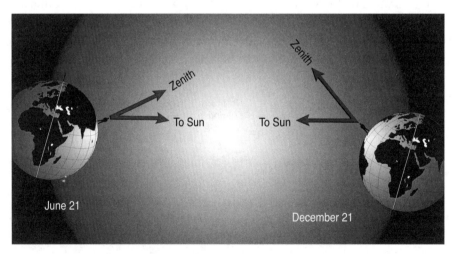

FIGURE 13.15 Altitude of the Sun at midday on the two solstices. Note that on June 21 the Sun is seen closer to the zenith (overhead position, 90° above the horizon) in the Northern Hemisphere. Source: *Let's Review: Earth Science—The Physical Setting*, 2nd Ed., Edward J. Denecke, Jr., Barron's Educational Series, Inc., 2002.

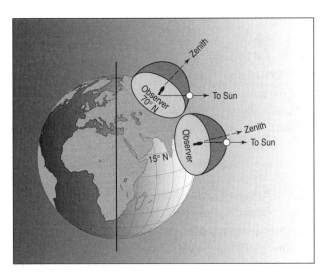

FIGURE 13.16 Midday altitude of the Sun as observed at different latitudes. Since Earth is essentially a sphere, the Sun will appear at different altitudes (solar elevations) as latitude varies. An observer in a tropical location (e.g., 15° N) will see the midday Sun near the zenith, whereas an observer at a polar location will see the midday Sun closer to the horizon. Source: *Let's Review: Earth Science—The Physical Setting*, 2nd Ed., Edward J. Denecke, Jr., Barron's Educational Series, Inc., 2002.

Figure 13.16 illustrates the altitude of the Sun at different latitudes. Keep in mind that the greater the solar elevation, the greater the intensity of insolation. Figure 13.17 shows the daily and annual cycles of solar radiation intensity. Figure 13.18 provides an additional perspective; the most intense insolation is received in the tropics, where solar elevations are greatest year-round.

Since the two hemispheres (Northern and Southern) cannot be tilted toward the Sun simultaneously, the seasons observed are opposite each other. Thus, summer in the Northern Hemisphere is winter in the Southern Hemisphere. The day when the summer solstice occurs in the Northern Hemisphere marks the winter solstice in the Southern Hemisphere. When the Northern Hemisphere is tilted toward the Sun, the Southern Hemisphere is tilted away from it. Daylight hours are short, and solar elevations are relatively low. For the 6 months that the North Pole is bathed in sunlight, the South Pole is shrouded in darkness.

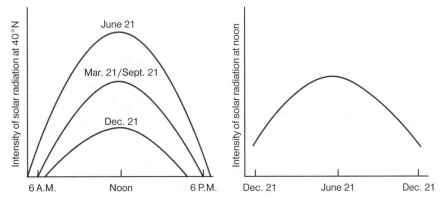

FIGURE 13.17 Intensity of incoming solar radiation (insolation) varies from season to season as well as throughout the day. The closer the Sun is to the zenith (the greater the angle of solar elevation), the greater the intensity of insolation. Source: *Let's Review: Earth Science—The Physical Setting*, 2nd Ed., Edward J. Denecke, Jr., Barron's Educational Series, Inc., 2002.

FIGURE 13.18 Insolation near the Equator and the North Pole. Note that the two beams of sunlight approaching Earth are identical, but the angle at which they strike Earth's surface causes insolation received near the Equator to be more concentrated than that received near the poles. The result is greater heating at the Equator. Source: *Let's Review: Earth Science—The Physical Setting*, 2nd Ed., Edward J. Denecke, Jr., Barron's Educational Series, Inc., 2002.

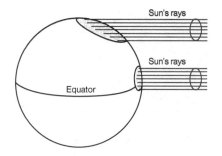

Six months later, on or about December 21, the Northern Hemisphere reaches its winter solstice. At this time, most midlatitude regions, including much of the United States, are experiencing cold weather as daylight hours are short, and low solar elevations provide for indirect or low-intensity insolation. Figure 13.19 shows variations in the lengths of days as a result of Earth's rotation.

The *equinoxes* occur midway between the solstices (Figure 13.20). On the day of each equinox, the Sun appears directly overhead at the Equator. The vernal equinox occurs on or about March 21, and the autumnal equinox on or about September 21. On these dates, daylight and darkness occur for approximately 12 hours at all locations across the planet.

Each of these four dates marks the beginning of a season. Summer begins with the summer solstice; autumn, with the autumnal equinox; winter, with the winter solstice; and spring, with the vernal equinox.

Despite the extent of insolation received throughout the midlatitudes on the summer solstice, this is not the warmest day of the summer. The hottest part of the summer for much of this region occurs some 5 to 6 weeks later, in late July and early August. This phenomenon, known as *seasonal lag*, occurs because solar elevation and intensity of insolation are not the only factors that produce heat. All through June and July the oceans and the landmass continue to heat up, thus causing temperatures to rise through early August. By that time, days are beginning to shorten, and solar elevations are diminishing to the point where surface temperatures across much of North America, Europe, and Asia begin to decrease.

The seasonal lag is again observed when the coldest winter temperatures typically

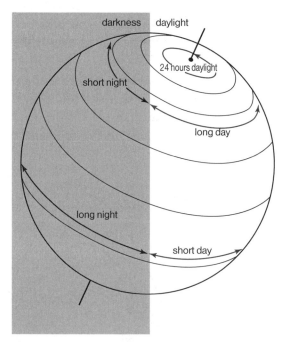

FIGURE 13.19 Variations in the lengths of day and night as a result of Earth's tilted axis of rotation. Source: *Let's Review: Earth Science—The Physical Setting*, 2nd Ed., Edward J. Denecke, Jr., Barron's Educational Series, Inc., 2002.

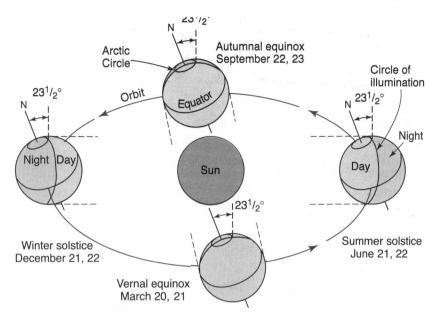

FIGURE 13.20 Earth's changing position in relation to the Sun. Note Earth's position at the beginning of each season. Source: *Let's Review: Earth Science—The Physical Setting*, 2nd Ed., Edward J. Denecke, Jr., Barron's Educational Series, Inc., 2002.

occur in late January and early February, despite the winter solstice's having occurred in late December. Even though days grow longer after the winter solstice, the pool of cold air over the Arctic intensifies in the relative darkness and impacts much of the midlatitudes in the Northern Hemisphere until well into February.

NOTES:
A "daily lag" is also observed on most days. Although the greatest solar elevations are observed midday at "solar noon," the warmest temperatures usually occur a few hours later.

Circulation of the Atmosphere—Global Convection Cells

Unequal heating across the globe produces convection currents that carry energy from one place to another. In the atmosphere this occurs by means of *global convection cells*. Refer to Figure 12.1b, which illustrates the entire global atmospheric circulation as a complete system. The illustrations that follow break the system down into its component parts and help to clarify how the global circulation shown in this diagram develops.

Working through the global atmospheric circulation piece by piece, we start with the tropical regions. Located near the Equator, these regions experience an excess of

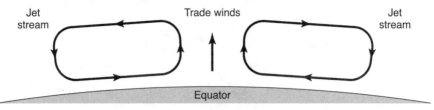

FIGURE 13.21 Upward atmospheric motion caused by intense heating along the Equator.

insolation, which causes rising motion in the atmosphere, as illustrated in Figure 13.21. Air is induced to rise by intense heating. The warm air becomes more buoyant, thus tending to rise still higher. Rising air at the Equator sets off a chain of events that ultimately produce three relatively distinct global convection cells.

In 1735, George Hadley, a British meteorologist, was trying to develop a model of atmospheric circulation. His model correctly identified rising air at the Equator and sinking or subsiding air at the poles. He thought, however, that there was only one large circulation between the Equator and the poles. More modern models incorporate the understanding that moving so much air directly from the Equator to the North Pole is not practicable, and that the rising air at the Equator must subside somewhere between that latitude (0°) and the North Pole.

Modern atmospheric circulation models include a *Hadley cell* (named in honor of George Hadley) that includes rising air at the Equator and subsiding air at the horse latitudes (30° from the Equator). The flow of subsiding air then splits at the surface; part turns toward the poles, carrying warm air in their directions, and part returns to the Equator. Figure 13.22 illustrates this movement.

As noted, the Equator is a zone of rising air. The rising motion is enhanced by the return at the surface of air from the horse latitudes in the form of the southeasterly trade winds (north of the Equator) and the northeasterly trade winds (south of the Equator). Where these winds converge, usually within 5° of the Equator, the intertropical convergence zone (ITCZ) forms. This is a zone of rising air, cloudiness, and frequent thunderstorms.

Figures 13.23–13.26 describe some of the activity occurring along or near the Equator.

When large parcels of air rise, particularly if they are laden with moisture, the inevitable result is clouds and precipitation. The mechanism works as follows: Envision an imaginary parcel of rising air at the Equator that begins to expand as it moves vertically upward into a region with less atmospheric pressure. As the parcel expands, the temperature in this region falls. Any moisture (water vapor) carried with the air from the surface begins to condense, as cooler air cannot hold as much water vapor. As the water vapor condenses, clouds form and eventually precipitation falls. Many thunderstorms occur along the Equator as a result of the persistent upward motion of the atmosphere (Figure 13.24).

Once the tropopause is reached, the rising air turns poleward. As noted earlier, the air movement begins to subside at 30° north and south of the Equator in the region known as the *horse latitudes*. Here, at the surface, is a zone of weak wind flow where, legend has it, Spanish sailors carried horses in ships toward the West Indies. In this zone, the

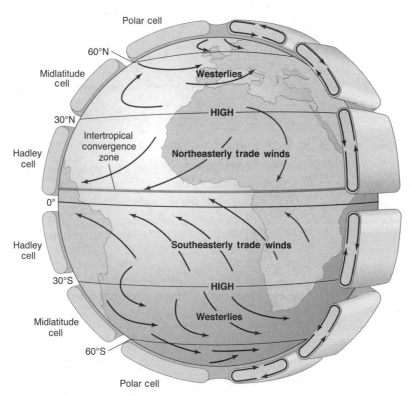

FIGURE 13.22 Global atmospheric circulation. Air rising at the Equator travels toward the poles but sinks at or about the 30th parallel (30° latitude). This sinking air then splits (diverges), with part of the flow returning to the Equator and part again heading toward the poles along the surface. The tropopause caps the upward motion of the air along the Equator.

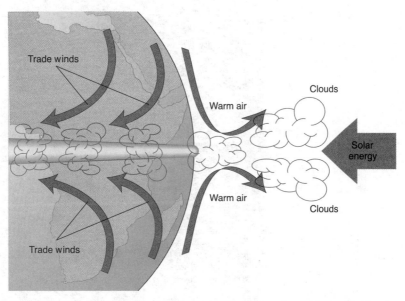

FIGURE 13.23 Trade winds carry subsiding (sinking) air at the 30th parallel back toward the Equator (bold line) where the northern and southern air streams converge. This convergence adds to the rising motion near the Equator, as seen along the right side of the diagram.

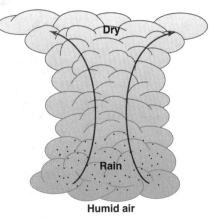

FIGURE 13.24 Rising air along the Equator is laden with moisture (water vapor), which, upon cooling, condenses and forms a relatively permanent band of clouds just north of the Equator called the intertropical convergence zone (ITCZ). This zone is the site of most of the world's thunderstorms, and this figure depicts the formation of those thunderstorms.

FIGURE 13.25 The intertropical convergence zone (ITCZ). The bright band parallel to and just north of the Equator represents a region of very heavy rainfall, caused by showers and thunderstorms within the ITCZ. Source: NOAA.

Three-Year TRMM Climatology

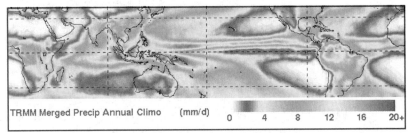

January 1998–December 2000

FIGURE 13.26 This spectacular photo, taken by satellite, shows the ITCZ cloud band as seen from space. The ITCZ, or band of tropical moisture, appears at the base of this photograph, south of Mexico and most of Central America. Source: NOAA.

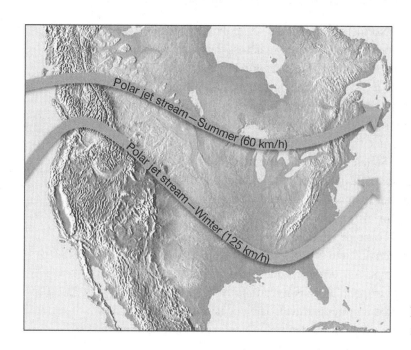

FIGURE 13.27 Seasonal jet stream— idealized positions.

weak winds slowed the sailing ships' journey. According to the story, they threw the horses overboard as water and supplies became limited—hence the name horse latitudes.

In addition to weak wind flow, the horse latitudes feature many of the world's deserts. As air flow subsides in this region, our imaginary parcel of air is compressed and warmed. Any condensation evaporates, and the cloudless sky produces minimal precipitation. The result is a region where the Sun is still quite strong, and since evaporation greatly exceeds precipitation, deserts result.

The region surrounding the Gulf of Mexico, however, despite its location in the horse latitudes, is not a desert. The Gulf of Mexico, a body of very warm water throughout much of the year, seems to be the saving grace as it supplies abundant amounts of moisture, thus keeping the U.S. Southeast from experiencing the climatic patterns typical of the U.S. Southwest.

Turning our attention toward the midlatitudes, we can characterize the region, metaphorically, as a battle zone between warm air transported northward along the surface from the horse latitudes and cold air transported southward from the poles. The United States experiences wild temperature swings as these vastly different air masses compete for geographical position. The transitional seasons of spring and autumn offer the most spectacular examples. Residents of the Dakotas can regale visitors with tales of a "blue norther," a cold front carrying arctic air that can drop temperatures up to 40C° in a matter of hours.

The Coriolis effect impacts both streams of air, cold and warm, as they move toward the midlatitude convergence zone. The polar easterlies are created by air flowing from the North Pole toward the Equator. The Coriolis effect acts to curve the moving air to the right of its original path, thus deflecting the flow to the west. The same effect is observed at the South Pole as a left curvature causes a similar belt of polar easterlies to prevail. These winds ideally occur between the poles and 60° latitude. In reality, however, especially during winter in the Northern Hemisphere, the polar easterlies dip much farther south as the jet stream, or polar front, rides more across the northern United States and southern Canada than at the idealized latitude reported by most texts.

In the midlatitudes, between the subsidence zone at the thirtieth parallel and the polar front, the prevailing winds flow from the southwest as poleward-moving air is deflected toward the right (in the Northern Hemisphere), thus creating the prevailing westerlies. These winds guarantee that most weather systems move in general from west to east across the United States. Airline passengers who travel coast to coast are familiar with the prevailing westerlies because a flight bound for New York from Los Angeles is about an hour shorter than a flight headed westbound.

As these opposite-flowing air masses converge, another zone of rising air is found. Models often depict this at 60° latitude. In the Northern Hemisphere, however, this zone more often occurs between 35° and 50° N—in other words, right over the United States! Where the convergence occurs, a polar front forms. This polar front acts to separate polar air masses from tropical air masses, and provides a zone along which midlatitude or extra-tropical storm systems travel. The occasional full-day rain or snowstorms experienced across much of the United States, particularly during the fall and winter, tend to form and travel along the polar front.

Finally, as we turn our attention toward the poles, the cold air subsides to create a relatively dry environment. Although precipitation is as limited as in the deserts, conditions near the poles are not considered to be desert because evaporation is minimal. To have desert conditions, evaporation must exceed precipitation. What little snow falls near the poles either fails to melt or melts quite slowly.

Review Exercises for Chapter 13

WORD-STUDY CONNECTION

absorb	equinox	microwave
aerosols	evaporation	midlatitude
albedo	extra-tropical	millibar
allotrope	gamma rays	nitrogen dioxide
autumn	greenhouse effect	oxidizing
carbon dioxide	Hadley cell	oxygen
chlorofluorocarbons (CFCs)	horse latitudes	ozone
clouds	infrared	ozone layer
condensation	insolation	pollutants
convection	ITCZ	precipitation
desert	long-wave radiation	radiation
electromagnetic radiation	mesosphere	radio waves
spectrum	methane	reducing

reflect   stratosphere   ultraviolet
scatter   stratospheric ozone   Van Allen radiation belt
season   subsiding   visible
short-wave radiation   summer   water vapor
solstice   thermosphere   winter
spring   tropopause   X-rays
stratopause   troposphere

SELF-TEST CONNECTION

PART A. Completion. Write in the word or words that correctly completes the sentence.

1. _____ is the process whereby water vapor changes phase to become water droplets.

2. The _____ are a region of subsiding air and most of the world's deserts.

3. The Hadley cell is an example of a _____ cell.

4. The _____ is the boundary between the troposphere and the stratosphere.

5. Infrared radiation emitted by Earth's surface is known as _____.

6. The ozone layer is located within the _____.

7. Chlorofluorocarbons are known to destroy _____ in the stratosphere.

8. Collectively, airborne dust, pollen, sea salts, microorganisms, and human-induced pollutants are known as _____.

9. _____ describes the percentage of incoming solar radiation that is reflected by a planet.

10. When insolation encounters Earth's atmosphere, this energy is either scattered, reflected, or _____.

11. The _____ is the uppermost layer of the atmosphere.

12. _____ radiation is a form of the electromagnetic radiation spectrum that the human eye is sensitive to.

13. _____ are greenhouse gases with no known source.

14. Temperatures tend to _____ with height within the stratosphere.

15. _____ air is likely to produce clear skies and dry conditions.

PART B. Multiple Choice. Circle the letter of the item that correctly completes the statement.

1. The strongest radiation in the electromagnetic radiation spectrum consists of
   (a) gamma rays
   (b) X-rays
   (c) ultraviolet rays
   (d) infrared rays
   (e) microwaves

2. The portion of the electromagnetic radiation spectrum felt as heat consists of
   (a) gamma rays
   (b) X-rays
   (c) ultraviolet rays
   (d) infrared rays
   (e) microwaves

3. An important naturally occurring gas that traps long-wave radiation is
   (a) oxygen
   (b) nitrogen
   (c) carbon dioxide
   (d) chlorofluorocarbons
   (e) hydrogen

4. Upward motion of air occurs along
   (a) the ITCZ
   (b) the horse latitudes
   (c) the polar front
   (d) both a and b
   (e) both a and c

5. The greenhouse effect is
   (a) a natural effect that results in warming of the planet
   (b) a natural effect that results in cooling of the planet
   (c) a human-induced effect only that warms the planet
   (d) a human-induced effect only that cools the planet
   (e) a natural effect that has an impact on the United States only

6. The vernal equinox occurs on the first day of
   (a) spring
   (b) summer
   (c) autumn
   (d) winter
   (e) both a and c

7. Humans and all other life-forms live within Earth's
   - (a) troposphere
   - (b) stratosphere
   - (c) mesosphere
   - (d) thermosphere
   - (e) atmosphere

8. The gas, found in both the troposphere and the stratosphere, that is a pollutant in the former and vital to the biosphere in the latter is
   - (a) oxygen
   - (b) nitrogen
   - (c) ozone
   - (d) water vapor
   - (e) methane

9. The ozone layer is located within the
   - (a) troposphere
   - (b) stratosphere
   - (c) mesosphere
   - (d) thermosphere
   - (e) atmosphere

10. A desert is best defined as a
   - (a) very dry region
   - (b) place with lots of cactus
   - (c) region where precipitation exceeds evaporation
   - (d) region where evaporation exceeds precipitation
   - (e) place with lots of sand

PART C. Modified True/False. *If a statement is true, write "true" for your answer. If a statement is incorrect, change the <u>underlined</u> expression to one that will make the statement true.*

1. <u>Solstices</u> mark the dates when the Sun appears directly over the Equator at midday.

2. The <u>greenhouse effect</u> causes Earth to be warmer than it otherwise would be.

3. <u>Ultraviolet</u> radiation can cause your skin to tan.

4. The summer solstice occurs on or about <u>July 21</u>.

5. The <u>midlatitudes</u> are where most of the world's deserts are found.

6. <u>Carbon dioxide</u> is an important greenhouse gas.

7. <u>Nitrogen</u> is an allotrope of oxygen.

8. Aerosols play an important role in <u>scattering</u> incoming solar radiation.

9. Six months of daylight at the North Pole begin with the first day of <u>spring</u>.

10. The <u>ITCZ</u> is a zone of rising air where polar and tropical air masses converge.

CONNECTING TO CONCEPTS

1. Why are the world's temperature records usually set in the horse latitudes instead of near the Equator?

2. Explain how ozone is both a gas that is vital to the survival of all life on the surface and also a dangerous air pollutant.

3. Why is it not possible for Earth to absorb long-wave radiation and emit short-wave radiation?

4. If Earth's surface emitted short-wave instead of long-wave radiation, how would conditions on the planet differ from those that currently exist?

5. If Earth's albedo changed from 30 to 40 percent, what other changes might occur?

6. Where did most of the oxygen in our atmosphere come from?

7. One major global environmental issue faced by all humanity is the depletion of stratospheric ozone. What are the consequences of its continued depletion?

8. Base your answers to a–c on the graph below, which shows the average daily temperatures and the duration of insolation for a location in the midlatitudes of the Northern Hemisphere during a year.

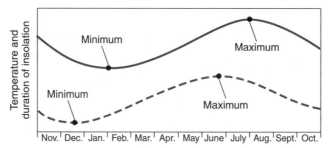

Surface Temperature (—) versus Duration of Insolation (- - -)

a. State the month of the year during which insolation reaches its minimum at this location.

   b. State how the relationship between surface temperature and duration of insolation during the months of January and July differs from this relationship during most other months of the year.

   c. Explain why surface temperatures continue to increase between late June and early August even though the duration of insolation is decreasing during this time period.

9. Why does Earth not increase in temperature even though it is constantly absorbing solar energy?

10. Base your answers to a and b on the diagram below, which shows identical solar electric cells placed on the roofs of two adjacent houses, *A* and *B*, in Chicago.

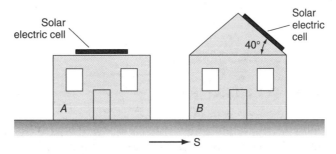

   a. On which house will the roof-mounted solar cell produce more electricity on March 21 at midday? Why?

   b. How would the output of house *B*'s solar cell change if it was relocated to the north side of the roof?

CONNECTING TO LIFE/JOB SKILLS

This chapter introduces some of the issues studied in atmospheric science. From the open-ended questions in the preceding section, it is obvious that many "what if" scenarios exist. Atmospheric scientists strive to understand the workings of the atmosphere and its interactions with all other major systems. Many of the significant environmental issues faced by society today, including ozone depletion, ground-level ozone pollution, and global warming, are being studied by atmospheric scientists. A strong background in chemistry is also helpful in understanding these issues.

ANSWERS
SELF-TEST CONNECTION

Part A

1. Condensation
2. horse latitudes
3. convection
4. tropopause
5. long-wave radiation
6. stratosphere

7. ozone
8. condensation nuclei
9. Albedo
10. absorbed
11. thermosphere
12. Visible
13. Chlorofluorocarbons
14. increase
15. Subsiding

Part B

1. **(a)**
2. **(d)**
3. **(c)**
4. **(e)**
5. **(a)**
6. **(a)**
7. **(a)**
8. **(c)**
9. **(b)**
10. **(d)**

Part C

1. False; Equinoxes
2. True
3. True
4. False; June 21
5. False; horse latitudes
6. True
7. False; Ozone
8. True
9. True
10. False; polar front

CONNECTING TO CONCEPTS

1. The horse latitudes are a zone of subsiding air, which tends to compress and warm. Additionally, subsiding air suppresses cloud formation; thus sunshine is common. Although solar elevations are not as direct year-round at the horse latitudes as on the Equator, the combination of subsiding air and sunshine produces the world's highest temperatures.

2. In both cases, ozone is ozone; that is, it's an allotrope of oxygen (three atoms instead of two in the molecule). The difference lies in where the ozone is located. Ozone in the stratosphere acts to absorb, or effectively block, incoming ultraviolet radiation from the Sun, thereby making life possible on Earth's surface. In contrast, ground-level ozone is located where animal life, including humans, comes into direct contact with it. Ozone is a corrosive gas that poses a health threat to life-forms in direct contact with it for prolonged periods of time.

3. Short-wave radiation is higher energy radiation than long-wave radiation. It is not possible to absorb a lower form of energy and then emit a higher form.

4. The atmosphere contains gases that absorb long-wave radiation. These gases, known collectively as greenhouse gases, include carbon dioxide, methane, and water vapor. Greenhouse gases act to keep surface temperatures in a habitable range. If Earth emitted short-wave radiation, it is possible that the lower atmosphere would not be warmed and conditions on the planet's surface would be significantly colder.

5. Albedo is a measure of reflectivity. The more energy reflected, the less is absorbed by Earth's surface. Reflected solar energy does little to warm the planet

because the atmosphere is largely "transparent" to this energy. Therefore, reflected radiant energy returns to space without interacting with the atmosphere, and an increase in albedo would have little or no effect.

6. Probably the first free oxygen came from primitive photosynthetic life-forms such as phytoplankton and bacteria.

7. Stratospheric ozone absorbs harmful high-energy radiation emitted by the Sun. If the ozone layer is eventually destroyed, life on Earth's surface will be threatened.

8. a. Insolation reaches a minimum in December at the location in question.
   b. During January, duration of insolation is increasing, yet temperature is still decreasing. By early February, incoming energy exceeds outgoing energy, and temperatures respond by gradually rising. During July, duration of insolation is decreasing, yet temperature is still increasing. In August, outgoing energy begins to exceed incoming energy, and temperatures fall until late January.
   c. Temperatures begin to decrease when outgoing energy exceeds incoming energy. After June 21 and until late July or early August, incoming energy still exceeds outgoing energy; hence average temperatures continue to rise. Although the duration of insolation is greatest on June 21 or thereabouts, solar elevations in the Northern Hemisphere continue to remain high, and daylight is still much longer than darkness through July and into early August.

9. Globally, the amount of energy absorbed equals the amount of energy emitted; thus Earth is neither warming nor cooling over time.

10. a. The solar electric cell on house *B* will produce more energy because the solar panel on this house is angled toward the Sun in such a manner that, at midday, the Sun and the solar panel form a right angle.
    b. If the solar panel were relocated to the north side of the roof, the output would diminish greatly. In Chicago, and for that matter in all cities north of the Tropic of Cancer, the Sun will always appear to be seen in the southern sky. As a result, the solar panel would not receive direct sunlight at any time of the day if the panel were relocated.

Atmospheric Moisture, Pressure, and Winds

WHAT YOU WILL LEARN

This chapter focuses on humidity, air pressure, and wind. In this chapter you will learn

- what relative humidity is and how it is determined;
- why dew point is a more useful measurement of how we feel outside;
- what clouds are and how they form;
- what types of precipitation occur and the conditions required for each;
- how air pressure is measured and the kinds of weather conditions associated with low- and high-pressure systems.

SECTIONS IN THIS CHAPTER

- Moisture
- Clouds and Fog
- Precipitation
- Air Pressure
- Winds
- Review Exercises for Chapter 14

Moisture

RELATIVE HUMIDITY

Atmospheric moisture is an important topic because we humans are very sensitive to the amount of moisture or humidity in our environment. Air that is too dry or humid is often perceived as uncomfortable. Atmospheric moisture is also important to plant life, as certain species require water vapor in order to thrive.

There are many ways of measuring moisture in the atmosphere. This section deals with atmospheric moisture and the various ways of measuring it. Radio and television weather reports often include a measurement called relative humidity. Relative humidity can be a useful statistic, but the public at large often misunderstands what the term actually means.

Relative humidity is defined as the ratio of the amount of moisture or water vapor (in grams) in 1 kilogram of air per (divided by) the maximum capacity (in grams) that 1 kilogram of air can hold *at that temperature*. This relationship can be expressed as a mathematical equation:

$$\text{Relative humidity} = \frac{\text{Water vapor in atmosphere}}{\text{Water vapor in atmosphere at equilibrium}} \times 100\%$$

The phrase "at that temperature" is the key to understanding relative humidity. Cold air can hold only a fraction of the water vapor per unit volume (or unit mass) of air that warm air can hold. For example, at 0°C saturation is reached with just 3.5 grams of water vapor in every kilogram of air, while at 30°C a kilogram of air can hold up to 26.5 grams of water vapor before saturation is reached. Refer to Table 14.1 for supporting data.

Saturation is the condition that is said to exist when the atmosphere is holding, or contains, all the water vapor that it can. More technically, at saturation the rate of

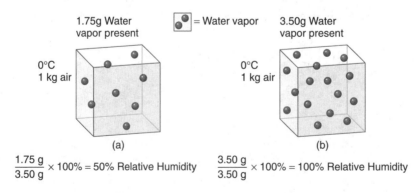

Two 1 kg parcels of air at 0°C, each with a capacity to hold 3.50g of water vapor. Note the actual water vapor present in each parcel and the relative humidity in each.

TABLE 14.1
MASS OF WATER VAPOR REQUIRED TO SATURATE
THE ATMOSPHERE AT VARIOUS TEMPERATURES

Temperature (°C)	Temperature (°F)	Mass of Water Vapor per 1 kg of Air
−20	−4	0.75
−10	14	2
0	32	3.5
10	50	7
20	68	14
30	86	26.5
40	104	47

evaporation, the process of liquid water changing phase to water vapor, equals the rate of condensation, the process of water vapor changing phase to liquid water. Saturation, or a state of equilibrium (between the rates of evaporation and condensation), and its relationship to air temperature are illustrated in Figure 14.1.

The data from Table 14.1 and the equation given above can be used to calculate the relative humidity. For example, if 3.5 grams of water vapor is actually present in every kilogram of air at 0°C:

$$\frac{3.5\,\text{g of water vapor in atmosphere}}{3.5\,\text{g of water vapor capacity}} \times 100\% = 100\%$$

If the temperature were to rise to 30°C, and the amount of water vapor in the atmosphere did not change, the relative humidity would fall to approximately 13 percent. Note that the amount of water vapor present did not change.

In essence, the word *relative* in *relative humidity* should be read as a measurement of atmospheric moisture that is "relative to the current temperature." In practical terms, it is not unusual to wake up to a very high relative humidity most winter mornings, particularly in the eastern two-thirds of the United States. Despite this high relative humidity, however, very little water vapor is actually present in the atmosphere because cold air can hold only a minimal amount of water vapor before saturation is reached.

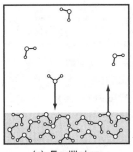

(a) Equilibrium
at low temperature

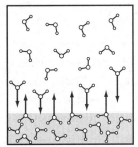

(b) Equilibrium
at high temperature

FIGURE 14.1 The amount of water vapor present in the atmosphere is controlled in part by the air temperature. Limited water vapor is present in a cold atmosphere (a). A state of equilibrium or saturation, in which the rate of evaporation equals the rate of condensation, is reached with a minimal amount of water vapor. In a warm atmosphere (b), equilibrium or saturation is achieved with much larger amounts of water vapor in the atmosphere. Source: *Let's Review: Earth Science—The Physical Setting*, 2nd Ed., Edward J. Denecke, Jr., Barron's Educational Series, Inc., 2002.

In contrast, while much of the eastern half of the United States experiences very humid conditions throughout much of the summer, with air temperatures well above 30°C, relative humidity values during midday hours remain well below 50 percent. This mathematical quirk occurs because warm air can hold large amounts of water vapor.

In conclusion, relative humidity can be a misleading statistic as high relative humidities readily occur in cold air and low relative humidities are common on hot summer afternoons. Another way to view this situation is presented in Figure 14.2.

When the atmosphere is saturated, the relative humidity must equal 100 percent since the amount of water vapor in the atmosphere equals the atmosphere's capacity to hold water vapor (at that temperature). When the numerator of the relative humidity equation equals the denominator, and you multiply by 100%, the final result must be 100 percent, as shown in the example on page 323.

Saturation can be recognized by the presence of fog at ground level. Rain or any other form of precipitation need not be falling. Rain can occur with humidity values well below 100 percent as long as a saturated state exists in the clouds where the rain is forming. Occasionally, rain falls from clouds where the atmosphere is saturated into a lower layer of the atmosphere that is quite dry. Under these circumstances, the rain may evaporate before reaching the ground, thus creating a

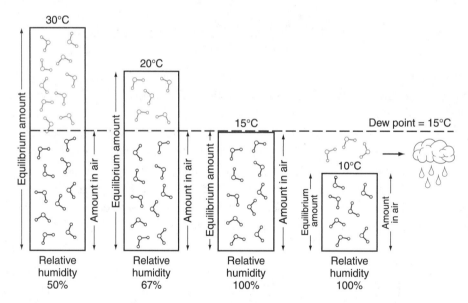

FIGURE 14.2 As the atmosphere cools, the amount of water vapor present at equilibrium or saturation diminishes. The atmosphere's capacity to hold water vapor diminishes as temperatures fall. Relative humidity must increase as air temperature falls. The third sketch illustrates saturation. Additional reductions in temperature (from the third to the fourth sketch) will result in the forced condensation of water vapor and the subsequent formation of clouds and precipitation—or dew or frost if this process occurs near the surface. Source: *Let's Review: Earth Science—The Physical Setting,* 2nd Ed., Edward J. Denecke, Jr., Barron's Educational Series, Inc., 2002.

FIGURE 14.3 Photograph of virga—precipitation falling from clouds but evaporating before it reaches the ground.

condition called *virga*. This is particularly common in the southwestern states, where dry air near the ground prevents rain falling from developing showers and thunderstorms from reaching the ground. Spectacular views are created when virga occurs; Figure 14.3 is a photograph taken by the author in northern Arizona.

DEW POINT

Dew point offers an alternative method for reporting atmospheric moisture. *Dew point* is defined as the temperature at which saturation is reached. In other words, it is the temperature at which the relative humidity will reach 100 percent. For example, if the air temperature is 25°C, and the relative humidity is 50 percent, it can be determined from statistical tables similar to Table 14.1 that 10 grams of water vapor is present in every kilogram of air. If the air temperature were to fall to 15°C, that sample of air would reach saturation (and the relative humidity would rise to 100%). Since in this example saturation occurs at 15°C, the dew point is reported as 15°C. If less moisture were present, the atmosphere would have to cool to a lower temperature to reach saturation.

As you can see, dew point is dependent upon the amount of moisture in the atmosphere; in fact, dew point can be considered to be a good gauge of how much moisture is present in the atmosphere at any given time. Since humans are quite sensitive to the amount of water vapor in the atmosphere, dew point can be a very useful measurement. When it falls below 0°C, the air is quite dry and some people need to use humidifiers to add moisture to the air in their homes. At the other end of the scale, when dew points exceed 20°C, the resulting conditions are generally considered to be uncomfortably humid.

In the midwestern and eastern states, where moisture from the Gulf of Mexico advects (travels laterally across the nation) during the summer months, residents are wise to watch dew point levels. Conditions can change from comfortable (dew points between 5 and 15°C) to uncomfortable (higher dew points) in a matter of hours.

MEASURING DEW POINT AND RELATIVE HUMIDITY

Both dew point and relative humidity can be measured by using a type of hygrometer (humidity meter) called a *sling psychrometer* (see diagram on page 327). A sling psychrometer is an instrument in which two thermometers are attached to a wooden rod that can be rotated rapidly. One thermometer, the dry bulb, is uncovered. The other thermometer has its bulb covered by a "sock," which is wet with water before the instrument is put into motion. To obtain a reading, the operator "slings," or swings, the psychrometer; as the instrument rotates, the spinning motion causes the water on the

TABLE 14.2
RELATIVE HUMIDITY (%)

Dry-Bulb Tempera- ture (°C)	Difference Between Wet-Bulb and Dry-Bulb Temperatures (C°)															
	0	1	2	3	4	5	6	7	8	9	10	11	12	13	14	15
−20	100	28														
−18	100	40														
−16	100	48	0													
−14	100	55	11													
−12	100	61	23													
−10	100	66	33	0												
−8	100	71	41	13												
−6	100	73	48	20	0											
−4	100	77	54	32	11											
−2	100	79	58	37	20	1										
0	100	81	63	45	28	11										
2	100	83	67	51	36	20	6									
4	100	85	70	56	42	27	14									
6	100	86	72	59	46	35	22	10	0							
8	100	87	74	62	51	39	28	17	6							
10	100	88	76	65	54	43	33	24	13	4						
12	100	88	78	67	57	48	38	28	19	10	2					
14	100	89	79	69	60	50	41	33	25	16	8	1				
16	100	90	80	71	62	54	45	37	29	21	14	7	1			
18	100	91	81	72	64	56	48	40	33	26	19	12	6	0		
20	100	91	82	74	66	58	51	44	36	30	23	17	11	5	0	
22	100	92	83	75	68	60	53	46	40	33	27	21	15	10	4	0
24	100	92	84	76	69	62	55	49	42	36	30	25	20	14	9	4
26	100	92	85	77	70	64	57	51	45	39	34	28	23	18	13	9
28	100	93	86	78	71	65	59	53	47	42	36	31	26	21	17	12
30	100	93	86	79	72	66	61	55	49	44	39	34	29	25	20	16

Source: The State Education Department, *Earth Science Reference Tables*. 2001 ed. (Albany, New York; The University of the State of New York).

TABLE 14.3
DEW POINT TEMPERATURES (°C)

Dry-Bulb Temperature (°C)	Difference Between Wet-Bulb and Dry-Bulb Temperatures (C°)														
	0	1	2	3	4	5	6	7	8	9	10	11	12	13	14	15
−20	−20	−33														
−18	−18	−28														
−16	−16	−24														
−14	−14	−21	−36													
−12	−12	−18	−28													
−10	−10	−14	−22													
−8	−8	−12	−18	−29												
−6	−6	−10	−14	−22												
−4	−4	−7	−12	−17	−29											
−2	−2	−5	−8	−13	−20											
0	0	−3	−6	−9	−15	−24										
2	2	−1	−3	−6	−11	−17										
4	4	1	−1	−4	−7	−11	−19									
6	6	4	1	−1	−4	−7	−13	−21								
8	8	6	3	1	−2	−5	−9	−14								
10	10	8	6	4	1	−2	−5	−9	−14	−28						
12	12	10	8	6	4	1	−2	−5	−9	−16						
14	14	12	11	9	6	4	1	−2	−5	−10	−17					
16	16	14	13	11	9	7	4	1	−1	−6	−10	−17				
18	18	16	15	13	11	9	7	4	2	−2	−5	−10	−19			
20	20	19	17	15	14	12	10	7	4	2	−2	−5	−10	−19		
22	22	21	19	17	16	14	12	10	8	5	3	−1	−5	−10	−19	
24	24	23	21	20	18	16	14	12	10	8	6	2	−1	−5	−10	−18
26	26	25	23	22	20	18	17	15	13	11	9	6	3	0	−4	−9
28	28	27	25	24	22	21	19	17	16	14	11	9	7	4	1	−3
30	30	29	27	26	24	23	21	19	18	16	14	12	10	8	5	1

Source: The State Education Department, *Earth Science Reference Tables.* 2001 ed. (Albany, New York; The University of the State of New York).

A basic sling psychrometer.

sock to evaporate. The less moisture present in the air, the faster the water evaporates off the sock. Since water requires energy to evaporate, the temperature of the wet-bulb thermometer falls as the evaporating water draws energy from the thermometer bulb. Water evaporates more rapidly in dry air, thereby producing a greater drop in the wet bulb's temperature. After a minute or so of slinging the psychrometer, the operator reads the wet- and dry-bulb temperatures and then applies the values to tables from which dew point and relative humidity can be determined. Table 14.2 is used to determine relative humidity, and Table 14.3 to determine dew point. For example, if the dry-bulb reading is 10°C and the "depression" (the difference between the dry- and wet-bulb temperatures) is 5°C, then, according to Table 14.2, the relative humidity is 43 percent, and according to Table 14.3, the dew point is −2°C.

LATENT HEAT

Meteorologists use humidity measurements to determine the likelihood of condensation and evaporation. These processes are important as they pertain to latent heat. *Latent heat* is the energy absorbed during evaporation (from the solid or liquid state) and released during condensation or at the time when freezing occurs.

Thunderstorms, for example, feed off the latent heat released during the process of condensation that occurs as air rises in the towering clouds. During the winter, latent heat is absorbed as some of the precipitation falling from clouds evaporates. Evaporation cools the atmosphere, sometimes just enough to ensure freezing or frozen precipitation at the ground instead of just rain. Meteorologists refer to this process as *evaporational cooling*.

Clouds and Fog

COMPOSITION

Clouds are a mixture of water droplets, condensation nuclei, and in some cases ice crystals. Condensation nuclei are usually aerosols, that is, tiny airborne particles, that provide a surface for water vapor to condense on. The water droplets, which are about one-hundredth the diameter, or less, of a raindrop, act as a suspension, remaining in the cloud instead of falling to the surface. Under the right conditions, water droplets will *coalesce*, or combine, to form raindrops of sufficient size to fall from the clouds. A raindrop contains about 1 million times more water than a cloud droplet by volume. Clouds that form at ground level are referred to as "fog."

Ice crystals, also found in clouds, can form at any temperature below 0°C. Water droplets, however, may remain in their liquid form until a temperature of −40°C is reached. Water existing under these conditions is said to be *supercooled*.

CLOUD FORMATION

Clouds form when rising air cools, achieves saturation, and then forces the water vapor it contains to condense. There are three mechanisms in the atmosphere that can cause this rising motion. The first is *global convection*, which causes air to rise at the Intertropical Convergence Zone (ITCZ) and along the polar front. Both of these occurrences are discussed in some detail in Chapter 13. The second mechanism that can cause rising motion in the atmosphere is an *orographic feature*, specifically a mountain, as illustrated in Figure 16.13. When a parcel of air encounters a mountain, it is forced to rise, and when condensation occurs, clouds form. You may have noticed that mountains are often enshrouded in clouds.

The final mechanism that can cause lifting in the atmosphere is a front. A *front* is a boundary between two different types of air masses. If one air mass is trying to overtake another, the less dense air mass may "overrun" the denser air mass. Overrunning produces clouds and, occasionally, prolonged periods of light to moderate precipitation in the midlatitudes. This topic is discussed more fully in Chapter 15.

TYPES OF CLOUDS

Studying clouds (Figure 15.3) can help both the amateur and professional meteorologist to forecast the weather. Clouds can be classified into three broad categories: cumulus, cirrus, and stratus. The type of clouds that form depends upon the elevation at which air is rising, the elevation in the atmosphere at which saturation is reached, and the rate at which the lifting (regardless of the lifting mechanism) is occurring.

Cumulus clouds, often perceived as puffy, fair-weather clouds, form when air is rising rapidly and saturation occurs at relatively low levels in the troposphere. Cumulus clouds are so named because of their vertical development; they often exhibit a stacked appearance in the sky. If the atmosphere is unstable (see the next section for clarification of this term), cumulonimbus clouds will form. These clouds, better known as thunderheads, are spectacular, vertically developed clouds that can produce severe thunderstorms with hail and frequent lightning, and in extreme cases tornadoes. Cumulonimbus clouds are easily recognized by their flat, dark bases and their distinctive tall, anvil-shaped tops. In extreme cases, the tops of these clouds can stretch to the tropopause. In fact, it is the high-velocity winds in the upper troposphere and lower stratosphere that tend to shear off the tops of these giant cloud masses, creating the signature anvil shapes.

When air is rising, but is doing so more gradually, clouds still form, but they lack the vertical development exhibited by cumulus clouds. Both cirrus and stratus clouds are horizontally rather than vertically developed. *Cirrus clouds* are high, thin, wispy clouds that contain only small amounts of ice crystals because of the limited moisture supply at their high elevation. Cirrus clouds are usually thin enough for sunlight to penetrate easily. Many varieties of these clouds exist. To cite just two examples, cirrocumulus clouds exhibit a small degree of vertical development, and cirrostratus clouds are lower and thicker than other cirrus clouds.

Although cirrus clouds do not produce precipitation, they may precede lower, more moisture-laden stratus clouds. *Stratus clouds*, easily identified as low, gray, layered clouds, can produce a prolonged period of light to moderate precipitation that can take the form of rain, freezing rain, ice pellets, or snow. What falls from the clouds and reaches the ground is determined by the temperature profile of the atmosphere between the base of the clouds and the surface. There is a more extensive discussion on forms of precipitation in a later section of this chapter.

Stratus clouds are particularly common throughout the Pacific Northwest, west of the Cascade Mountains, throughout much of the year. In the eastern half of the nation, midlatitude or extra-tropical storms typically occur in the autumn or winter. These storms create a widespread blanket of stratus clouds, usually ahead of a warm front,

> **REMEMBER**
> Clouds form when air rises. The more rapidly and violently air rises, the more vertically developed the clouds become. Cumulonimbus clouds form in the most violently rising air.

that produce steel-gray skies for 2 or 3 days at a time. Stratus clouds that are particularly laden with moisture are classified as nimbostratus. Nimbostratus clouds are quite low and dark, often obtaining abundant moisture from the Gulf of Mexico or the Atlantic Ocean and producing extended periods of moderate to heavy precipitation.

ATMOSPHERIC STABILITY

Regardless of what causes the upward motion of a parcel of air, rising air eventually reaches a level where it is as buoyant as the surrounding air. When this occurs, the parcel of air ceases its upward motion. If saturation occurs, condensation takes place, clouds form, and if enough moisture is available, precipitation will fall from them. The buoyancy of a parcel of air is related to its temperature in comparison to the temperature of the surrounding air. If a parcel of air that is initially warmer than the air surrounding it, and hence more buoyant than its surroundings, is lifted, it will continue to rise until its temperature is the same as that of the surrounding air at the same level.

Stable air tends to resist upward vertical movement, and any rising air will form cirrus or stratus clouds at best. Cumulus clouds, particularly cumulonimbus clouds, form in an unstable atmosphere. Under these conditions, air will tend to rise for quite a while before achieving equal buoyancy with the surrounding atmosphere. If the upward motion occurs with enough vertical velocity and reaches a sufficient height, towering thunderheads or cumulonimbus clouds develop.

Precipitation

HOW PRECIPITATION FORMS

As explained above, clouds are composed of water droplets and/or ice crystals that are too small to fall as precipitation. Condensation nuclei are also present. *Condensation nuclei*, as previously explained, are dust-sized or smaller, solid particles on which water droplets and ice crystals can coalesce. Essentially, condensation nuclei provide surfaces for the water droplets and ice crystals to collect upon. Through a *collision-coalescence* process, water droplets or ice crystals collide "constructively" with smaller particles and grow into larger particles as the process continues. Eventually the particles (raindrops, snowflakes, or ice pellets) grow to a sufficient size to fall from the cloud to the ground. This process as described occurs most commonly in the tropics.

An additional process, more common to the midlatitudes, called the *Bergeron process* in honor of its discoverer, builds upon two interesting aspects of water droplet behavior in clouds: supercooling and supersaturation. Supercooling, as noted earlier, is the ability of water to remain liquid until nearly −40°C. At subfreezing temperatures, common in clouds even during the summer, water droplets and ice crystals can coexist. The atmosphere, however, may be saturated with respect to water droplets and, at the same time, be supersaturated with respect to ice crystals. The Bergeron process forces the ice crystals to grow, thus removing moisture from the cloud. Eventually the ice crystals grow large enough to fall from the cloud as snowflakes. If temperatures in the air through which they fall are warm enough, the snow will melt and fall as rain. It is thought that many rainfalls reaching the surface in the middle of summer start as snow in the clouds.

FORMS OF PRECIPITATION

Major forms of precipitation include rain, snow, ice pellets (also known as sleet), freezing rain, drizzle, freezing drizzle (a personal favorite!), and hail. Fog is not considered to be a form of precipitation since fog is essentially a ground-level cloud. The water droplets or ice crystals in clouds are much smaller than the drops of water, snowflakes, or ice pellets that fall from clouds as precipitation. Precipitation forms in clouds as water droplets or ice crystals coalesce (join together) to make larger and larger particles until the force of gravity takes over and the particles fall to the ground. This section will describe each form of precipitation and explain the conditions under which each may occur. Table 14.4 lists each form of precipitation and the atmospheric conditions that enable it to form.

Snow is perhaps the easiest to describe. Snowflakes start out in a cloud as ice crystals, which must coalesce or grow until they are large enough to fall from the cloud.

TABLE 14.4
TYPES OF PRECIPITATION AND ORIGINS

Name	Description	Origin
Rain	Droplets of water up to 4 mm in diameter	Cloud droplets coalesce when they collide.
Drizzle	Very fine droplets falling slowly and close together	Cloud droplets coalesce when they collide.
Sleet	Clear pellets of ice	Raindrops freeze as they fall through layers of air at below-freezing temperatures.
Glaze	Rain that forms a layer of ice on surfaces it touches	Supercooled raindrops freeze as soon as they come in contact with below-freezing surfaces.
Snow	Hexagonal crystals of ice, or needlelike crystals at very low temperatures	Water vapor sublimes, forming ice crystals on condensation nuclei at temperatures below freezing.
Hail	Balls of ice ranging in size from small pellets to as large as a softball, with an internal structure of concentric layers of ice and snow	Again and again hailstones are hurled up by updrafts in thunderstorms and then fall through layers of air that alternate above and below freezing. Each cycle adds a layer to the hailstone. The more violent the updrafts, the larger and heavier the hailstone can become before falling.

When this occurs, the column of air through which the snowflakes fall must remain near or below freezing. If temperatures exceed 3°C for more than a few hundred meters, the snow will melt and fall as rain. All snowflakes are hexagonal in shape. This property results from the shape of the water molecules from which the ice crystals formed and the interactions between these molecules.

Rain may begin as snow, but if the air column is too warm, as described above, the snow melts and rain is experienced at the surface. Precipitation is often divided into liquid, freezing, and frozen forms. The latter two forms of precipitation are often confused. *Freezing rain* is classified as freezing precipitation; in contrast, *ice pellets* are classified as frozen precipitation. Freezing rain actually consists of rain droplets that fall into a shallow, subfreezing layer of air at and very near Earth's surface. Since temperatures are below freezing at the surface, often between −1°C and −4°C, the rain freezes upon making contact with the surface and quickly forms a slick glaze. Conditions become quite hazardous as the ice coats everything in sight. *Drizzle* and *freezing drizzle* form in a similar manner to rain and freezing rain; however, the particle sizes are smaller.

Ice storms are all too common throughout the eastern third of the nation from the inland southeastern states north to eastern Canada. They often occur when shallow arctic air masses become entrenched at the surface, thus creating subfreezing conditions as far south as central Alabama and Mississippi. At the same time, warm, maritime tropical air from the Gulf of Mexico overruns the arctic air, setting up conditions for an ice storm. Ice storms can cause millions of dollars in damage; multivehicle accidents frequently occur, tree limbs are broken, and power lines are knocked down.

If the layer of arctic air near the surface is a bit thicker, the rain may freeze as it falls into small, hard pellets of ice. *Ice pellets*, also known as sleet, are easily recognized as balls of ice, but are sometimes mistaken for hailstones.

Hail forms under circumstances completely different from those described for other forms of precipitation (see Figure 14.4). Hail forms in cumulonimbus clouds only; all of the other forms of precipitation typically form in stratus clouds. Hail, being associated with thunderstorms, is a spring or summer event, whereas sleet occurs almost exclusively during cold-weather months.

The process that causes hail to form involves the carrying of water droplets by the strong updrafts present within thunderstorms to a height above the freezing level, which is often around 3000–5000 meters above the surface during the summer months. Raindrops carried above this level freeze into ice pellets and eventually fall. Sometimes the partially melted ice pellet picks up more water droplets, is then recaptured by an updraft, and carried above the freezing level again. The ice pellet, now larger than before, refreezes and again falls below the freezing level. As this process repeats itself several times, the ice pellet grows, layer by layer, into a hailstone. Eventually, the hailstone is too heavy to be supported by even the strongest updraft and falls to the ground. The largest hailstones to reach the surface are about the size of a grapefruit.

> **REMEMBER**
> Frozen precipitation (sleet and snow) require a layer of below-freezing air to exist above Earth's surface.

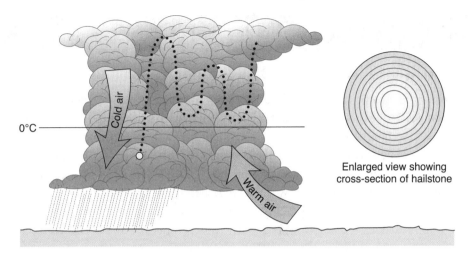

FIGURE 14.4 Hail formation.

These objects can cause considerable damage if they hit people, cars, or other objects when they hit the ground.

MEASURING PRECIPITATION

Various methods are employed to measure precipitation. All forms of precipitation are ultimately reported in the "melted equivalent." For rain, already in liquid form, this is an easy task; all that is required is a *standard rain gauge*. This device has a circular opening about 20 centimeters in diameter that funnels into a cylindrical tube with a cross-sectional area of 2 centimeters. This device measures rainfall accurately to the hundredth of a centimeter. To be used effectively (and according to National Weather Service guidelines), a rain gauge must be in an open area, unobstructed by man-made structures or natural features.

Snow is often measured by the depth of accumulation. The problem here is finding a representative site at which to measure depth. Snow readily drifts, even when the wind is relatively light, so most observers select a few measurement sites and take an average of the readings. After every 5 centimeters (2 in) of accumulated snow, the measurement site must be cleaned off so that compaction does not occur. If snow is allowed to accumulate to a sufficient depth, the lower layers begin to compact and will not yield an accurate measurement of the actual accumulation.

Once the total accumulation is known, an estimate of the water equivalent can be made. An average snowfall has a 10:1 ratio between the accumulation and the water equivalent; that is, if 5.0 centimeters of snow accumulates, this is equivalent to 0.50 centimeter of rainfall. It is important to keep track of winter snows, particularly in the mountains, so that, when the spring melt comes, the amount of water stored in the snow is known. In many cases, moderate spring rainfalls, when combined with snowmelt, have produced significant flooding.

Air Pressure

MEASURING AIR PRESSURE

Air pressure is created by the weight of the column of air above. At sea level, air pressure averages 14.7 pounds per square inch. Evangelista Torricelli, a student of Galileo, constructed the first barometer in 1643. The mercury barometer illustrated in Figure 14.5a is designed to measure air pressure. The barometer consists of a long, evacuated tube with a bowl of mercury below it. Pressure caused by the surrounding air forces the mercury up the tube to an average height of 760 millimeters (30.00 in.) at sea level. More and more commonly, weather maps are reporting pressure in units called millibars. The *millibar* (mb) is a metric unit of pressure and is related to the more familiar unit, inches of mercury (Hg), as follows: 30.00 in. Hg = 1013.2 mb.

It is important to understand that pressure varies with altitude. At higher elevations, a column of air above the observer has less weight. The atmosphere thins out quite quickly with elevation; pressures respond accordingly and drop rapidly with altitude. Average air pressures in Denver, for example, at an elevation of about 1600 meters (1 mi), are nearly 20 percent lower than those experienced at sea level. To standardize the reporting of air pressures across the nation, all pressures are corrected to sea level. In other words, actual air pressures recorded at altitudes are converted to sea-level pressures in order to determine where areas of highest and lowest air pressures are located.

The aneroid barometer shown in Figure 14.5b is a more recent invention for measuring air pressure and is much more convenient to use than the mercury barometer with its long glass tube partially filled with toxic mercury! The aneroid barometer

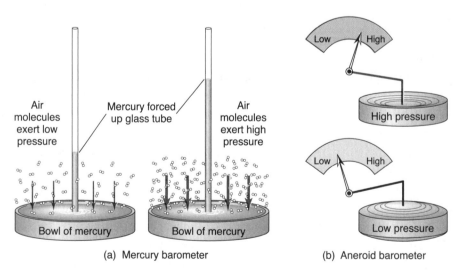

FIGURE 14.5 Two commonly used barometers. Source: *Let's Review: Earth Science—The Physical Setting,* 2nd Ed., Edward J. Denecke, Jr., Barron's Educational Series, Inc., 2002.

contains an internal spring that contracts or expands with changes in air pressure. The instrument reports the pressure via a dial and pointer or through a digital readout.

LOWS AND HIGHS: SYNOPTIC-SCALE WEATHER SYSTEMS

One of the first discoveries made by Torricelli while using his barometer was that air pressure changes from day to day, even from hour to hour. Through observations made over time, he determined that lower pressure is generally associated with stormy weather and higher pressure with fair weather. These basic observations were extended when Benjamin Franklin noted that stormy weather seems to be associated with "storm systems" that travel across the surface of the planet.

With modern technology, even a cursory analysis of a satellite photograph (see Figure 14.6) reveals the presence of swirling cloud patterns that cover regions

FIGURE 14.6 Hemispheric satellite image. Note the storm systems in the top center and bottom center portions of the image. Each cyclone has a characteristic comma-shaped cloud associated with it. Contrast these Northern and Southern Hemisphere midlatitude cyclones with the tropical cloud masses located near the Equator. The tropical cloud masses lack the curved cloud patterns exhibited by the midlatitude cyclones. The curvature of the cloud patterns is because of the Coriolis effect, caused by changes in Earth's orbital velocity with latitude, along with global convection. Source: NOAA.

a few hundred kilometers across. Areas of generally clear skies, again often spanning hundreds of kilometers, are evident between the cloud patterns. When compared to a map depicting the centers of lowest and highest pressure, the swirling clouds are easily seen to be synoptic-scale low-pressure "storm" systems; the regions experiencing clear skies, or at least a lack of "organized" clouds, are under the influence of synoptic-scale high pressure. The term *synoptic scale* is explained below.

There are four primary scales for weather systems or phenomena on our planet. The largest are *global scale*; examples are the global wind belts discussed in Chapter 13. Since global winds are relatively permanent features, a flight from New York City to Los Angeles will always take longer than an eastbound flight because global winds called the westerlies are always flowing in some manner from west to east.

Large-scale storm systems and air masses are classified as *synoptic scale*. These weather systems, which affect large geographic regions for periods of a day or more, may be considered to be embedded within the global winds that drive them. The essential job of a synoptic meteorologist is to predict the movement of these systems in order to provide accurate weather forecasts for several days or a week.

Mesoscale weather systems or phenomena span just a few kilometers. Thunderstorms are classic examples of mesoscale phenomena. Smaller scale systems, such as thunderstorms, are much shorter lived than synoptic-scale systems, which can travel across the globe for days before breaking down or dissipating. Thunderstorms, in contrast, exist only for an hour or so before collapsing. The shorter life span of these systems makes accurate forecasting of them more problematic.

Microscale phenomena are the smallest systems of all. Tornadoes, for example, are microscale phenomena. Their span is measured on the order of meters, and they can form and dissipate with little or no advance warning. Consider this: Residents of coastal communities generally receive advance warning for days of an approaching hurricane, a synoptic-scale phenomenon. By contrast, Midwest residents often receive less than 10 minutes' warning of an approaching tornado, a microscale phenomenon.

Only global- and synoptic-scale systems are impacted by the Coriolis effect. Refer to Figure 14.6 and note the curvature to the right, as expected in the Northern Hemisphere.

Low-pressure systems, or extra-tropical cyclones, develop as air rises more rapidly through their centers than air can flow in to replace the rising air (Figure 14.7). Thus, over time, the central pressure of the cyclone falls. The rising air, as described in

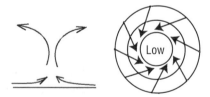

FIGURE 14.7 Rising motion as air flows into a cyclone. Source: *Let's Review: Earth Science—The Physical Setting,* 2nd Ed., Edward J. Denecke, Jr., Barron's Educational Series, Inc., 2002.

other sections, results in condensation, clouds, and precipitation.

In contrast, regions of strong high pressure, also known as anticyclones or air masses, are characterized by relatively tranquil and usually fair weather. Whether an air mass is warm or cold, air movement in an anticyclone tends to be outward and down. This subsidence, or subsiding motion, prevents cloud formation. As the air moves outward from the center of highest pressure, the Coriolis effect causes the moving air to spiral toward a cyclone. In the Northern Hemisphere, circulation is clockwise in a high-pressure system and counterclockwise in a low-pressure system. Figure 14.8 illustrates anticyclone motion. Air masses can develop as cold air builds up in a stable region, such as northern Canada. The pressure is high because of the weight of the cold, dense air. Warm air masses also exist. In regions where these weather systems form, the air pressure is not especially high, but it *is* higher than in any areas of low pressure in the same part of the world.

Figure 14.9 depicts the various types of air masses and their "source regions," that is, the areas where they tend to form. Table 14.5 details the characteristics of each air mass. Air masses are classified according to their temperatures and humidity levels. It is not surprising that maritime air masses (air masses that form over the ocean) have high humidity levels, whereas continental air masses are often quite dry. Air masses of arctic or polar origin are generally cold (or frigid), and when they advect into the midlatitudes, they bring normal temperatures for periods of up to several days at a time. Tropical air masses, which form over the lower latitudes, are characteristically warm. In the summer, maritime tropical air masses that form over the Gulf of Mexico tend to advect northward into the upper Great Plains, Midwest, and Northeast. There they can cause extended periods of above-normal temperatures, more commonly referred to as "heat waves" by residents of these regions.

Air masses can modify their characteristics when they move to a new geographic region. For example, if a continental polar air mass, initially containing cold, dry air, moves southeastward to a position off the eastern seaboard, it will begin to "pump" warm, moist air northward into the same region that was much colder a day or two earlier.

FIGURE 14.8 Air subsides as it flows into an anticyclone. Source: *Let's Review: Earth Science—The Physical Setting,* 2nd Ed., Edward J. Denecke, Jr., Barron's Educational Series, Inc., 2002.

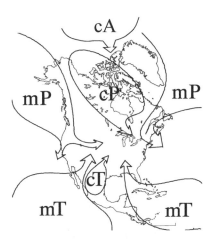

FIGURE 14.9 Air masses that influence weather patterns in the United States and their source regions. Source: *Let's Review: Earth Science—The Physical Setting,* 2nd Ed., Edward J. Denecke, Jr., Barron's Educational Series, Inc., 2002.

TABLE 14.5
AIR-MASS NAMES AND SOURCES

	Arctic A	Polar P	Tropical T
	Formed over extremely cold, ice-covered regions	Formed over regions at high latitudes where temperatures are relatively low	Formed over regions at low latitudes where temperatures are relatively high
maritime m Formed over water, moist		mP—cold, moist Formed over North Atlantic, North Pacific	mT—warm, moist Formed over Gulf of Mexico, middle Atlantic, Caribbean, Pacific south of California
continental c Formed over land, dry	cA—dry, frigid Formed north of Canada	cP—cold, dry Formed over northern and central Canada	cT—warm, dry Formed over southwestern United States in summer

As noted earlier, most synoptic-scale weather systems are driven by the global winds within which they are embedded. Since the westerlies blow across most of the United States, most weather systems move in some way from west to east. Figure 14.10 illustrates a "typical" cyclone (low-pressure storm system) and anticyclone (air mass) centered over Ohio and Utah.

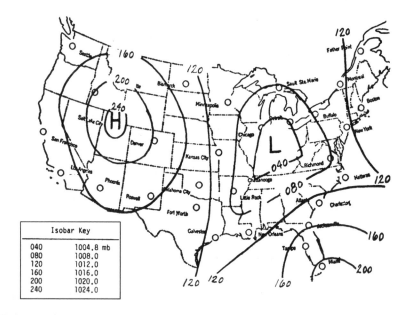

Isobar Key	
040	1004.8 mb
080	1008.0
120	1012.0
160	1016.0
200	1020.0
240	1024.0

FIGURE 14.10 Synoptic-scale weather systems. Meteorologists use wind patterns in the upper troposphere to predict the movements of these weather systems. Here, areas near the center of the low in Ohio are likely experiencing stormy weather, while the high-pressure system in the West is probably bringing fair weather to the entire Rocky Mountain region. The key shows pressure units measured in millibars (mb); 1013.2 mb is equivalent to 30.00 inches of mercury.

Winds

RELATIONSHIP BETWEEN AIR PRESSURE AND WIND

As stated above, in the Northern Hemisphere, air spirals clockwise out of a high-pressure, fair-weather system and counter-clockwise into a low-pressure storm system. The pattern is different in the Southern Hemisphere; air still flows from high to low pressure, but in the opposite direction. In the Northern Hemisphere, as a high-pressure system moves toward or builds into a region, winds often blow from the northwest (see Figure 14.11).

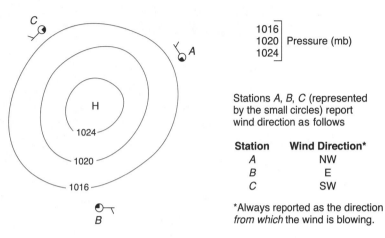

1016
1020 Pressure (mb)
1024

Stations *A*, *B*, *C* (represented by the small circles) report wind direction as follows

Station	Wind Direction*
A	NW
B	E
C	SW

*Always reported as the direction *from which* the wind is blowing.

FIGURE 14.11 Wind flow around an anticyclone, or high-pressure system.

The eastern United States experiences a well-known type of storm system called a *nor' easter*. The name, which Benjamin Franklin is credited with coining, describes the direction from which the wind blows as the storm approaches the region. Study Figure 14.12 and imagine the storm system depicted, currently centered over North Carolina, moving northeastward up along and parallel to the eastern seaboard. The counterclockwise circulation around the cyclone or storm system assures that winds ahead, or winds that precede the passage of the

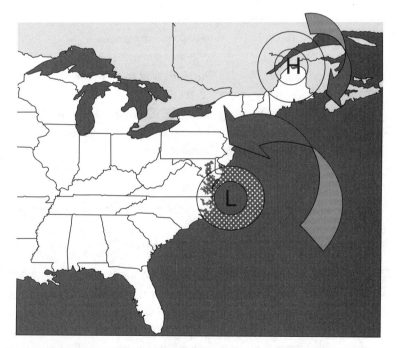

FIGURE 14.12 Coastal storm pattern along the U.S. eastern seaboard. The clockwise circulation around the high-pressure center (H) and the counterclockwise circulation around the low-pressure storm center combine to produce winds from the northeast in the region between these two storm systems.

storm's center, blow from the northeast. These storms often become quite strong; they are fed by the temperature contrast between the polar or arctic air masses that are frequently situated over the United States during the winter months and the warm waters of the Gulf Stream, located just offshore (see Figure 14.13). When a nor'easter impacts the eastern seaboard, heavy rains often fall along the coast, with heavy snows inland. High winds whip up the ocean, causing beach erosion and, under the right conditions, coastal flooding. Refer to Chapter 10, where astronomical high tides are discussed, to learn when a nor'easter is most likely to cause these types of environmental damage.

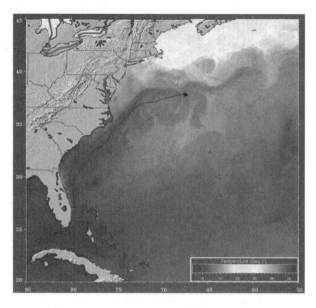

FIGURE 14.13 AVHRR satellite image of the Gulf Stream as it meanders northeastward, roughly paralleling the U.S. eastern seaboard. Source: *http://oceanexplorer.noaa.gov/explorations/05coralbanks/ background/vaults_of_history/media/fig1_600.jpg.*

Most of the greatest storms on record in the New York City area have been nor'easters. Included are the blizzard of 1888, the massive, paralyzing snowstorm of December 1947, the "superstorm" of March 1993, and the blizzard of January 1996. Some locales in the Northeast and even in the Mid-Atlantic interior received upward of 150 centimeters (4–5 ft) of snow in as little as 2 days!

Look again at Figure 14.12. Can you determine from what direction the winds will blow when the storm has passed by? The winds behind the storm often carry with them polar air from Canada.

Incidentally, in a midlatitude cyclone, sometimes called an extra-tropical storm, most if not all of the stormy weather precedes the arrival of the center of lowest pressure, and clearing takes place shortly after this center passes by.

FACTORS AFFECTING WINDS

You may have noted that some days are quite windy, whereas others are relatively calm. What causes high winds to occur one day and calm conditions to prevail the next? The answer is that the atmosphere is continually in a state of flux as regions of high and low pressure develop. Once these systems form, the atmosphere tries to return to a state of equilibrium; air in regions of higher pressure flows outward toward centers of lower pressure. This process is illustrated in Figure 14.14. The voracity with which wind blows depends upon the relative strengths of the high- and low-pressure weather systems and the proximity of these systems to each other. If a very strong center of high pressure, an anticyclone, is centered near a strong cyclone that has very low pressure at its center, the wind flow between the anticyclone and the cyclone will be quite strong. Strong winds are more common in the winter, when weather systems fueled by temperature contrasts between the cold interior of the continent and the warm coastal regions may well facilitate the development of strong areas of high and low pressure. The cold, dense air common across the interior of the northern United States and Canada naturally tends to exhibit high pressure.

Once a strong cyclone exists, the pressure difference between the cyclone and the anticyclone creates a *pressure gradient*, that is, a large change in air pressure over a relatively small distance that generates high-velocity winds between the two. This scenario can produce very windy conditions after a storm system with very low central pressure has passed by. Winds will continue to blow strongly until the air pressure approaches the levels in the anticyclone that develops after the passage of a cyclone.

Strong low-pressure systems can have central pressures below 950 millibars (29.00 in.). When a strong high-pressure system with a central pressure in excess of 1040 millibars (about 30.80 in.) approaches a region after the passage of a strong cyclone, high winds are an easy call for the weather forecaster.

FIGURE 14.14 Air pressure and wind. Winds blow from areas of high to low pressure. High-pressure air exerts more force than low-pressure air and pushes the low-pressure air ahead of it. The result is the horizontal movement of air commonly known as wind. Source: *Let's Review: Earth Science—The Physical Setting,* 2nd Ed., Edward J. Denecke, Jr., Barron's Educational Series, Inc., 2002.

In contrast, if the centers of a weak anticyclone and a weak cyclone are 1500 kilometers distant from each other, the wind flow between the two will be weak. This situation is more common during the summer when weather systems are generally less powerful than in the winter.

Review Exercises for Chapter 14

WORD-STUDY CONNECTION

aerosol	cirrocumulus	condensation nuclei
air pressure	cirrostratus	convection
aneroid barometer	cirrus	Coriolis effect
anticyclone	cloud	cumulonimbus
barometer	coalesce	cumulus
Bergeron	condensation	cyclone

dew point	mercury barometer	sling psychrometer
drizzle	mesoscale	snow
dry bulb	microscale	stable
evaporation	nimbostratus	stratus
fog	orographic	supercooled
freezing drizzle	overrunning	supersaturated
freezing rain	precipitation	suspension
global scale	pressure gradient	synoptic
hail	rain	thunderstorm
hygrometer	rain gauge	tornado
ice pellets	relative humidity	unstable
ITCZ	saturation	water vapor
latent heat	sleet	wet bulb
lightning		

SELF-TEST CONNECTION

PART A. Completion. *Write in the word or words that correctly complete the statement.*

1. Atmospheric moisture is most often reported in weather reports as _____.

2. The temperature at which the atmosphere reaches saturation is called the _____.

3. According to Table 14.1, for every 10C° temperature increase, the atmosphere's ability to hold water vapor roughly _____.

4. When fog is observed at ground level, the atmosphere is most likely _____.

5. _____ is precipitation that evaporates before reaching the ground.

6. A _____ is an instrument that can be used to determine both relative humidity and dew point.

7. Evaporational cooling occurs in a "dry" atmosphere because water droplets _____ energy as they evaporate.

8. The Intertropical Convergence Zone and the polar front are both sites of _____ air where clouds tend to form.

9. _____ clouds are easily recognized as vertically developed, "puffy" clouds in the sky.

10. _____ clouds are often the forerunners of a prolonged period of light to moderate precipitation.

11. _____ air tends to resist upward vertical movement.

12. Water vapor requires _____ to provide surfaces on which cloud droplets or ice crystals can form.

13. _____ is a form of precipitation that falls as rain, but freezes on contact with cold surfaces.

14. Snow will be observed at ground level as long as the temperature of the column of air between the clouds and the surface remains near or below _____ °C.

15. If snow melts as it falls, only to refreeze in a layer of cold air near Earth's surface, it will reach the ground as _____.

PART B. Multiple Choice. *Circle the letter of the item that correctly completes the statement.*

1. Relative humidity is affected by changes in
   (a) temperature
   (b) the amount of water vapor in the atmosphere
   (c) the amount of water vapor required to saturate the atmosphere
   (d) both a and b
   (e) a, b, and c

2. Using Table 14.1, if 0.5 gram of water vapor per kilogram of air is measured, and the temperature is −10°C, the relative humidity is
   (a) 0.5%
   (b) 2%
   (c) 25%
   (d) 50%
   (e) 100%

3. According to Table 14.2, when the dry-bulb temperature is 20°C and the wet bulb temperature is 17°C, the relative humidity is
   (a) 20%
   (b) 37%
   (c) 55%
   (d) 74%
   (e) 98%

4. According to Table 14.3, when the dry-bulb temperature is 14°C and the wet-bulb temperature is 10°C, the dew point is
   (a) 0°C
   (b) 5°C
   (c) 6°C
   (d) 7°C
   (e) 10°C

5. Latent heat, released as water vapor condenses, can provide the energy needed to fuel
   (a) thunderstorms
   (b) snowstorms
   (c) icestorms
   (d) evaporational cooling
   (e) prolonged rainfalls

6. Air tends to rise
   (a) along the ITCZ
   (b) when it encounters a mountain
   (c) along fronts
   (d) both a and b
   (e) a, b, and c

7. Clouds that yield prolonged periods of light to moderate precipitation are
   (a) cumulus
   (b) cumulonimbus
   (c) cirrus
   (d) cirrostratus
   (e) stratus

8. The form of precipitation that is generally not associated with winter storms or stratus clouds is
   (a) rain
   (b) snow
   (c) sleet
   (d) hail
   (e) freezing rain

9. Cold fronts tend to produce
   (a) rain
   (b) sleet
   (c) brief periods of heavy precipitation
   (d) long periods of light precipitation
   (e) no precipitation

10. Cyclones and anticyclones are examples of _____ scale systems.
    (a) global
    (b) synoptic
    (c) meso
    (d) micro
    (e) variable

PART C. Modified True/False. *If a statement is true, write "true" for your answer. If a statement is incorrect, change the <u>underlined</u> expression to one that will make the statement true.*

1. Cyclones spin <u>counterclockwise</u> in the Northern Hemisphere.

2. Latent heat releases energy as <u>evaporation</u> occurs.

3. <u>Thunderstorms</u> are classic examples of a mesoscale weather system.

4. Weather systems generally move <u>westward</u> across the United States.

5. When a large difference in pressure exists between two locations, the region in between can be quite <u>windy</u>.

6. A cyclone with a central pressure of 935 millibars is classified as a <u>strong</u> cyclone.

7. As temperature decreases and atmospheric moisture remains constant, relative humidity must <u>decrease</u>.

8. As temperature decreases and atmospheric moisture remains constant, the dew point will <u>decrease</u>. (Assume that the atmosphere remains unsaturated.)

9. <u>Subsidence</u> is common in anticyclones.

10. <u>Advection</u> is the horizontal transport of air masses across the face of the planet.

CONNECTING TO CONCEPTS

1. Explain why relative humidity can be a misleading measurement of humidity.

2. Explain the conditions that can cause virga to occur.

3. Explain the principles governing the way a sling psychrometer works. What controls the wet-bulb temperature?

4. Compare the conditions that produce snow, sleet, and freezing rain. Discuss in terms of a hypothetical vertical column of air that exists between the clouds in the sky and Earth's surface.

5. Explain the difference in the conditions that produce stratus and cumulus clouds.

6. Base your answers to parts (a) and (b) on the data table below, which shows the air temperatures and dew points over a 24-hour period for a particular location in New York State.

Time of Day	Air Temperature (°C)	Dew Point (°C)
12:00 midnight	19	12
2:00 A.M.	17	11
4:00 A.M.	14	10
6:00 A.M.	13	12
8:00 A.M.	15	11
10:00 A.M.	17	10
12:00 noon	18	9
2:00 P.M.	21	7
4:00 P.M.	23	6
6:00 P.M.	21	8
8:00 P.M.	19	10
10:00 P.M.	18	12
12:00 midnight	17	13

a. Use the data to construct a graph on the grid, following the directions below.
   1. Mark an appropriate scale on the axis labeled "Temperature."
   2. Plot a line graph for air temperature and label the line "Air temperature."
   3. Plot a line for the dew point and label the line "Dew point."

b. Based on your graph, state the hour of the day when the relative humidity was lowest.

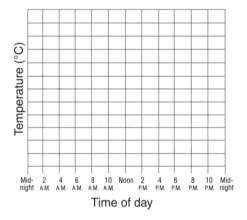

CONNECTING TO LIFE/JOB SKILLS

This chapter concentrates on several core topics in meteorology. The study of atmospheric moisture, pressure, and winds helps meteorologists to produce forecasts that can be relied upon by individuals, industry, and government alike. Severe weather events and storm systems, if unanticipated, can cause extensive loss of life and property damage. The accurate forecasting of these events and improvements upon current methods are major goals of the National Weather Service.

New instruments and methods of remote-sensing the atmosphere are being developed to improve forecasting accuracy. Many opportunities exist with universities and corporations that are actively researching and developing these tools and techniques. Students wishing to pursue careers in this area are encouraged to enroll in as many math and science courses as possible.

ANSWERS
SELF-TEST CONNECTION
Part A

1. relative humidity
2. dew point
3. doubles
4. saturated
5. Virga
6. sling psychrometer
7. absorb
8. rising
9. Cumulus
10. Cirrus
11. Stable
12. condensation nuclei
13. Freezing rain
14. 0
15. sleet or ice pellets

Part B

1. (e)
2. (c)
3. (d)
4. (c)
5. (a)
6. (e)
7. (e)
8. (d)
9. (d)
10. (b)

Part C

1. True
2. False; condensation
3. True
4. False; eastward
5. True
6. True
7. False; increase
8. False; remain constant
9. True
10. True

CONNECTING TO CONCEPTS

1. Warm air has a great capacity to hold water vapor without becoming saturated. Since relative humidity is a ratio of the water vapor in a given volume of air to the atmosphere's capacity to hold water vapor at that temperature, and since the atmosphere's capacity is high when the air is warm, relative humidity will remain quite low even though a significant amount of water vapor is in the atmosphere.

2. Virga occurs when precipitation falls from clouds but evaporates before reaching the ground. This occurs when the lower atmosphere, into which the precipitation is falling, is quite dry.

3. A sling psychrometer consists of two thermometers attached to a rod. One thermometer, called the wet bulb, has a sock attached to it. This sock is soaked with water, and then the instrument is "slung" or twirled for about a minute. As this

action takes place, the dry-bulb thermometer, which is directly in contact with the atmosphere, measures the actual air temperature. The wet bulb records a temperature cooler than the actual one because water is evaporating from the sock and absorbs energy as it evaporates. The less moisture present in the atmosphere, the faster the evaporation from the wet bulb and, consequently, the lower the wet-bulb temperature. Hence a "dry" atmosphere will produce a wet-bulb temperature significantly lower than the dry-bulb temperature.

4. If air temperatures are below freezing throughout a hypothetical column of air that stretches from the clouds to the ground, snow will fall. If temperatures in and near the clouds are below freezing, but warm to above freezing near the ground, the falling snow will melt and rain will be observed at ground level. The same is true if the column of air is above freezing throughout.

   Occasionally, an inversion can occur where warmer air overlies cold air. When cold air is "trapped" near the ground, freezing rain and/or sleet can result. Freezing rain occurs when rain falls into a shallow layer of subfreezing air. Sleet occurs if the subfreezing layer is a bit thicker, thus allowing the rain to freeze into ice pellets (also known as sleet) as it falls toward the ground.

5. Stratus clouds are horizontally oriented clouds; cumulus clouds are vertically developed. Vertical (cumulus) clouds form where air is rising rapidly. Stratus clouds develop where upward motion is quite gentle and gradual.

6. a.

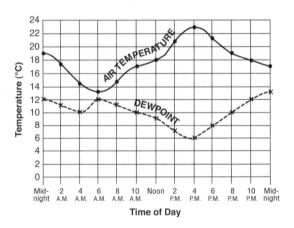

   b. Relative humidity was lowest at 4:00 P.M. At this hour, temperature and dew point were separated by 17C°. Furthermore, the dew point was at its low for the 24-hour period. This supports the answer, as dew point is an effective measure of atmospheric moisture.

15

Weather and
Weather Systems

WHAT YOU WILL LEARN

This chapter focuses on weather and weather systems. In this chapter you will learn

- the difference between cyclones and anticyclones;
- how fronts form and the various kinds of weather fronts;
- cloud patterns associated with cold and warm fronts;
- what severe weather is and how it forms;
- what tropical weather systems are and how they differ from extra-tropical systems;
- how meteorologists prepare a weather forecast.

SECTIONS IN THIS CHAPTER

- Cyclones and Anticyclones
- Other Forms of Severe Weather
- Tropical Weather Systems
- Weather Forecasting
- Related Internet Resources for Great Images and Information
- Review Exercises for Chapter 15

Cyclones and Anticyclones

AIR MASSES AND THEIR CHARACTERISTICS

Chapter 14 introduced the concept of synoptic-scale high- and low-pressure systems. This chapter explores these systems in somewhat greater depth and, in doing so, reviews some of the material covered earlier.

High-pressure systems, or anticyclones, are also known as *air masses*. Air masses develop over regions where temperature and humidity are similar. The strongest air masses form during the winter in northern Canada, Alaska, and Siberia as cold, dense pools of air that cover regions up to and over 1000 kilometers in all directions from the center of highest pressure. These cold, and air masses are classified as *continental polar* (*cP*) or *Arctic* (*A*), the only difference being the nature of the cold air. Arctic air masses are extremely cold, and polar air masses are somewhat less frigid. When these air masses travel into the United States, virtually under their own weight, and with the help of the prevailing winds, weather conditions can become bitterly cold across much of the nation, particularly east of the Continental Divide. Arctic air masses are responsible for the −40°C (−40°F) mornings that can occur in the Dakotas and Minnesota. Continental polar air also brings cold weather, but not quite as cold, to the midlatitudes.

Maritime polar (*mP*) air masses form over cool ocean waters and therefore exist as cool, moist high-pressure systems. Typical source regions include the northern Pacific Ocean and the northern Atlantic Ocean. Maritime polar air masses bring cool, damp weather to the Pacific Northwest much of the year. Although the prevailing winds limit the frequency of this event in the eastern states, occasionally a maritime polar air mass will slide southwestward from just off the Canadian Maritime Provinces to a position over the northeastern United States. When this happens, "Seattle-like" conditions occur from Maine to New Jersey, including the major cities of Boston and New York. These weather patterns, accompanied by drizzle and chilly northeasterly winds, occur most frequently during the autumn and winter months.

Air masses, also known as high-pressure systems, are often thought of as fair-weather systems. However, the cool, damp nature of a maritime polar air mass, when combined with orographic lifting caused by the prevailing flow of moist air flowing inland from the Pacific Ocean and up the windward side of the Cascades, causes overcast skies and periodic drizzle or light rain. Although the weather is not stormy per se, it certainly is not the typical fair weather, with mostly sunny skies, that is created by most air masses, particularly those of continental origin, across much of the nation.

Thus far, all three air masses discussed have had polar characteristics, that is, they are associated with cool or cold weather, the greatest difference among the three being the amount of moisture supplied by the source region. Arctic and continental polar air masses form over land or snow and ice and consequently are dry and cold; maritime polar air masses form over water and have high moisture levels.

Maritime tropical and *continental tropical* air masses form over warm regions, generally in the horse latitudes, located near the thirtieth parallel, or even nearer the Equator. As their name suggests, maritime tropical air masses form over warm water bodies; in terms of U.S. climate and weather, the most important of these is the Gulf of Mexico. Many maritime tropical air masses that form over the Gulf of Mexico eventually move northward with the prevailing winds. When maritime tropical air masses advect toward the Midwest, the Ohio valley, or the Northeast, conditions in these regions usually become quite warm and humid. Dew points often exceed 20°C if the advection occurs during the summer months.

In the United States, continental tropical air masses are confined to the Desert Southwest. This region is geographically cut off from any warm water sources; thus very little atmospheric moisture is available for any air mass forming over the area. Furthermore, the Desert Southwest is within or near the belt of global subsidence that occurs near the horse latitudes. Consequently, hot, dry conditions prevail, with Phoenix, for example, experiencing daytime temperatures in the summer well in excess of 40°C.

If a continental tropical air mass migrates eastward out of the Desert Southwest and into the southeastern states, it quickly modifies as it picks up moisture from the Gulf of Mexico and becomes, in effect, a maritime tropical air mass.

FRONTS

Fronts separate air masses with differing characteristics. A buoyant, warm, humid air mass tends to be less dense than a cold, dry air mass. Air masses with different densities do not mix well, and therefore a boundary, known as a *front*, forms between the two air masses. Midlatitude or extra-tropical storms often develop, clouds form, and precipitation falls along fronts.

Four different types of fronts are known to exist: stationary, warm, cold, and occluded. A *stationary* front is a boundary between two air masses that are temporarily motionless. Neither air mass is attempting to displace the other.

If, however, a warm air mass begins to move into a region occupied by cold air, the front separating them becomes a *warm* front. When a warm front passes through a region, the warm air displaces the cold air and the weather becomes warmer. In the eastern United States, warm air often advects northward from the Gulf of Mexico to a position just west of the Appalachian Mountains. Under certain conditions, temperatures in parts of western Pennsylvania and New York State can exceed 20°C (68°F), while at the same time, in cities such as New York and Boston, temperatures in a maritime polar air mass are hovering near 5°C (40°F). In this case, the Appalachian Mountains act as a physical block, preventing the warm air from moving eastward. This phenomenon, known to the region as "Appalachian Mountain damming," can occur east of the Appalachian Mountains as far south as South Carolina in extreme cases, although it is much more common to the northern Mid-Atlantic states and in the Northeast. Occasionally, as a cold front approaches the entire region, the warm front, which precedes the cold front, is pushed through Philadelphia and Baltimore, and rapid, though temporary, increases in temperature occur.

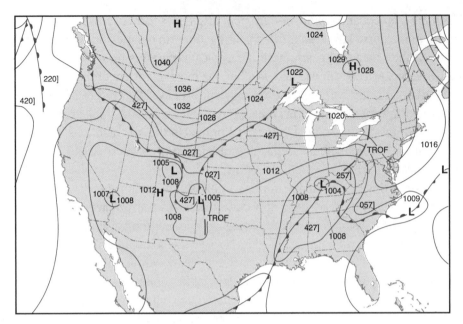

FIGURE 15.1 A typical weather map depicting frontal positions, along with high- and low-pressure centers across the United States. Note that a cold front stretches from northwestern Wisconsin to eastern Colorado. A warm front can be seen approaching the West Coast, just west of Washington and Oregon. A stationary front stretches from Georgia to South Carolina to a position just off the Mid-Atlantic coastline. Source: NOAA's National Weather Service.

A front must be thought of as a three-dimensional phenomenon. We can study fronts by looking at them on weather maps (see Figure 15.1), but this does not yield any real insight as to what is happening to the atmosphere as a front moves. When warm air attempts to displace cold air, the less dense warm air tends to overrun the dense, cold air, as illustrated in Figure 15.2. The overrunning may start at the point where the warm air first meets the cold air at the surface and may continue for a distance of about 1000 kilometers. Over this entire stretch, air is gently rising on a gradual slope. The rising air produces various types of clouds, beginning with high, thin cirrus clouds well ahead of the actual surface boundary. As the warm front moves closer to a given region, the clouds become lower and thicker.

The cirrus clouds are replaced in time by cirrostratus, stratus, and eventually nimbostratus clouds (see Figure 15.3). Fair weather may be replaced by a steady period of precipitation that often occurs as snow well ahead of the front, but may turn to rain by the time the front arrives (see Figure 15.4).

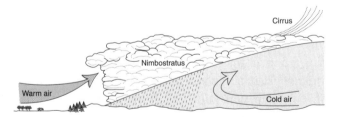

FIGURE 15.2 Structure of a typical warm front. Source: *Let's Review: Earth Science—The Physical Setting*, 2nd Ed., Edward J. Denecke, Jr., Barron's Educational Series, Inc., 2002.

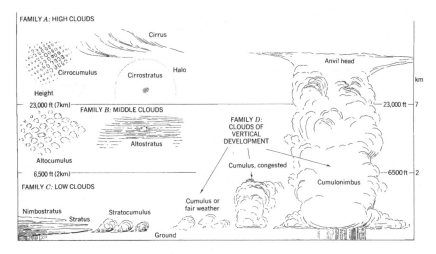

FIGURE 15.3 Cloud classification chart. Clouds are classified into families according to their altitude and degree of vertical development. Source: Arthur N. Strahler, *The Earth Sciences*, Harper & Row, 1971. Study figures 15.4 and 15.5 to better understand the relationship between cloud type and the frontal surface along which it forms.

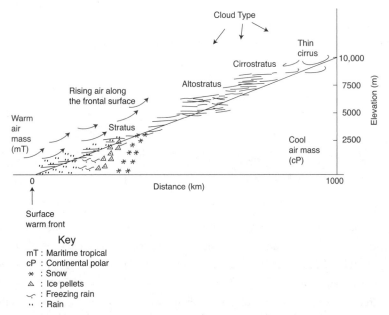

FIGURE 15.4 Cutaway view of a warm frontal surface. Warm air overruns the cold, dense air for distances of up to 1000 kilometers before the passage of the actual surface warm front. Before a warm frontal passage, a sequence of events occurs that begins with the observation of high, thin, "wispy" cirrus clouds. A series of lower and thicker clouds cover the sky as the surface warm front approaches. Closer to the surface front, stratus clouds produce precipitation. Snow falls where the atmosphere is colder, and temperatures are below freezing throughout. As the atmosphere warms, snow is often replaced by ice pellets (sleet), freezing rain, and finally rain. After the frontal passage, skies often clear and temperatures rise.

A *cold* front produces an entirely different set of conditions along its boundary, and the cold frontal surface has a different shape from that of a warm frontal surface. As cold air tries to displace warm air, the denser cold air begins to burrow under the warm air mass, as illustrated in Figure 15.5. This action forces the warm air to rise strongly or even, occasionally, violently and produces a relatively narrow band of clouds that exhibit vertical development, in other words, cumulus

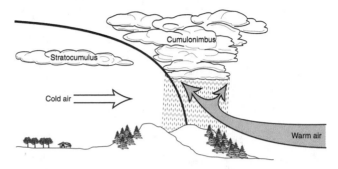

FIGURE 15.5 Structure of a typical cold front. Source: *Let's Review: Earth Science—The Physical Setting*, 2nd Ed., Edward J. Denecke, Jr., Barron's Educational Series, Inc., 2002.

clouds (refer to Figure 15.3). The strong vertical motion can produce thunderstorms, frequent lightning, hail, and, in extreme cases, tornadoes.

Cold fronts, with their dense, cold air, typically travel faster than warm fronts, and an *occluded* front (Figure 15.6) forms when a cold front catches up to a warm front. When this occurs, the warm air, or warm sector, is lifted aloft, and the boundary that exists at the surface is between two fairly similar air masses. This event marks the beginning of *frontolysis*, a stage where fronts break down or dissipate.

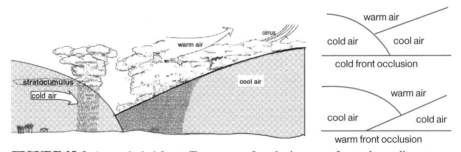

FIGURE 15.6 An occluded front. Two types of occlusions can form, depending upon whether the advancing cold air mass is warmer (less dense) or colder (more dense) than the cold air mass ahead of the warm air mass. When the advancing cold air is colder than the air mass ahead of the warm air mass, a cold occlusion forms as the advancing cold air "undercuts" both air masses that it is advancing toward. General movement in this diagram is to the right. Source: *Let's Review: Earth Science—The Physical Setting*, 2nd Ed., Edward J. Denecke, Jr., Barron's Educational Series, Inc., 2002.

CYCLOGENESIS—FORMATION OF A MIDLATITUDE STORM SYSTEM

Cyclones, or *extra-tropical storms*, form when air masses of differing characteristics clash. Although the initial conditions separating two dissimilar air masses may result in the formation of a stationary front, after a brief period of time a wave forms along that front where each air mass is trying to "invade" the other's territory (see Figure 15.7). *Cyclogenesis*, or the birth of a cyclone, occurs when the frontal wave forms. Cyclogenesis is the process of initiating a circulation about a central point, as illustrated in Figure 15.7.

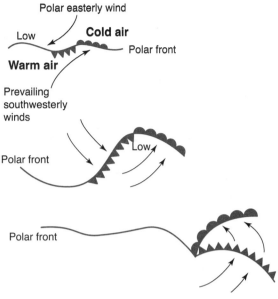

FIGURE 15.7 Formation of a wave cyclone. Source: *Let's Review: Earth Science—The Physical Setting*, 2nd Ed., Edward J. Denecke, Jr., Barron's Educational Series, Inc., 2002.

DEVELOPMENT OF A CYCLONE

If conditions are right, the cyclone will continue to strengthen as it forms well-defined cold and warm fronts and a cyclonic or counterclockwise circulation develops about the center of lowest pressure. The storm then engages in two distinct motions. Forward movement, often to the east or northeast, occurs as the cyclone is carried by the prevailing winds, and a counterclockwise, rotational movement occurs about the center of deepening low pressure. At this stage, the cyclone is approaching its maximum strength. Air is rising near the center faster than the surface circulation can replace it; hence air pressure is falling near the center (see Figure 15.8).

PRECIPITATION PATTERNS ASSOCIATED WITH A CYCLONE

The rising air associated with a cyclone is most significant along both fronts, the cold and the warm; hence the stormy weather typically occurs along and ahead of each (see Figure 15.9). Since the warm frontal surface is a gradual one that extends for hundreds of kilometers as illustrated

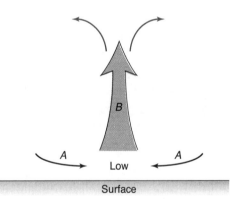

FIGURE 15.8 Air flows upward (*B*) and out of the cyclone's center faster than air along the surface (*A*) flows in to replace the rising air. The net result is a drop in air pressure at the point marked "Low."

in Figure 15.10, a prolonged period of rain or snow will precede the arrival of the center of lowest pressure. The cold front's zone of stormy weather is more narrow, but is often more intense.

Take note of the two diagonal lines drawn across the map in Figure 15.11. If the storm passes to the north and west of the observer's location along line 1, two frontal

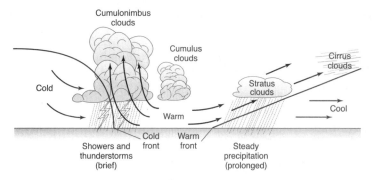

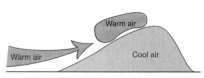

Warm air is forced upward by wind.

FIGURE 15.10 Cutaway diagram of a warm frontal surface. Source: Adapted from *Let's Review: Earth Science—The Physical Setting*, 2nd Ed., Edward J. Denecke, Jr., Barron's Educational Series, Inc., 2002.

FIGURE 15.9 Stormy weather is associated with each front. Note the broad band of prolonged precipitation that occurs ahead of the warm front. The cold front is accompanied by a brief period of intense precipitation (showers and thunderstorms) as warm air is forced to rise violently along the steep frontal surface.

passages can be expected. As mentioned above, a prolonged period of steady precipitation will precede the arrival of the warm front. After the passage of this front, however, there is no longer a surface for the warm air to rise over. Skies may clear, and conditions may be relatively tranquil and warm in the so-called warm sector of the storm. The relative calm is abruptly shattered by the approach of the cold front, perhaps with a squall line of thunderstorms with it or ahead of it. A *squall line* is a nar-

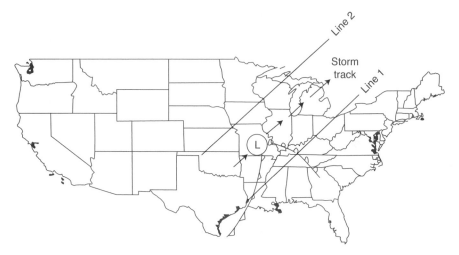

FIGURE 15.11 The weather that occurs with the passage of a cyclone depends largely on whether the storm center passes to the north or south of your location. Line 1 depicts the storm tracking to the north and west; two frontal passages will occur. In between, the warm sector will pass through the region and temperatures will temporarily rise along with a period of clearing skies before the arrival of the cold front. Line 2 depicts the storm passing to the south and east. Residents along this line will experience a swath of clouds and precipitation and will remain in the cold sector of the storm throughout the event. During the winter, a snowstorm is most likely to occur along line 2.

row zone of severe weather that precedes a cold front, usually by 100–200 kilometers. Cities along line 1 may experience some snow, but as the warm front approaches, the snow changes to rain.

Line 2 depicts the conditions experienced by a location that has the storm center pass to its south and east. No frontal passages are experienced by observers living along this line, and the conditions found in the warm sector never occur in this location. This is the region that in the winter usually receives a heavy snowfall when a cyclone passes by.

> **REMEMBER**
> The East Coast's greatest winter (snow) storms occur when a cyclone forms along the Gulf Coast or just off the coast of the Carolinas and then tracks northeastward along, but to the immediate east of, the eastern seaboard of the United States.

DISSIPATION OF A CYCLONE

The conditions that existed to create the cyclone begin to break down as the cold front eventually catches up with the warm front. When this occurs, an *occlusion* forms (refer to Figure 15.6) and forces the warm air aloft, where it begins to cool. Once the thermal differences that initiated the entire event become minimal, the storm begins to "fill" and its central pressure rises. Upward movement of air ceases, and the clouds and precipitation dissipate, marking the end of the cyclone.

Here's an interesting comparison. On Earth, storm systems typically exist for a week or two. In stark contrast, the Great Red Spot on Jupiter, a storm system into which 12 Earths could fit, has lasted for hundreds of years! Certainly, planetary meteorology represents an entirely new challenge for today's meteorologist or atmospheric scientist!

Other Forms of Severe Weather

THUNDERSTORMS AND LIGHTNING

Severe weather is the subject of much study in the field of atmospheric science, as many lives are still unfortunately lost each year because of lightning strikes and tornadoes. Forecast techniques in current use provide as little as 10 minutes', or even less, warning of an impending tornado. Also, it is nearly impossible to predict exactly where lightning will strike when thunderstorms are present. New, state-of-the-art forecasting techniques are improving the "lead time," and thus increasing the advance-warning time, provided to residents in areas threatened by severe weather. The other variable, however, is whether people will heed the warnings issued.

Thunderstorms are mesoscale phenomena that occur throughout much of the United States but reach a peak in the spring across the Great Plains and Midwest. Florida actually experiences the most thunderstorms in the nation, but they tend not to be of the severe variety. The clash of warm, moist Gulf of Mexico air and cool, dry continental polar air in April and May creates conditions that spawn thunderstorms quite often in the regions cited above. The United States is the only midlatitude nation on the planet without a major east-west mountain range. The relatively flat terrain allows tropical and polar air masses that have traveled in some cases over great distances to

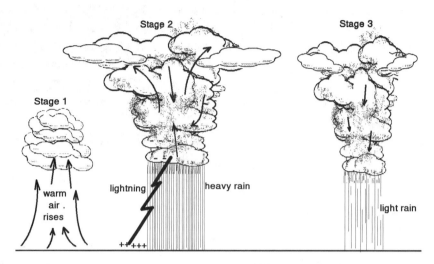

FIGURE 15.12 Formation of a thunderstorm. Stage 1: Intense heating causes warm air to rise rapidly, forming tall, vertically developed clouds. Stage 2: Strong updrafts keep precipitation aloft until it is heavy. Falling rain creates downdrafts, and internal friction with updrafts causes buildup of electrostatic charge, leading to lightning and then thunder. Stage 3: Downdrafts cause air to cool and updrafts subside, so rain becomes lighter. Source: *Let's Review: Earth Science— The Physical Setting*, 2nd Ed., Edward J. Denecke, Jr., Barron's Educational Series, Inc., 2002.

clash with very little interference. The fronts that form between these air masses can produce highly unstable atmospheric conditions.

Thunderstorms result when unstable warm air rises, in some instances to the tropopause (see Figure 15.12). As this air rises, towering thunderheads, or cumulonimbus clouds, form. When the condensed water vapor, or cloud tops, reach the tropopause, high-velocity winds shear off the thunderhead into a characteristic anvil shape. The cirrus clouds that result from this action can travel many miles ahead of the main cloud mass and can serve as an early warning of what is approaching a region.

To achieve "severe" status, a thunderstorm must exhibit winds in excess of 90 kilometers (55 mi) per hour, and/or frequent lightning, and/or hail and tornadic development, which is explained in the next section. If any combination of the conditions named above is deemed possible, the National Weather Service will issue a severe thunderstorm watch or a tornado watch. A *watch* means that the potential for severe weather exists over a fairly large region for a number of hours. Since severe thunderstorms are most common in the mid to late afternoon and early evening, residents in the watch area should observe the skies and keep tuned to weather information during those hours. A watch, however, does not guarantee the occurrence of severe weather. It is very possible, if not likely, that most, or all, of the watch area on a given day will not see any severe weather and may not even experience a shower!

Once severe thunderstorms actually form and are detected by radar and/or surface observations, the National Weather Service will issue a severe thunderstorm warning. The *warning* will report the current location and path of any storms that currently exist. Residents of locations in the path of the storms usually have an hour or less to prepare for the arrival of the severe weather.

The updrafts and downdrafts in thunderstorms are responsible for their production of hail and can act to spawn tornadoes. Wind currents are also the reason why many pilots avoid these violent storms at all costs.

Thunderstorms can be deadly. Lightning kills an average of 62 people each year across the United States. As noted in Figure 15.12, the buildup of electrostatic charge leads to the occurrence of lightning. Lightning is a static electricity discharge. Lightning also causes thunder. A lightning strike can heat the air it passes through to a temperature of about 30,000 °C. This causes rapid expansion and the resulting shock wave is heard as thunder.

TORNADOES

A *tornado* is basically a very small (microscale) and very intense vortex of spinning air (see Figure 15.13). Most tornadoes have diameters of only 200–300 meters (600–900 ft); however, wind velocities are thought to exceed 500 kilometers (300 mi) per hour in extreme cases. Precise measurements of these extreme winds do not yet exist. The air pressure within the tornado may be as low as 880 millibars (26.50 in.).

The United States, in particular the Great Plains and Midwest, experience most of the 1000 tornadoes that occur annually across the globe. The reasons, as described earlier, include the absence of mountains between the Arctic Ocean, or region, and the Gulf of Mexico, and the warm and abundant moisture that the Gulf of Mexico represents. The peak month for tornadic activity is May. During that month, solar elevations, and hence surface heating, are very great, and polar air masses can penetrate as far south as Oklahoma and Texas.

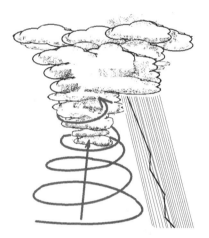

FIGURE 15.13 Formation of a tornado. Source: *Let's Review: Earth Science—The Physical Setting*, 2nd Ed., Edward J. Denecke, Jr., Barron's Educational Series, Inc., 2002.

Tornadoes are rated according to the Fujita intensity scale (see Table 15.1). Named for its developer, T. Theodore Fujita, an atmospheric scientist, the scale is based upon the damage caused by a tornado, which is usually assessed after the event. The scale ranges from F0, where "light damage" has resulted, to F5, which indicates "incredible damage."

F5 tornadoes are rare and often make news for years after their occurrence. In 1974, Xenia, Ohio, experienced several F5 tornadoes; more recently, in the spring of 2002, an F5 actually touched down in eastern Maryland, a region and state unaccustomed to such strong tornadoes.

Modern atmospheric models attempt to duplicate the formation of a tornado. Current thinking rests on an interaction among the updrafts, downdrafts, and outside winds that help to form the vertically oriented vortex and drive it toward the ground. The problem with forecasting tornadoes is the same as that experienced with the forecasting of severe

TABLE 15.1
THE FUJITA SCALE FOR TORNADOES

Rating	Wind Speed (mph)	Damage
F0	<73	Light damage. Some damage to chimneys; branches broken off trees; shallow-rooted trees pushed over; sign boards damaged.
F1	73–112	Moderate damage. Surfaces peeled off roofs; mobile homes pushed off foundations or overturned; moving autos blown off roads.
F2	113–157	Considerable damage. Roofs torn off frame houses; mobile homes demolished; boxcars overturned; large trees snapped or uprooted; light-object missiles generated; cars lifted off ground.
F3	158–206	Severe damage. Roofs and some walls torn off well-constructed houses; trains overturned; most trees in forest uprooted; heavy cars lifted off the ground and thrown.
F4	207–260	Devastating damage. Well-constructed houses leveled; structures with weak foundations blown away some distance; cars thrown and large-object missiles generated.
F5	261+	Incredible damage. Strong frame houses leveled off foundations and swept away; automobile-sized missiles fly through the air in excess of 100 meters (109 yd); trees debarked; incredible phenomena occur.

thunderstorms. The National Weather Service tends to issue a tornado watch for a wide region for a several-hour period and to issue a tornado warning only when a tornado is actually sighted or the "signature" or mark of a tornado is observed on Doppler radar.

Doppler radar has been a significant step forward in adding lead time to the issuance of a tornado warning, as a direct sighting is no longer necessary. Doppler radar helps meteorologists to measure the rate at which particles (raindrops) are moving toward and away from the radar. Since tornadoes feature a small, high-velocity circulation, the radar will show "reflections" moving rapidly toward the radar viewer or screen and, immediately adjacent, reflections moving rapidly away. Meteorologists use false color images to enhance the view. The use of Doppler radar has improved warning times for tornadoes from less than 10 minutes to between 20 and 30 minutes. This may seem like a small improvement, but it can mean the difference between life and death for people affected by these storms.

In summary, tornadoes are fascinating phenomena, generally associated with severe thunderstorms. These systems can also be quite deadly. Much work remains to be done on generating more accurate forecasts with greater lead times to warn those who may

> **REMEMBER**
> An excellent description of Doppler radar is available at this *USA Today* page:
> *www.usatoday.com/weather/wdoppler.htm.*

be in harm's way. The National Weather Service's Storm Prediction Center in Norman, Oklahoma (*www.spc.noaa.gov/*), is the agency most directly responsible for researching tornadoes and severe thunderstorms and their associated weather.

Tropical Weather Systems

TROPICAL STORMS AND HURRICANES

Tropical storms, or tropical cyclones, are quite different from their midlatitude or extra-tropical counterparts. Midlatitude cyclones derive their energy from the temperature contrast between, and the mixing of, the tropical and polar air-mass boundaries along which they are born. Often called "cold core" storms, they "live" as long as there is sufficient contrast between the two air masses that they border. When the circulation created by one of these storms achieves sufficient mixing of warm and cold air, the storm ceases to exist.

Tropical storms, in contrast, form over warm, tropical waters. They derive their energy exclusively from an abundant supply of water vapor, which, as it rises near the storm center and condenses, releases latent heat of condensation. A strong tropical storm also tends to develop a well-defined wall of clouds because the center, or eye, of the storm forms a narrow, circular region at the storm's core, where air is actually subsiding.

This eye is characterized by clear to partly cloudy skies and relatively light winds. The break in storm conditions may last up to an hour, but rarely longer. A thick, dark wall of clouds known as the *eye wall* surrounds the eye. Once the eye passes, the eye wall moves through the affected region and storm conditions resume. Figure 15.14 illustrates the structure of a tropical cyclone, including the eye (*A*) and eye wall.

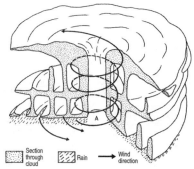

FIGURE 15.14 Cross section of a hurricane. Source: Earth Science on File. © Facts on File, Inc., 1988.

A tropical system begins as a tropical wave or a region of building thunderstorms in warm equatorial waters, usually between 5° and 15° latitude. At these latitudes, the tropical wave is carried westward with the prevailing trade winds. Tropical cyclones do not form any closer to the Equator because the Coriolis effect is too weak.

If the conditions are right, the tropical wave will strengthen into a tropical depression, which is characterized by increasing winds of up to 61 kilometers (38 mi) per hour. Further strengthening produces a tropical storm, which begins to have an organized circulation on radar and satellite images. Tropical storms have winds that range between 61 and 119 kilometers (38 and 74 mi) per hour.

When a tropical depression becomes a tropical storm, it acquires a name that has been preselected by the National Hurricane Center, located in Miami, Florida. Each year, names are generated randomly, alternating between male and female, well in advance of the hurricane season, which runs from June 1 through November 1. There are separate lists of names each season for storms that form in the Atlantic, western Pacific, and eastern Pacific. Lists of names can be found on the National Hurricane Center Web site: *www.nhc.noaa.gov/*.

When winds near the storm's center exceed tropical storm levels, the tropical cyclone has reached *hurricane* strength. Although the wind speeds of hurricanes are not as great as those of tornadoes, hurricanes are much larger storms and can produce much greater damage over a widespread region. Hurricanes are rated according to the Saffir-Simpson scale (Table 15.2) from one to five. A Category One hurricane has winds that do not exceed 153 kilometers (95 mi) per hour and will usually produce minimal damage if it makes landfall—something that most do not. An average season in the Atlantic Ocean, for example, will produce approximately eleven named storms, all of which must have reached at least tropical storm status. Of these eleven, typically

TABLE 15.2
THE SAFFIR-SIMPSON SCALE FOR HURRICANES

Category	Wind Speed (mph)	Damage
One	74–95	No real damage to buildings. Damage primarily to unanchored mobile homes, shrubbery, and trees. Also, some coastal road flooding and minor pier damage.
Two	96–110	Some roofing material, door, and window damage to buildings. Considerable damage to vegetation, mobile homes, and piers. Coastal and low-lying escape routes flood 2–4 hours before arrival of center. Small craft in unprotected anchorages break moorings.
Three	111–130	Some structural damage to small residences and utility buildings with a minor amount of curtainwall failures. Mobile homes destroyed. Flooding near the coast destroys smaller structures, with larger structures damaged by floating debris. Terrain continuously lower than 5 feet ASL* may be flooded inland for 8 miles or more.
Four	131–155	More extensive curtainwall failures with some complete roof structure failure on small residences. Major erosion of beach. Major damage to lower floors of structures near the shore. Terrain continuously lower than 10 feet ASL may be flooded, requiring massive evacuation of residential areas inland as far as 6 miles.
Five	Greater than 155	Complete roof failure on many residences and industrial buildings. Some complete building failures with small utility buildings blown over or away. Major damage to lower floors of all structures located less than 15 feet ASL and within 500 yards of shoreline. Massive evacuation of residential areas on low ground within 5–10 miles of shoreline may be required.

*Above sea level.

only two or three will make landfall somewhere in the Caribbean Islands, in Mexico, or on the Gulf or East Coast of the United States.

Category Five hurricanes with winds exceeding 250 kilometers per hour are rare. The United States has been impacted by only a few in the past century. Most recently, Katrina hit the U.S. mainland in August 2005. Katrina impacted the Louisiana and Mississippi coastlines and forced the near total evacuation of New Orleans. Katrina was the most costly and the third-deadliest hurricane in U.S. history. The strongest storm (as measured by a central pressure of 26.35 inches!) was dubbed the Great Labor Day Storm, as it formed before storms were named. More than 400 people died when it made landfall on September 2, 1935. Hurricanes Camille (1969) and Andrew (1992) were notable Category Five storms as well.

> **REMEMBER**
> A complete list of major hurricanes that have impacted the United States may be found at *www.epicdisasters.com/index.php/site/ comments/the_ten_strongest_hurricanes/.*

PATHS OF DESTRUCTION

In the Northern Hemisphere, hurricanes tend to follow curved paths that eventually take them north of the trade winds into the zone of the global wind belt known as the prevailing westerlies. Since the westerlies flow in the opposite direction to the trade winds, a hurricane's path often looks like a large hook, with the initial path to the west-northwest and the final track toward the northeast (see Figure 15.15). This path spares the eastern United States if a hurricane makes its turn while it is still out at sea.

Hurricanes, as noted earlier, feed off warm water, generally in excess of 26°C (80°F). When a storm encounters cooler waters or makes landfall, it loses its source of energy from latent heat and rapidly weakens. Friction caused by land surfaces also acts to quickly lower wind velocity; as a result, inland locations rarely experience serious damage from high winds. These regions are more often impacted by excessive rainfall, which can exceed 30 centimeters (1 ft) in a brief period of time.

The Atlantic Ocean and the Gulf of Mexico are ideal places for hurricanes to develop and gain strength as the waters are quite warm, especially in the late summer. Seawater temperatures of nearly 26°C are typically observed as far north as the waters off Long Island, New York, in the late summer and early autumn.

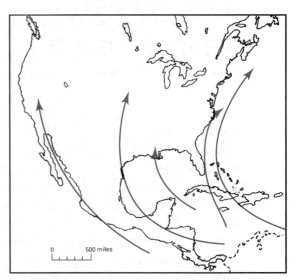

FIGURE 15.15 Typical tracks of tropical cyclones-hurricanes near the United States. Source: *Let's Review: Earth Science—The Physical Setting*, 2nd Ed., Edward J. Denecke, Jr., Barron's Educational Series, Inc., 2002.

The eastern Pacific is cooler as a result of the currents that run southward along the West Coast of the United States. Tropical storms that develop in the Pacific Ocean off the coast of Mexico usually head northwest and, upon reaching cooler waters, quickly dissipate. Tropical storms that form in the western Pacific, however, have dynamics similar to those that form in the Atlantic Ocean, and populated regions such as the Philippines, southeastern Asia, Japan, and Korea are periodically threatened by these superstorms. In these regions, hurricanes are known as tropical storms, tropical cyclones, or typhoons, and in Australia they are called willy-willies!

STORM SURGE

One of the more destructive features of hurricanes is the storm surge. As mentioned earlier, winds create waves, so it should come as no surprise that the wave action in a hurricane can be quite spectacular. In addition, the persistent winds pile water up on the northern side of the storm, where onshore winds can create literally a wall of water upward of 7 meters (21 ft) high in some Category Five storms. This *storm surge* arrives just as the hurricane's eye comes onshore and can inundate an entire coastal region.

Since many coastal areas in the eastern United States and the Gulf Coast region are only 2 or 3 meters above sea level even a kilometer or two inland, the storm surge can be quite devastating and destructive. If it occurs at or around the time of high tide, flooding will be enhanced.

> **REMEMBER**
> Storm surges are often the most deadly feature of a tropical cyclone.

Weather Forecasting

IS IT SCIENCE OR ART?

Meteorologists today have an arsenal of data and computer forecast models to study in an effort to refine and improve their forecasts. The public is often frustrated when a forecast of "partly cloudy" for the afternoon is promptly followed by 10 centimeters of snow! The bottom line is that we humans do not fully understand the workings of our atmosphere, and the computers that crunch data, model the atmosphere, and produce forecasts as guidance for forecasters are virtually useless for weather predictions beyond a few days from the time the data were collected.

In other words, short-term forecasts, those for a day or two in the future, are reasonably, sometimes very, accurate for most regions. Longer-term forecasts, however, are still quite problematic in many regions, especially when a storm system is approaching.

Ideally, when a meteorologist issues a forecast, he or she is combining computer-generated information and models with personal knowledge of the local region and experience. Specific weather information about a local area and work experience in that region for many years can greatly enhance the accuracy of the computer modeling that the National Weather Service provides.

MODERN METHODS OF FORECASTING

Today the National Weather Service has hundreds of reporting stations across the nation and many more beyond U.S. borders that share information in an effort to model the atmosphere. In the United States, most reporting stations are located at airports, each of which is identified by the same three call letters used to tag passengers' luggage. Some of these stations use radiosonde packages to measure conditions aloft, in addition to monitoring surface conditions. *Radiosonde* is a miniature transmitter carried by balloon that measures temperature, humidity, and wind speed and direction at various heights above the observing station. These observations are usually made twice per day at 0 and 1200 Greenwich Mean Time (GMT). This is the standardized time zone used by all weather service offices to avoid the confusion of local time zones and conventions such as daylight savings time.

All the data collected by National Weather Service reporting stations are sent electronically to the National Center for Environmental Prediction (NCEP) in Camp Springs, Maryland. The NCEP is a supercomputing center where all the National Weather Service forecasts are issued for the entire nation. These forecasts are based in part upon computer models with initialism names such as MRF (Medium Range Forecast), GFS (Global Forecast System), and WRF (Weather Research and Forecasting Model). Each of these models is somewhat unique in its algorithms (the computational procedure a model follows in generating its forecast, which may include the geographic, oceanographic, or atmospheric information provided to the model). Also, resolution (the detail with which it models the atmosphere) varies from model to model. Some are graphical (maps), and some are numerical (charts of data). As computers become more powerful, the resolution of the model can be improved.

Each shift of meteorologists studies a set of these models and then issues a forecast that reflects the "thinking" of the models, along with their own knowledge and experience. With improvements in the processing power of computers, new satellites that can observe the atmosphere in greater detail and more often than radiosonde, and new models that reflect the real atmosphere more accurately, forecasts, both short-term and long-term, should gradually improve in coming years.

Related Internet Resources for Great Images and Information

www.intellicast.com

The Intellicast Web site provides access to real-time satellite images, radar images, and surface observations. The information is provided in a well-organized and understandable format.

http://weather.unisys.com

The Unisys Corporation maintains this very comprehensive site that provides access to current and archived weather data and images. Guides are scattered throughout the Web site to help the visitor to understand how to interpret the wealth of material available here.

http://ww2010.atmos.uiuc.edu/(Gh)/home.rxml

The "World Weather 2010 Project," developed at the University of Illinois, provides an extensive bank of weather-related information along with a number of real-time products, including surface observations, satellite images, and radar images. The user can control the level of analysis that is provided on each map.

www.nws.noaa.gov

The National Weather Service has many regional and local offices, along with a wealth of data. Virtually all of these data are made available to the public.

www.ametsoc.org/dstreme

The Datastreme Web site provides access to real-time maps and images. This site is an excellent educational tool because images can be selected that allow the user to perform certain analyses, such as an isobar analysis on a pressure chart or an isotherm analysis on a temperature chart.

REVIEW EXERCISES FOR CHAPTER 15

WORD-STUDY CONNECTION

advect	dew point	maritime polar
advection	Doppler radar	maritime tropical
air mass	downdraft	mesoscale
anticyclone	Equator	microscale
arctic	eye	nimbostratus
cirrostratus	eye wall	occluded front
cirrus	front	occlusion
cold front	frontal wave	overrunning
Continental Divide	frontolysis	precipitation
continental polar	Fujita intensity scale	radar
continental tropical	hail	radiosonde
cumulus	horse latitudes	Saffir-Simpson scale
cyclogenesis	hurricane	satellite
cyclone	lightning	squall line

stationary front	thunderstorm	typhoon
storm surge	tornado	unstable
stratus	trade winds	updraft
subsidence	tropical cyclone	vortex
synoptic	tropical storm	warm front

SELF-TEST CONNECTION

PART A. Completion. *Write in the word or words that correctly complete the statement.*

1. _____ is the form of precipitation most often associated with severe thunderstorms.

2. _____ air masses tend to produce damp, cool weather. These conditions are common to the Pacific Northwest.

3. Tornadoes are an example of a _____ scale weather phenomenon.

4. Typhoons are more commonly called _____ in the United States.

5. Overrunning tends to occur along the gradually sloping surface of a _____ front.

6. When a cold front "catches" a warm front, an _____ forms.

7. Low, moisture-laden clouds are called _____ clouds.

8. A _____ is a narrow zone of severe weather that may precede a strong cold front.

9. The eye of a hurricane is surrounded by an _____, which is the zone of highest winds.

10. Thunderstorms often contain strong updrafts and _____, making them a hazard to aviators.

11. _____ is the process whereby a cyclone forms.

12. The coldest air masses to affect the United States are known as _____ air masses.

13. The strength of a tornado is rated by the _____.

14. The _____ that occurs within a hurricane's eye ensures fair to partly cloudy weather in the region.

15. Hot, dry air masses that form over the Desert Southwest are known as _____.

PART B. Multiple Choice. *Circle the letter of the item that correctly completes the statement.*

1. During the months of September and October, cool, dry air masses begin to advect southward from Canada to the central and eastern United States. These masses are best characterized as
   (a) continental polar
   (b) continental tropical
   (c) maritime polar
   (d) maritime tropical
   (e) arctic

2. An air mass that forms over the Gulf of Mexico and advects northward to bring warm, humid conditions to most locales east of the Continental Divide during the months of July and August is known as
   (a) continental polar
   (b) continental tropical
   (c) maritime polar
   (d) maritime tropical
   (e) arctic

3. Warm fronts have a gradually sloping surface that can extend for hundreds of kilometers. As a warm front approaches, the sequence of increasing clouds that ground-based observers will see is
   (a) cirrus, cumulus, stratus
   (b) cumulus, cirrus, stratus
   (c) cirrus, cirrostratus, stratus
   (d) cirrus, stratus, cumulus
   (e) stratus, cirrus, cumulus

4. Frontolysis generally occurs after a
   (a) cold front catches a stationary front, forming a warm front
   (b) warm front catches a cold front, forming an occluded front
   (c) warm front catches a cold front, forming a stationary front
   (d) cold front catches a warm front, forming an occluded front
   (e) stationary front turns into an occluded front

5. As a cyclone moves toward a region, the longest period of steady precipitation occurs ahead of the
   (a) warm front
   (b) cold front
   (c) occluded front
   (d) stationary front
   (e) none of the above

6. A squall line may be accompanied by
   (a) severe thunderstorms
   (b) frequent lightning
   (c) hail
   (d) tornadoes
   (e) all of the above

7. The best tool for studying and identifying tornadoes is
   (a) a satellite image
   (b) a radar image
   (c) a Doppler radar image
   (d) surface observations
   (e) long-term climatic data

8. The number of named storms that an average hurricane season produces in the Atlantic Ocean basin, including the Gulf of Mexico and the Carribean is
   (a) 9
   (b) 11
   (c) 13
   (d) 15
   (e) 17

9. As soon as a hurricane makes landfall,
   (a) winds begin to diminish
   (b) the rainfall ends
   (c) skies clear
   (d) the hurricane turns around and heads back to sea
   (e) wind velocity increases

10. As a warm front approaches, the most likely precipitation sequence is
    (a) rain may turn into sleet and then snow
    (b) snow may turn into sleet and then rain
    (c) snow may turn into rain and then sleet
    (d) sleet may turn into snow and then rain
    (e) rain may turn into sleet and then hail

PART C. Modified True/False. *If a statement is true, write "true" for your answer. If a statement is incorrect, change the <u>underlined</u> expression to one that will make the statement true.*

1. Cyclones and anticyclones are <u>global</u>-scale phenomena.

2. <u>Continental</u> tropical air masses are warm and moist.

3. Seattle, Washington, experiences <u>maritime polar</u> air masses through much of the year.

4. Cold fronts have <u>steeper</u> surfaces than warm fronts.

5. <u>Stratus</u> clouds eventually replace cirrus clouds as a warm front approaches.

6. Cumulonimbus clouds are most often associated with <u>stationary</u> fronts.

7. Lightning, hail, and tornadoes are all associated with <u>cirrus</u> clouds.

8. Cyclones are often entering the dissipation stage when an <u>occluded</u> front forms.

9. A <u>squall line</u> is a narrow zone of severe weather that can precede a cold front.

10. Thunderstorms tend to form in <u>stable</u> air.

CONNECTING TO CONCEPTS

1. Describe the changes that can occur to a continental tropical air mass that forms over the Desert Southwest and later drifts eastward toward a position over the southeastern states and then off the coast into the nearby Atlantic Ocean.

2. Explain the factors that contribute toward creating a dry, hot climate through the interior of southern California and much of Arizona.

3. Why is the precipitation along a cold front usually shorter in duration but more intense than the precipitation that precedes a warm front?

4. a. After Studying Figure 15.4, explain why snow will likely change to sleet and then rain before the passage of the warm front.

   b. Study Figure 15.11, and determine the wind direction that would most likely be observed at the surface both before and after the passage of the warm front.

5. Explain the series of events that result in precipitation when air is forced to rise.

6. If global warming becomes more pronounced, how might the balance between maritime tropical and continental polar air masses be changed? Also, since the winter storm track is often along the boundary between these two dissimilar air masses, how might precipitation patterns across the United States be affected during the winter?

7. Currently, New England depends greatly upon annual snowmelt for much of its water during the summer. If global warming becomes more pronounced, will New England benefit or suffer in this regard?

8. Base your answers to parts a–d on the diagram below, which represents a vertical cross section of a frontal system moving from west to east across the nation. Temperatures for six weather stations, labeled *A* through *G*, are shown on the diagram.

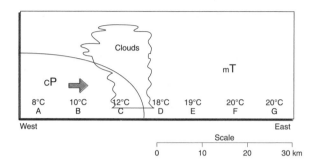

a. How will the cloud cover change at station *E* as the front approaches *and* after the front passes?
b. What barometric pressure change will occur as the front passes station *F*?
c. What change in temperature will occur after the front passes station *D*?
d. Describe the vertical movements of the air masses that cause clouds to form along the front.

9. Base your answers to parts a–e on the information and data provided in the narrative and table below.

   In August 1992, Hurricane Andrew, the most costly natural disaster in U.S. history, hit southern Florida. The data table below shows the locations and classifications of Hurricane Andrew on 7 days in August 1992.

DATA TABLE

Day	Latitude	Longitude	Storm Classification
August 18	13°N	46°W	Tropical storm
August 20	19°N	59°W	Tropical storm
August 22	25°N	66°W	Hurricane
August 24	25°N	78°W	Hurricane
August 26	28°N	90°W	Hurricane
August 27	32°N	91°W	Tropical storm
August 28	34°N	86°W	Tropical storm

a. On the hurricane tracking map provided, plot the location of Hurricane Andrew on each of the 7 days. Mark each location with an **X**, note the appropriate date on the map, and, finally, connect the **X**'s with a line to show the path of the storm from August 18 to 28.

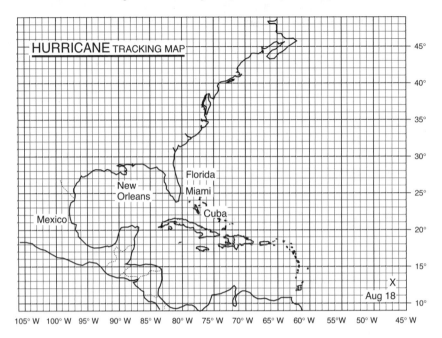

b. As Hurricane Andrew approached Miami, Florida, clouds increased and precipitation began to fall heavily. Describe how air pressure changed as the storm approached.

c. Between August 26 and 27, Andrew's winds diminished and the storm was downgraded from a hurricane to a tropical storm. What may have caused the wind velocity to decrease?

d. Describe at least one threat to human life as Hurricane Andrew made landfall in southern Florida.

e. What actions might a resident of southern Florida take to reduce his or her risk of injury when a hurricane warning is issued?

CONNECTING TO LIFE/JOB SKILLS

Weather forecasting, particularly during the threat of severe weather, is critical to the health and safety of the U.S. population. Despite advanced computer modeling, however, seasoned meteorologists often base their forecasts, at least in part, upon their knowledge of local weather patterns and past events in the region.

Today's meteorologist is a trained professional, usually possessing a master's degree or higher. The required coursework emphasizes calculus and the physical sciences, including thermodynamics and fluid dynamics.

ANSWERS
SELF-TEST CONNECTION

Part A

1. Hail	6. occlusion	11. Cyclogenesis
2. Maritime polar	7. nimbostratus	12. arctic
3. micro	8. squall line	13. Fujita intensity scale
4. hurricanes	9. eye wall	14. subsidence
5. warm	10. downdrafts	15. continental tropical

Part B

1. (a)	3. (c)	5. (a)	7. (c)	9. (a)
2. (d)	4. (d)	6. (e)	8. (b)	10. (b)

Part C

1. False; synoptic	5. True	8. True
2. False; maritime	6. False; cold	9. True
3. True	7. False; cumulonimbus	10. False; unstable
4. True		

CONNECTING TO CONCEPTS

1. A continental tropical air mass that drifts from the Desert Southwest to a position over the southeastern states near the Atlantic Ocean will pick up moisture from the Gulf of Mexico and the Atlantic Ocean. The air mass will, in short order, convert from continental tropical to maritime tropical.

2. Large areas of southern California and Arizona are located near the thirtieth-parallel zone of global subsidence. Furthermore, mountains located just to the west act to prevent Pacific moisture from reaching this region. When the two factors are combined, the result is a very hot, dry region.

3. Air rises more rapidly and vertically along a cold than along a warm front. The result is a geographically smaller, but more intense, region of stormy weather.

4. a. As a warm front approaches, snow often changes to sleet and rain as the cold air becomes more and more shallow. Notice in Figure 15.4 that the warm air is overrunning the cold air and, as the surface warm front is approached, the warm air is getting closer to the surface. When little cold air remains, snow will change to sleet and then rain as long as the temperature of the warmer air aloft is above freezing.

   b. When a warm front passes by, winds generally shift from the east or southeast to the south or southwest.

5. When a parcel of air is forced to rise, it expands (occupies greater volume), its temperature drops, and condensation occurs. Condensation can lead to precipitation.

6. Global warming will likely lead to a greater dominance of maritime tropical air masses over the United States, with continental polar air masses remaining farther north over northern Canada. If this happens, the winter storm track will shift farther north, taking the associated precipitation with it.

7. New England will likely suffer from the effects of global warming. Less winter snowfall will result in shortages of water in the spring and summer months.

8. a. Cumulus and cumulonimbus clouds will build at station *E* as the front approaches; then skies will clear after it passes.
    b. Air pressure will fall as the front approaches station *F* and then rise after the front passes.
    c. Temperatures fall after a cold front passes, so the temperature will be lower at station *D*.
    d. Cold air behind the front tries to "burrow" under the warm air ahead of the front. The warm air rises in response and thus creates cumulus or vertically developed clouds.

9. a.

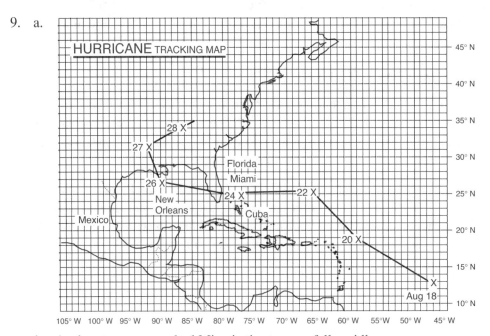

    b. As the storm approached Miami, air pressure fell rapidly.
    c. Andrew made landfall between August 26 and 27. When a hurricane makes landfall, wind speeds rapidly diminish.
    d. Threats to human life include high winds, flying debris, flooding, coastal storm surge, and tornadoes spawned by the hurricane.
    e. Many actions can be taken, starting with building a home designed to withstand the force of hurricane winds. Also, appropriate action should be taken when signaled by the National Hurricane Center, including evacuation if required. Taping and boarding windows can also minimize risk of injury and loss of property.

Climate and Climatic Zones

16

WHAT YOU WILL LEARN

This chapter focuses on climate and climatic patterns. In this chapter you will learn

- the difference between weather and climate;
- what climatic zones exist and factors that control climatic conditions;
- climatic patterns across our nation;
- how climate is changing over time and the possible role of humans in that change.

SECTIONS IN THIS CHAPTER

- Climate Defined
- Global Climatic Zones
- Climate Patterns Across the United States
- Climate Changes
- Internet Resources to Climatic Data and More
- Review Exercises for Chapter 16

Climate Defined

In this chapter, you will learn what climate is and what types of climates exist worldwide. Particular emphasis is placed upon climatic patterns and the reasons for the patterns found across the United States.

Climate is defined as the average and extremes of the long-term weather patterns that occur at a specific place or region. For example, New York City can best be described as a place that experiences cold, wet winters and warm, humid summers. To elaborate further, precipitation events (rain, sleet, or snow) during the winters often last for 12 hours or more; in comparison, brief showers and thunderstorms account for most of the rainfall experienced during the summer. In describing New York City's climate, both temperature and precipitation patterns have been addressed. Similarly, in describing the climate of a region, both of these parameters are generally included.

The following example will clarify the difference between the study of weather (meteorology) and the study of climate (climatology). The fact that New York City typically experiences a cold, wet climate during the winter season does not preclude the possibility that, on any given day during the winter, the weather can be warm and sunny. On any given day, the weather can diverge from the general climatic pattern.

GLOBAL CLIMATIC PATTERNS

Figure 16.1 depicts global climates. The climates illustrated can be organized into four general regions or zones ranging from the Equator to the poles. These climates are, in order from the Equator, tropical, subtropical, temperate, and polar. In general, as distance from the Equator increases, the climate becomes less and less tropical.

Tropical climates are characterized as being warm year-round. Often these locations experience ample rainfall in the form of brief, torrential showers on a daily basis.

Subtropical climates exhibit some variability from summer to winter. Summers are quite warm or hot. Depending upon other factors to be discussed later, they can be wet or dry. Winters are still quite mild, but some cooling during this season is apparent. Again depending upon other factors, a region that is dry in the summer may become wet in the winter, or vice-versa. The southernmost United States can be characterized as subtropical.

Most of the United States, as well as Europe and Asia, fall within the temperate midlatitude climate zone. *Temperate climates* often feature significant changes from summer to winter. In general, summers are warm and winters are cold. Exceptions exist, however, particularly along the west coasts of the United States and Europe. Reasons for these exceptions are discussed later in this chapter. As for precipitation, some areas are quite wet and others quite dry. This aspect of midlatitude climates is discussed more fully in the section headed "Climate Patterns Across the United States."

Polar climates are characteristic of regions closest to the poles, where temperatures are naturally quite low year-round. Precipitation tends to vary according to location, as is the case in the other climatic zones.

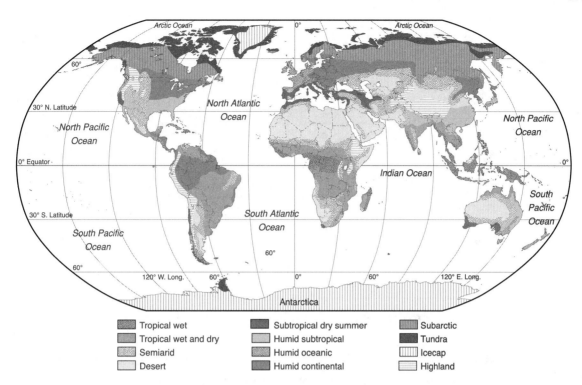

FIGURE 16.1 Global climates.

Legend:
- Tropical wet
- Tropical wet and dry
- Semiarid
- Desert
- Subtropical dry summer
- Humid subtropical
- Humid oceanic
- Humid continental
- Subarctic
- Tundra
- Icecap
- Highland

FACTORS AFFECTING CLIMATE

As the preceding discussion indicates, the basic factor that determines a region's climate is latitude. As discussed in Chapter 13, solar elevation varies according to latitude, and the most intense insolation (incoming solar radiation) is received annually near the Equator. This fact should and often does produce a pattern in which temperature gradually decreases from the Equator to the poles.

Many other factors, however, impact the rate at which this expected change in temperature occurs. These other factors that affect climate include proximity to global convection cell boundaries, as these tend to be locations where air is forced either to rise or to sink; proximity to bodies of water and ocean currents; and elevation and proximity to mountains.

IMPORTANCE OF VERTICAL MOTION

As stated in Chapter 13, the troposphere is the lowest layer of the atmosphere, where all weather occurs. The troposphere exhibits both horizontal and vertical air currents. Both types of motion can influence the kind of climate experienced by a particular region. Most important, however, vertical movements have a pronounced impact upon how much precipitation is experienced in the region.

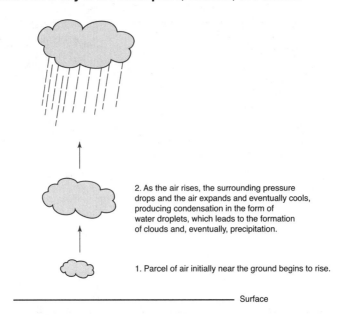

2. As the air rises, the surrounding pressure drops and the air expands and eventually cools, producing condensation in the form of water droplets, which leads to the formation of clouds and, eventually, precipitation.

1. Parcel of air initially near the ground begins to rise.

Surface

FIGURE 16.2 Upward movement of a parcel of air.

To understand the importance of vertical air currents, we will consider a parcel of air, initially near the ground, that begins to rise. This process is illustrated in Figure 16.2. For the moment, we will not concern ourselves with what causes the parcel to rise.

Instead, let's analyze what happens to our parcel of air near the ground as it begins to rise. Since pressure decreases as one moves higher in the atmosphere, the parcel of air will tend to expand. The expansion causes the temperature of the air within the parcel to drop because of a phenomenon known as *adiabatic cooling.*

Adiabatic cooling can be understood by viewing the parcel of air as being separate or isolated from the air surrounding it. In that case, as the parcel rises, the quantity of heat within it does not change; however, the parcel is growing larger (increasing in volume). Since our parcel has neither gained nor lost any heat (energy), the original quantity of heat is now distributed over a larger volume. The result is a lower average kinetic energy at any particular location within the parcel—in basic terms, the temperature of the air within the parcel drops.

The reduction in temperature will eventually produce condensation (assuming the presence of condensation nuclei, usually dust or other airborne particles), because, as the temperature of the parcel drops, the air can't hold as much water vapor. Here's a simple example of what happens. If a bottle of soda pop is removed from the refrigerator during the summer, a layer of condensed water vapor, or water droplets, often forms on the outside of the bottle in short order. This occurs as the layer of air near the chilled bottle is cooled, thus forcing the condensation of water vapor into water droplets, which subsequently collect on the bottle.

When condensation occurs in the atmosphere, clouds form; and, if the air continues to rise and enough moisture is present, precipitation is inevitable. In summary, we may conclude that the cooling associated with rising air can produce precipitation.

The opposite argument applies when subsidence (sinking air) occurs. Subsidence produces a net warming of our parcel of air, forcing any water droplets present to evaporate and resulting in clear skies over the region. Thus, where air is subsiding, little rainfall, if any, will occur.

Global Climatic Zones

THE INTERTROPICAL CONVERGENCE ZONE

Study Figure 16.3a to locate the *Intertropical Convergence Zone*, or ITCZ, discussed also in Chapter 13. This region, centered near the Equator, is so named because of the convergence of the trade winds, which meet after carrying air from latitudes 30°N and 30°S to the Equator. The convergence in this region produces a zone of rising air along the Equator. For reasons discussed in a preceding section, the inevitable result is considerable precipitation. Because of the temperatures along the Equator, virtually all of this precipitation falls in the form of rain (see Figure 16.3b).

Upward vertical movement of the atmosphere in this region is further enhanced by the strong tropical Sun, which warms the air, making it more buoyant. This rising motion acts to build towering cumulonimbus clouds. The inevitable result is thunderstorms that

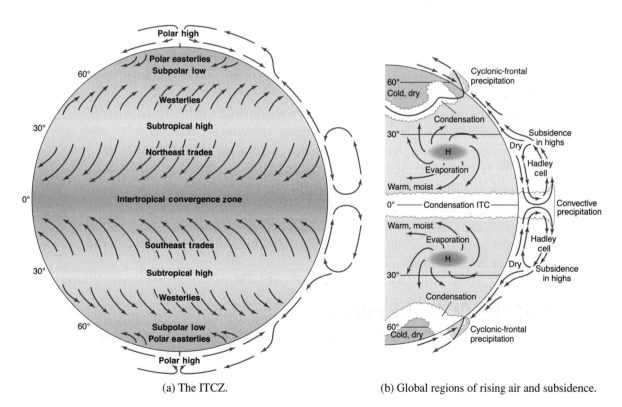

(a) The ITCZ. (b) Global regions of rising air and subsidence.

FIGURE 16.3 Global atmospheric circulation.

reach toward the top of the troposphere. In fact, the majority of the world's daily 25,000 thunderstorms occur within this region, producing torrential rainfalls.

THE HORSE LATITUDES

The *horse latitudes*, located in each hemisphere near 30° latitude, are zones of subsiding air (refer to Figure 16.3b). Air subsides along and near these latitudes as poleward flowing air cools and becomes denser. These subsidence zones should and often do produce regions where very dry conditions are observed. The result is a desert climate across much (but not all) of the global belts. Refer to Figure 16.1 to confirm that all the world's great deserts are located along or near the thirtieth parallels.

THE POLAR STORM TRACK WESTERLIES

The *polar storm track* (embedded within the Westerlies) is created by another convergence zone, produced by cold air returning toward the Equator after subsiding at the poles and by warm air traveling toward the poles after subsiding near the thirtieth parallels. On an "idealized earth" (refer to Figure 16.3a), this convergence zone is located near the sixtieth parallels. In reality, however, the boundary is quite mobile and can be found as far south as below the fortieth parallel and as far north (or south) as near the Arctic Circle (or Antarctic Circle). The position of the convergence zone varies according to the time of year, reaching its southernmost location during the winter and northernmost location in the summer. It is often easy to locate, particularly during the winter, as the jet stream, a belt of high-velocity winds that often divides the cold air toward the poles from warmer air toward the Equator, transports storms from west to east across the United States.

An important difference between this convergence zone and the ITCZ is the fact that warm air from lower latitudes is converging with cold air from the poles, as opposed to warm air converging from both directions along the ITCZ (refer to Figures 16.3a and 16.3b). Since warm air is less dense than cold air, the warm air tends to ride over (overrun) the cold air. This causes fronts to form, namely, boundaries between different air masses (the cold air to the north and the warm air to the south), resulting in much precipitation along the (polar) front. Since temperatures can be cold enough for snow at these latitudes, the precipitation that falls can make a region quite wet or white!

SUBSIDENCE AT THE POLES

It should come as no surprise that the North and South poles are quite cold; however, you may be surprised to learn that they are quite dry as well. The reason is the cold air, which is heavy and dense and subsides at these high latitudes. Nevertheless, although snowfall is not abundant, the year-round near to below freezing temperatures serve to retain the snow pack, which has accumulated for thousands of years in these regions.

Climate Patterns Across the United States

A THREE-CITY STUDY OF CLIMATIC DATA FROM SELECTED LOCATIONS

We begin our study of climatic patterns across the United States by analyzing the graphs of average monthly temperature and precipitation for three cities across the nation—New York, New York; Lincoln, Nebraska; and Eureka, California (see Figures 16.4 and 16.5).

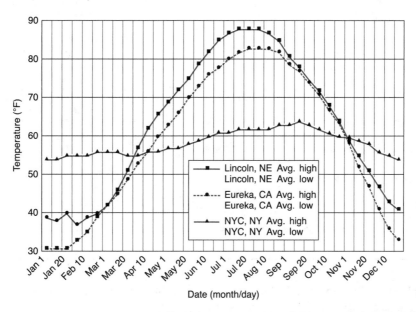

FIGURE 16.4 Average high temperatures for three selected cities.

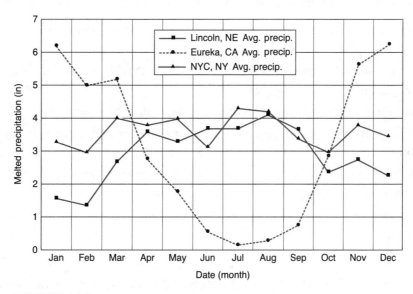

FIGURE 16.5 Average precipitation patterns for three selected cities.

Why were these three sites selected? All three cities are located near 41°N latitude and are situated at relatively low elevations above sea level. When latitude and elevation are eliminated as variables, other factors may be considered in addressing the significant climatic differences among these locations. Figure 16.6 shows the geographic locations of the three cities; note that one is situated on the West Coast, one along the East Coast, and one near the geographic center of our nation.

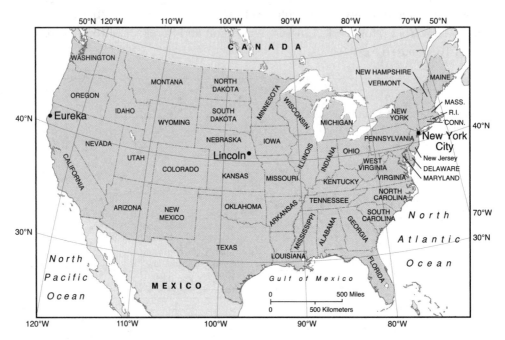

FIGURE 16.6 Map depicting the location of the three cities in our study. Note the similar latitudes of New York, NY; Lincoln, NE; and Eureka, CA.

Climatic Differences Among Eureka, CA; Lincoln, NE; and NY, NY

- New York City and Lincoln both experience large annual temperature ranges, with cold winters and warm summers.
- Eureka's annual temperature range is quite small with mild winters and cool summers.
- Lincoln has a greater annual temperature range than New York City.
- Throughout much of the year, New York City and Lincoln experience moderate amounts of precipitation each month (3–4 in. liquid equivalent per month).
- Precipitation in Lincoln drops off significantly during the autumn and winter months.
- Eureka experiences a very wet winter and a pronounced dry season during the summer.

To identify climatic differences among the three sites, refer to Figures 16.4 and 16.5. You are encouraged to construct your own list of observed differences and to compare it to the list below Figure 16.6.

From the observations reported in this list, it is evident that significant differences exist among the three locations despite their similar latitudes and elevations. A reasonable question to ask is, Why do these differences exist? For example, what keeps Eureka's temperatures so moderate throughout the year? Why does it have a pronounced dry season in the summer? Why do Lincoln's and New York City's temperature patterns vary so throughout the year?

We shall begin our study of U.S. climates along the West Coast and will then work our way eastward. Figure 16.7 summarizes the various types of climates found across the nation. It will be helpful to refer to this map as each of the climate zones is discussed in the following sections.

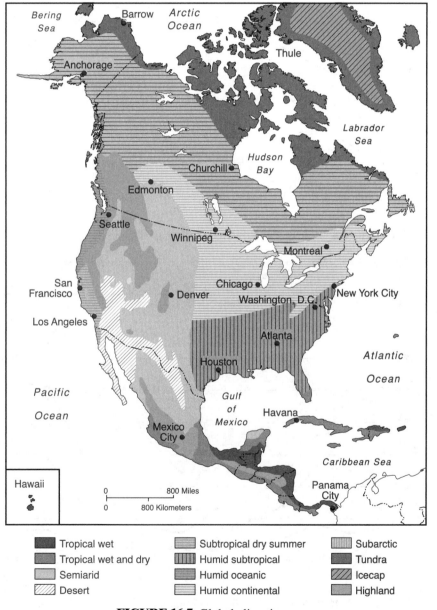

FIGURE 16.7 Global climatic zones.

AN ANALYSIS OF THE WESTERN SEABOARD

Figures 16.8 and 16.9 and Table 16.1 reveal very interesting patterns along the western seaboard of the United States. Using Table 16.1, we can see that the four sites selected for this analysis—Imperial Beach, California; Eureka, California; Astoria, Oregon; and Annette, Alaska—range from southern California to coastal Alaska. The distance

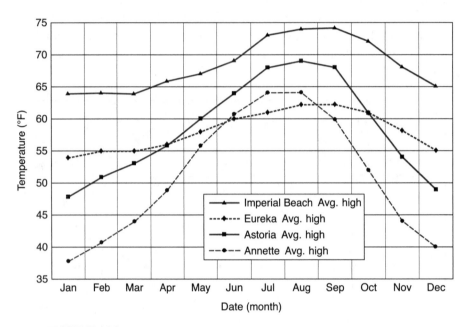

FIGURE 16.8 Average monthly high temperatures along the West Coast.

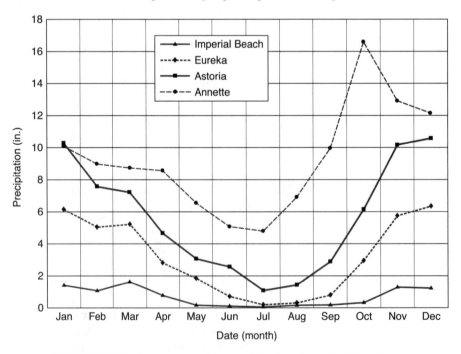

FIGURE 16.9 Average monthly precipitation along the West Coast.

TABLE 16.1
LATITUDE AND ELEVATION DATA FOR FOUR WEST COAST LOCATIONS

Location	Imperial Beach, CA	Eureka, CA	Astoria, OR	Annette, AK
Elevation	18 m	8 m	2 m	33 m
Latitude	32°34′ N	40°48′ N	46°09′ N	55°03′ N

from the southernmost to northernmost location is over 900 miles (1500 km). Despite this distance, temperature patterns (see Figure 16.8) all along the West Coast are quite similar in each season, with the possible exception of winter. For example, a typical day is no more than 15°F (9°C) cooler in coastal Alaska than in coastal southern California. The only season in which Alaska appears to become significantly cooler than its neighbors to the south is winter.

A similar study conducted over the same distance along the eastern seaboard would not reveal such a homogeneous temperature pattern. The reason for the lack of temperature variation from coastal Alaska to southern California is evident in Figure 16.3a, which illustrates the movement of the prevailing winds in the midlatitudes. Note that the prevailing winds at this latitude, known as the westerlies, blow from west to east. This pattern causes a persistent, prevailing on-shore wind along the entire western seaboard. Since ocean temperatures do not vary much throughout the year, a relatively cool ocean keeps coastal locations reasonably cool during the summer. During the winter, the ocean keeps these same locations from becoming extremely cold, especially when compared to areas a bit farther inland.

The East Coast, in comparison, is influenced by the same westerlies; this time, however, they are carrying air from the continent to the coast. Since the mainland can get quite hot during the summer and cold during the winter, climatic patterns along the East Coast are quite different from those experienced along the West Coast. The Atlantic Ocean, located immediately to the east of the eastern seaboard, thus has a much smaller impact upon climate along the East Coast than the Pacific Ocean has along the West Coast. A more detailed description of East Coast climates appears in the section headed "The Northeast."

Turning our attention to precipitation patterns along the West Coast, we can see that they vary considerably from Alaska to southern California. Figure 16.9 shows that all locations receive more precipitation in the winter than in the summer. The more southern locations overall, however, receive much less precipitation than the northern ones, and the summer dry season is much longer and more severe.

Climatologists have acknowledged the fundamental differences mentioned above between the northern and southern western seaboards and have identified two types of climates along the West Coast: the *humid oceanic* (sometimes called the marine West Coast) and the *subtropical dry* (refer to Figure 16.7).

Cool summers, mild winters, and adequate year-round precipitation characterize the humid oceanic climate. Autumn and winter in these regions can be quite wet as the jet stream, or midlatitude storm track, is strengthened and drops southward from the

Arctic. This type of climate occurs from approximately 40°N to about 55°N, or from northernmost California through Oregon, Washington, and coastal British Columbia, Canada, to the Alaskan coastline. Cities such as Portland and Seattle are famous for experiencing months on end of persistent cloud cover and periods of rainfall during the rainy season.

Farther south, in central and southern California (approx. 30°–40°N), the subtropical dry climate produces much less annual precipitation than is experienced in the humid oceanic climate. In this region, winter rains (and nearby mountain snows) are depended upon for summer water supplies, as summers tend to be extremely dry. Here, too, the jet stream helps to carry storms onshore during the winter to produce most of the precipitation experienced annually. In El Niño years, this pattern is modified in that storms are carried farther south more frequently; hence rainfall (and mountain snowfall) increases significantly.

Another pattern is illustrated by studying the precipitation trend between Eureka and Imperial Beach, both in California. Imperial Beach shows that, as one approaches the thirtieth parallel, annual precipitation becomes quite light. This is consistent, as discussed earlier in this chapter, with the global belt of subsidence observed along this latitude.

Temperatures in the subtropical dry zone are milder than those in the humid oceanic zone. In these regions, too, ocean winds keep annual temperature ranges to a minimum, particularly when compared to locations farther inland.

THE MOUNTAIN RAIN SHADOW EFFECT

To understand the types of climates found farther inland, we need to explore a phenomenon known as the *mountain rain shadow effect*. The cities used as examples in our study are Astoria, Oregon, located on the western or windward side of the Cascade Mountains, and Yakima, Washington, located on the eastern or leeward side of the Cascades (see Figure 16.10). Actually, any two cities, in the Far West with similar latitudes could have been chosen as long as one is on the windward side of the mountains and the other on the leeward side.

Figures 16.11 and 16.12 reveal that significant changes occur as we cross the mountains. Astoria, located near the Pacific Coast in the humid oceanic climate zone, experiences (as expected) moderate temperatures throughout the year, with the classic wet winter pattern discussed earlier. In contrast, Yakima has a much greater annual temperature range with cold winters and hot summers. Perhaps the greatest difference between the two is seen in the precipitation pattern; Yakima receives much less precipitation throughout the year.

To best understand why this occurs, study Figure 16.13. This diagram shows that, as moist Pacific air rises over the Cascade Mountains on the windward side, it cools and drops its moisture in the form of rain and snow. This is similar to the rising motion discussed in conjunction with global convergence zones. In this case, the mountain acts as a physical lift, causing air to rise with the same net result as that observed in global convergence zones.

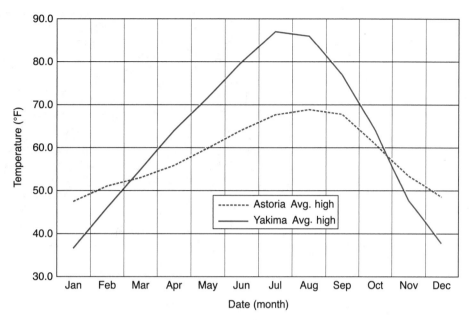

FIGURE 16.10 Average high temperatures for Yakima, WA, and Astoria, OR.

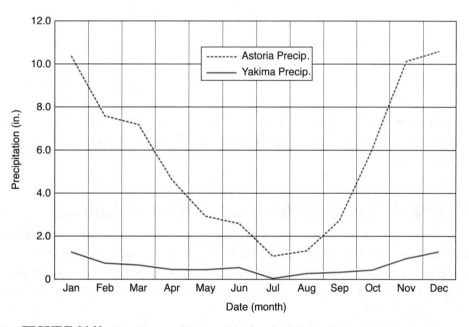

FIGURE 16.11 Average monthly precipitation for Yakima, WA, and Astoria, OR.

When the air reaches the leeward side of the mountains, it has lost its moisture. Subsiding air, particularly in the summer, can produce very warm conditions east of the mountains. When this warmth is combined with a lack of rainfall, a near-desert condition occurs.

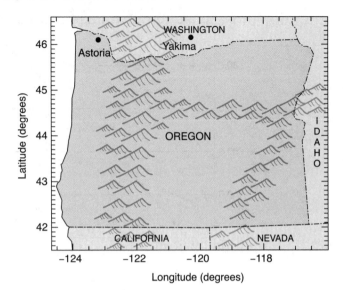

FIGURE 16.12 Map of the Cascades separating Astoria and Yakima.

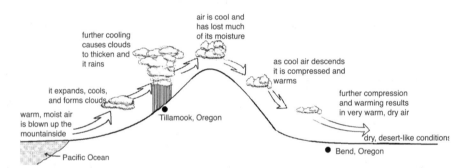

FIGURE 16.13 The mountain rain shadow effect. Source: *Let's Review: Earth Science—The Physical Setting*, 2nd Ed., Edward J. Denecke, Jr., Barron's Educational Series, Inc., 2002.

THE DESERT SOUTHWEST AND SEMIARID REGIONS

The U.S. West Coast is marked by significant north-south mountain ranges that effectively create a mountain rain shadow effect in all interior regions to the lee (east) of the mountains. Prevailing winds from the west, combined with a lack of a major moisture supply east of the mountains, ensure dry conditions from the interior of the Far West right into the Great Plains.

The most extreme of these dry conditions is experienced in the interior of southern California, much of Arizona, New Mexico, and parts of neighboring states. This region is known collectively as the *Desert Southwest*. A *desert* is a region where evaporation exceeds precipitation annually. In addition to the mountain rain shadow effect, this region's proximity to the global subsidence zone (near the thirtieth parallel) results in very little annual rainfall. The polar storm track rarely extends this far south; and since little moisture reaches over the coastal mountain ranges, precipitation in the desert occurs only in the form of brief, heavy showers. These showers are caused by a mild, monsoonal flow (reversal of the normal wind direction) that carries moisture into normally dry regions.

The dry conditions in the Desert Southwest allow for wide temperature swings daily as well as seasonally. Water vapor, present in more humid climates, limits the extent of the temperature range. For example, in a humid location such as New York City, if an early morning low temperature of 38°F occurs, the high temperature later that day will probably remain in the 50s. In contrast, Phoenix, Arizona, located in the desert with little water vapor in the air, often warms to the low 80s on a day with a similar morning low temperature. Desert regions have been known to record some of the highest temperatures ever observed on Earth. In southern Arizona, for example, average high temperatures regularly exceed 100°F (38°C) throughout the summer.

Farther to the north, through the interior of eastern Oregon and Washington (the interior Northwest), as well as to the east in the Great Plains, *semiarid* conditions prevail. Semiarid climates are best described as "not quite as hot or dry as desert conditions."

One such *semiarid region*, the interior Northwest, is far enough north to receive moisture from storms carried by the jet stream and its storm track, particularly during the winter. This factor, combined with lower temperatures than those farther to the south, in the Desert Southwest, allows this region to be considered to be semiarid.

The Great Plains region, once called the "Great American Desert," is also considered to be semiarid. This is now the region where much of our nation's (if not the world's) wheat is grown. Situated in the rain shadow of the Rocky Mountains, this region receives most of its moisture from the Gulf of Mexico, which accounts for the rain and snow that fall over much of the eastern half of our nation. Since the prevailing winds blow from the west, or more specifically from the southwest, however, the Great Plains region experiences only a small fraction of the moisture that regions to the east receive. In fact, the Great Plains have a history of receiving just enough precipitation to support the crops that are grown there! In the past 100 years or so, significant droughts have occurred in sites where conditions appear to have been much more desertlike than they are today.

MOUNTAIN CLIMATES

In regions with *mountain* (highland) climates, cool to cold conditions are experienced year-round. These conditions are created by the higher elevations, as air tends to cool about 3°C (5°F) for every 300 meters (1000 ft) of increase in height. Since air is forced to rise over mountains, rainfall and snowfall totals range from adequate to abundant. The mountain ranges in the Far West, such as the Sierra Nevada and the Cascades, get the full brunt (or benefit) of moisture from the Pacific. Winter snowfalls in this region frequently produce spectacular accumulations, often measured in feet! Farther east, the Wasatch Range in Utah and the Rocky Mountains receive progressively less Pacific moisture.

THE MIDWEST

The Midwest and neighboring regions to the east experience a *humid continental* climate, characterized by adequate monthly precipitation throughout the year, thanks to the Gulf of Mexico. The Gulf provides moisture to storm systems carried by the jet stream, producing extended periods of rain or snow in the winter, and periodic showers and thunderstorms in the summer. Winters in this region are generally cold and severe, and summers can be hot and humid.

THE GREAT LAKES EFFECT

The *Great Lakes effect* is an interesting phenomenon that affects specific regions in the humid continental climate zone. Areas to the lee (east and south) of the Great Lakes experience this effect frequently during the late autumn and early winter. The Great Lakes effect is produced by cold, dry air traveling southeastward from Canada that reaches the leeward side of the lakes after crossing over still relatively warm waters. These waters contribute a large amount of moisture to the cold air, and when the air is lifted, even slightly, over parts of Michigan, Ohio, Pennsylvania, and New York, copious amounts of snow can result (see Figure 16.14). Western, New York, located in a prime spot for this to occur, has received upward of 2 m (80 inches) of snow during a single "lake effect snow event." It is important to understand that the Great Lakes effect is a localized phenomenon and typically affects only the areas illustrated in Figure 16.15.

THE NORTHEAST

Climate maps often identify regions north of the fortieth parallel in the northeastern states as being within the humid continental climate zone. Identifying this entire region as humid continental, however, does not tell the whole story. The Appalachian Mountains, although not as impressive a mountain range as those in the West, can and do

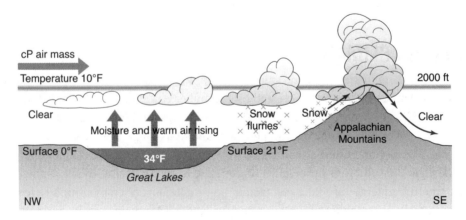

FIGURE 16.14 The production of lake effect snows.

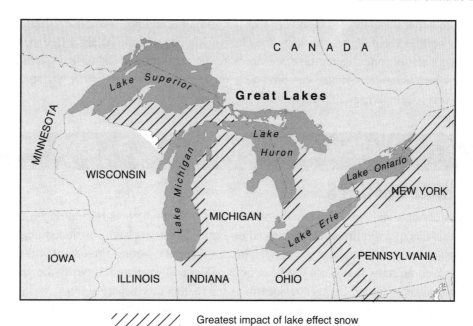

/////// Greatest impact of lake effect snow

FIGURE 16.15 Where lake effect snows occur. Cold, dry polar air, loaded with moisture from the Great Lakes, is given a slight "lift" by the topography (the hills) just east and south of the lakes. The result is lots of snow!

modify the climate experienced in the region. Residents living in or near the Appalachian Mountains, which stretch from the interior of Maine to northern Georgia, experience a cooler and wetter climate than is typical of locations immediately to the east or west.

The Appalachians also influence the climate of the regions east of the mountains. These regions, which include the coastal Northeast and coastal Mid-Atlantic, experience what is best described as a *modified continental* climate. Here, the Atlantic Ocean serves as the modifier, supplying additional moisture for this region, as well as keeping coastal areas from getting as cold during the winter or warm during the summer as those farther inland. This effect is muted, however, when compared with the situation on the West Coast, because prevailing winds act to minimize the impact of the Atlantic Ocean. The Appalachians often act as the dividing line between areas that experience some effects from the Atlantic and those that do not.

One of the local winds common east of the Appalachians is the summer "sea breeze." Cool ocean air flows inland for distances of up to 50 miles (although more commonly, much less) when the air over the mainland heats up and becomes more buoyant. This buoyant air rises, allowing the cooler, denser "ocean air" to flow inland, thus modifying the climate across much of the coastal plain.

THE SUBTROPICAL SOUTHEAST

The southeastern states generally fall within the *subtropical* climate zone. Subtropical climates are characterized by winters that are shorter and less severe than those in the humid continental zone. Summers are warmer and more humid than those experienced farther north. Precipitation, thanks to the Gulf of Mexico and the Atlantic Ocean, is still adequate, despite this region's proximity to the thirtieth-parallel global subsidence zone. An occasional hurricane can provide a significant percentage of the late summer—early autumn rainfall total in these southwestern states.

Climate Changes

NATURAL CHANGES

Studies of our planet, including studying ice cores and marsh cores for pollen grains, yield clues as to past climates experienced on Earth. Scientists have concluded that climate variation is a natural and continuous process. There is strong agreement within the scientific community that Earth has been both cooler and warmer at times in the past than it is today. Climate changes result from such factors as volcanic activity, variations in solar energy, and even the distribution of landmasses on our planet as continents have drifted over the eons. Volcanoes, for example, can release gases and particles that block incoming solar energy, causing global cooling that can last for years.

HUMAN IMPACT ON GLOBAL CLIMATE

In recent years, not only scientists and political leaders, but also average citizens, have begun to ask whether our species has impacted or has the potential to impact global climate. Those who believe that human activities can affect Earth's climate point to the increasing levels of carbon dioxide and other greenhouse gases in the atmosphere. Greenhouse gases have the ability to trap heat, and thus an increase in the atmosphere should make our planet warmer.

Scientists are still debating the degree to which global warming, observed in the past century or so, can be attributed to human activities. The most dramatic evidence of this warming is the increased rate at which glaciers are melting worldwide. Most climatologists who are studying this issue agree that the first signs of global warming will be increased variability in weather events. One has to wonder whether the unusually strong winter storms that occurred in recent years in the eastern states and the superhurricanes that formed in various locations worldwide during this same time period were the first signs of an ominous change in Earth's climate.

The Intergovernmental Panel on Climate Change (IPCC), a scientific intergovernmental body established in 1988 by the World Meteorological Organization and the United Nations Environmental Programme (both organizations of the United Nations), released its fourth assessment report on climate change in 2007. This report, assembled by some of the world's leading scientists, found the following: The climate system is warming; much of the warming is caused by anthropogenic (human) greenhouse gas contributions; the warming and subsequent sea level rise will continue even if greenhouse gas emissions are stabilized (as of the document's writing); the probability that this is caused by natural processes is less than 5 percent; world temperatures may rise between 1.1 and 6.4°C by the end of the twenty-first century. Agreement on these findings was not universal as of March 2009; more than 700 scientists, including some who wrote the IPCC report, had expressed their skepticism about them. The issue has become as much of a political debate as a matter of scientific research. Dissenting scientists point to peer-reviewed studies indicating that global warming has stopped or will stop by 2015 and that global temperatures are controlled to a greater degree than previously thought by solar activity. As computer models continue to become more sophisticated, their predictive power will improve. This should lead toward a true consensus in the coming years. In the interim, it is certainly safe to say that taking steps toward reducing the use of (and our dependence upon) fossil fuels will reduce the greenhouse gas emissions that form the core of concern within the IPCC report.

You can read the IPCC report at: *www.ipcc.ch/*.

You can read a report about the dissenting voices here: *http://epw.senate.gov/public/index.cfm?FuseAction=Minority.Blogs&ContentRecord_id=2674e64f-802a-23ad-490b-bd9faf4dcdb7*.

Global climate change is not limited to global warming. Acid rain deposition, ground-level ozone pollution, and stratospheric ozone depletion also threaten to affect many life-forms and their habitats. The acid rain problem stems from various gas emissions that combine with water vapor in the atmosphere to produce nitric and sulfuric acids. These strong acids are capable of increasing the acidity of surface waters and soils. Rain is naturally acidic, and carbon dioxide reacts with water to produce carbonic acid, a naturally occurring weak acid. Slight acidity is beneficial to many ecosystems, as aquatic life and plant life require mildly acidic conditions to thrive. However, too much of a good thing is not an improvement!

Higher acidity levels in soil and water negatively affect many systems. Aquatic life is impacted, and often entire ecosystems are destroyed. The extent of the damage depends upon the bedrock underlying the region; lakes bounded by granite are most affected. A lake whose pH (a measure of acidity) drops to levels near or below 5 is considered to be "dead"; little, if any, life can survive in its acidic waters. This problem is particularly acute in the northeastern United States and in southeastern Canada.

Acidified soil is an environmental problem also as minerals, particularly heavy metals found within the soil, are more easily dissolved and removed. Here, the effect is twofold: The soil is now low in the minerals necessary to support the various plants that have grown in that soil, and potentially dangerous heavy metals that would otherwise be limited in their ability to harm the environment are released into waters. Acidified soil tends to run off into lakes and streams, thus increasing their acidity levels.

The ozone problem has roots both in the troposphere and in the stratosphere. In the troposphere, ground-level ozone builds up in and near urban regions. A complex series of chemical reactions, starting with car exhausts and industrial emissions, results in the production of ozone near the surface. Ground-level ozone is a health hazard and is associated with respiratory distress for residents in the regions. The areas most affected include the entire northeastern megalopolis from north of Boston to south of Washington, D.C., the Houston metropolitan region, Denver, Phoenix, and most of southern California, particularly in and near Los Angeles.

Stratospheric ozone, in contrast, is potentially a global problem. Industrial emissions, largely of a family of gases known collectively as chlorofluorocarbons, migrate toward the stratosphere, where they act to break down the protective ozone layer. Stratospheric ozone absorbs much of the incoming high-energy radiation from the sun and therefore protects all life on Earth's surface. Without the protection afforded by the stratospheric ozone 30 kilometers above us, life as we know it would not be possible on this planet.

> **REMEMBER**
> Earth has a history of continuous and sometimes dramatic climate change.

In summary, each of these environmental changes has the potential to impact each of Earth's major systems in a variety of ways, many of which cannot be anticipated or predicted. It is these unforeseen effects that present the greatest concern for environmental scientists.

Related Internet Resources for Great Images and Information

www.cdc.noaa.gov/USclimate/

"U.S. Interactive Climate Pages" enables the creation of maps of monthly U.S. temperature and precipitation from 1895 to the present.

www.ncdc.noaa.gov/

The National Climatic Data Center is described as "the world's largest archive of weather data."

http://asp.usatoday.com/weather/weatherfront.aspx

USA Today has a very complete Web site that covers current weather, weather topics, and climate information.

www.wmo.ch/indexflash.html

The World Meteorological Organization has an extensive section on climate and weather.

REVIEW EXERCISES FOR CHAPTER 16

WORD-STUDY CONNECTION

adiabatic cooling
air pressure
Arctic Circle
climate
climatology
condensation
convection
desert
Equator
evaporation
front
Great Lakes effect
greenhouse gases
homogeneous
horse latitudes

humid continental
humid oceanic
insolation
ITCZ
jet stream
kinetic energy
latitude
leeward
marine West Coast
modified continental
mountain climate
mountain rain
 shadow effect
polar
precipitation

semiarid
subsidence
subtropical
subtropical dry
temperate
temperature
thirtieth parallel
trade winds
tropical
troposphere
water vapor
weather
westerlies
windward

SELF-TEST CONNECTION

PART A. Completion. *Write in the word or words that correctly complete the sentence.*

1. The science that studies long-term weather patterns is called
   _____.

2. _____ is moisture that falls to the ground in the form of rain
   or snow.

3. A climate zone where winters are cold, summers are warm, and adequate precipitation falls year-round is called _____.

4. The lowest layer of the atmosphere is the _____.

5. A zone of converging air along the Equator is called the _____.

6. The belt of winds that carries the polar storm track is called the
   _____.

7. An acronym for incoming solar radiation is _____.

8. Much of the United States is located within the _____ climate zone.

9. Locations above the Arctic Circle experience a _____ climate.

10. The western side of a mountain range in the midlatitudes is known as the _____ side.

11. This phenomenon, known as the _____, causes regions to the leeward side of a mountain range to be quite dry.

12. The term _____ describes a location's distance from the Equator.

13. The process by which water droplets turn into water vapor is called _____.

14. _____ may be responsible for global warming.

15. The subtropical region where global subsidence is occurring is the _____.

PART B. Multiple Choice. Circle the letter of the item that correctly completes the statement.

1. Condensation is the process whereby
   (a) water droplets change phase to water vapor
   (b) water vapor changes phase to water droplets
   (c) ice changes phase to water
   (d) water changes phase to ice

2. The global winds that blow from east to west across tropical latitudes from the Equator to approximately 30° latitude are the
   (a) polar easterlies
   (b) westerlies
   (c) trade winds
   (d) horse latitudes

3. A zone of convergence located along the Equator is called
   (a) the horse latitudes
   (b) the jet stream
   (c) subsidence
   (d) the ITCZ

4. A climatic zone characterized by mild winters, cool summers, and very wet winters is the
   (a) subtropical
   (b) modified continental
   (c) marine West Coast (humid oceanic)
   (d) desert

5. The climatic zone characterized as a region where evaporation exceeds precipitation is the
   (a) subtropical
   (b) modified continental
   (c) marine West Coast (humid oceanic)
   (d) desert

6. The climatic zone located along the immediate Atlantic coast, east of the Appalachians, is the
   (a) subtropical
   (b) modified continental
   (c) marine West Coast (humid oceanic)
   (d) desert

7. The windward side of a mountain often receives
   (a) more precipitation than the leeward side
   (b) less precipitation than the leeward side
   (c) the same precipitation as the leeward side
   (d) No consistent pattern has been found.

8. The phenomenon that can cause heavy snows to the lee of the Great Lakes is
   (a) the jet stream
   (b) adiabatic cooling
   (c) insolation
   (d) the Great Lakes effect

9. A semiarid climate is similar to, but not quite as dry as
   (a) a desert climate
   (b) a subtropical dry climate
   (c) a humid oceanic climate
   (d) a humid continental climate

10. A humid continental climate is similar to a _____ climate; however, winter and summer temperatures in the humid continental zone are more extreme.
    (a) modified continental
    (b) humid oceanic
    (c) semiarid
    (d) desert

PART C. Modified True/False. *If a statement is true, write "true" for your answer. If a statement is incorrect, change the <u>underlined</u> expression to one that will make the statement true.*

1. <u>Latitude</u> is a measurement that indicates how far north or south of the Equator a place is located.

2. <u>Polar</u> climates are characterized by warm weather year-round.

3. The westerlies are a belt of global winds that blow from <u>east to west</u> across the midlatitudes.

4. The horse latitudes are a zone of <u>rising</u> air that produces very dry conditions.

5. Much of the United States is located within the <u>temperate</u> climate zone.

6. When water vapor changes phase to water droplets, this process is known as <u>evaporation</u>.

7. Adiabatic cooling occurs when air <u>rises</u>.

8. Climate is the study of <u>short-term</u> weather patterns.

9. The mountain rain shadow effect causes areas east of the Cascade Mountains to experience <u>large</u> annual precipitation totals.

10. The <u>troposphere</u> is the lowest layer of the atmosphere.

CONNECTING TO CONCEPTS

1. What is the difference between a humid continental climate and a modified continental climate?

2. What factors are responsible for creating desert conditions in the southwestern United States?

3. Why are climatologists concerned about human influence on world climates?

4. What is the most significant difference between the humid oceanic and subtropical dry climate zones?

5. What conditions must be present for lake effect snows to fall?

6. Base your answers to parts a–d below upon the information provided and your knowledge of Earth science.

The climate of a region is determined by many factors, including elevation, latitude, and proximity to a large body of water. The effects of these variables upon climate can be shown on the following graph grids with the axes labeled as indicated below.

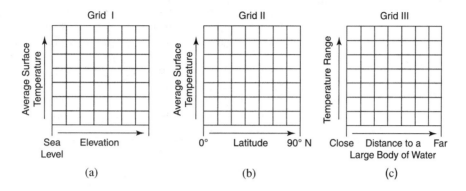

a. On grid I, draw a line that illustrates a relationship between elevation and average surface temperature.

b. On grid II, draw a line that illustrates a relationship between latitude and average surface temperature.

c. On grid III, draw a line that illustrates a relationship between distance to a large body of water and annual temperature range. (Assume that the wind is blowing from the water toward the point on land being considered. Assume also that the body of water does not freeze at any time throughout the year.)

d. The climate at most locations near the Equator is warm and humid. Explain why this is so.

7. Base your answers to parts a–d on the chart and map provided below. The chart gives the average high and low temperatures by month for four cities across the nation. The map shows the approximate locations of these four cities.

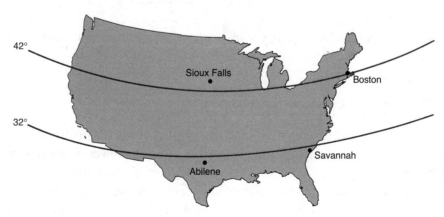

	Boston, MA		Sioux Falls, SD		Savannah, GA		Abilene, TX	
	High	Low	High	Low	High	Low	High	Low
Jan.	35.7	21.6	24.3	3.3	59.7	38.1	54.8	30.8
Feb.	37.5	23.0	29.6	9.7	62.4	41.1	59.7	35.1
Mar.	45.8	31.3	42.3	22.6	70.1	48.3	68.9	43.3
Apr.	55.9	40.2	59.0	34.8	77.5	54.5	77.8	52.9
May	66.6	49.8	70.7	45.9	84.0	62.9	84.4	61.1
June	76.3	59.1	80.5	56.1	88.8	69.2	91.4	68.9
July	81.8	65.1	86.3	62.3	91.1	72.4	95.2	72.7
Aug.	79.8	64.0	83.3	59.4	89.7	72.2	94.5	71.7
Sept.	72.8	56.8	73.1	48.7	85.2	67.8	86.7	65.3
Oct.	62.7	46.9	61.2	36.0	77.5	56.9	77.9	54.8
Nov.	52.2	38.3	43.4	22.6	70.0	48.1	66.3	43.4
Dec.	40.4	26.7	28.0	8.6	62.3	41.0	57.0	33.9
Avg. Yearly Precipitation (in)	41.51		23.86		49.22		24.40	

a.   Complete the table below.

	Boston	Sioux Falls	Savannah	Abilene
Highest summer temperature				
Lowest winter temperature				
Temperature range				

b.   On the basis of the map provided and the table you just completed, state the relationship between temperature range and distance from a large body of water.

c.   As the map indicates, Boston and Sioux Falls are located at approximately the same latitude. As a result, they receive roughly equal intensities and durations of insolation. Study the temperature pattern for each city, and explain why the patterns are slightly different.

d.   Why might Abilene experience higher temperatures year-round than Sioux Falls?

CONNECTING TO LIFE/JOB SKILLS

There is a growing need for climatologists to study climate and climate change. Universities and government research projects support many of these research positions. Advanced degrees in climatology, meteorology, and atmospheric science are often required to obtain leadership positions. Much of the research is done with advanced computer models, requiring a high degree of computer literacy. Most climatologists find their work to be rewarding and dynamic as cutting-edge ideas and issues are explored.

ANSWERS
SELF-TEST CONNECTION

Part A

1. climatology
2. Precipitation
3. humid continental
4. troposphere
5. ITCZ
6. jet stream
7. INSOLATION
8. temperate
9. polar
10. windward
11. mountain rain shadow effect
12. latitude
13. evaporation
14. Greenhouse gases
15. horse latitudes or thirtieth parallel

Part B

1. **(b)**
2. **(c)**
3. **(d)**
4. **(c)**
5. **(d)**
6. **(b)**
7. **(a)**
8. **(d)**
9. **(a)**
10. **(a)**

Part C

1. True
2. False; Tropical
3. False; west to east
4. False; subsiding or sinking
5. True
6. False; condensation
7. True
8. False; long-term
9. False; small
10. True

CONNECTING TO CONCEPTS

1. *Humid continental* is the term used to describe the climate across much of the Midwest and the northeastern United States. Summers are warm (or hot) and humid, and winters are cold and stormy.

   The only exceptions to this pattern are the locations sandwiched between the Atlantic Ocean and the Appalachian Mountains. Residents near the ocean are familiar with its impact upon the local climate. The ocean tends to prevent winters from being quite as cold as they are farther inland, and summers from being quite as warm. Hence, the term *modified continental* best describes this type of climate.

2. Desert conditions exist in the southwestern United States as a result of two major factors. First, this region is located near the thirtieth parallel, or zone of global subsidence. Second, the prevailing winds, which carry moisture inland from the Pacific Ocean, are largely blocked by impressive mountain ranges from precipitating that moisture farther inland.

3. Humans are the first species capable of influencing global climates. Any human-induced impact upon the climate system will almost certainly produce unintended and unwanted changes. For example, making the world a little warmer is not a change that will benefit all (or even most) regions across the globe.

4. The most significant difference in these climates is the amount of rainfall. As a whole, subtropical dry climates receive less precipitation than humid oceanic climates. Also, subtropical dry climates are, overall, somewhat milder than humid oceanic climates.

5. Lake effect snows occur when cold air is transported across the Great Lakes. Snowfall totals can be particularly impressive early in the winter season when lake-water temperatures are still relatively mild but air temperatures are quite cold.

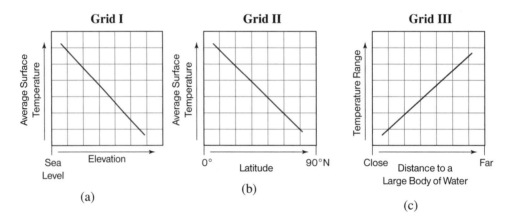

6. d. Equatorial regions can attribute their warmth and humidity to the fact that solar elevations are quite high year-round. High solar elevations contribute to strong surface heating. Equatorial waters are generally quite warm. Warm water has a great propensity to evaporate, thus raising the humidity in the lower atmosphere.

7. a.

	Boston	Sioux Falls	Savannah	Abilene
Highest summer temperature	81.8	86.3	91.1	95.2
Lowest winter temperature	21.6	3.3	38.1	30.8
Temperature range	60.2	83.0	53.0	64.4

b. The closer to a body of water, the smaller the annual temperature range.

c. Boston is a coastal city, and its climate is heavily influenced by the Atlantic Ocean. Although both Boston and Sioux Falls experience cold winters, conditions in Boston are best described as "not quite as cold." A similar moderating effect, again because of the Atlantic Ocean, occurs during the summer.

d. Abilene is located at a more southerly latitude than Sioux Falls. Therefore, insolation in Abilene is more direct year-round, causing temperatures to be warmer throughout the year.

Practice Exam

1. Ⓐ Ⓑ Ⓒ Ⓓ　　21. Ⓐ Ⓑ Ⓒ Ⓓ　　41. Ⓐ Ⓑ Ⓒ Ⓓ　　61. Ⓐ Ⓑ Ⓒ Ⓓ　　81. Ⓐ Ⓑ Ⓒ Ⓓ

2. Ⓐ Ⓑ Ⓒ Ⓓ　　22. Ⓐ Ⓑ Ⓒ Ⓓ　　42. Ⓐ Ⓑ Ⓒ Ⓓ　　62. Ⓐ Ⓑ Ⓒ Ⓓ　　82. Ⓐ Ⓑ Ⓒ Ⓓ

3. Ⓐ Ⓑ Ⓒ Ⓓ　　23. Ⓐ Ⓑ Ⓒ Ⓓ　　43. Ⓐ Ⓑ Ⓒ Ⓓ　　63. Ⓐ Ⓑ Ⓒ Ⓓ　　83. Ⓐ Ⓑ Ⓒ Ⓓ

4. Ⓐ Ⓑ Ⓒ Ⓓ　　24. Ⓐ Ⓑ Ⓒ Ⓓ　　44. Ⓐ Ⓑ Ⓒ Ⓓ　　64. Ⓐ Ⓑ Ⓒ Ⓓ　　84. Ⓐ Ⓑ Ⓒ Ⓓ

5. Ⓐ Ⓑ Ⓒ Ⓓ　　25. Ⓐ Ⓑ Ⓒ Ⓓ　　45. Ⓐ Ⓑ Ⓒ Ⓓ　　65. Ⓐ Ⓑ Ⓒ Ⓓ　　85. Ⓐ Ⓑ Ⓒ Ⓓ

6. Ⓐ Ⓑ Ⓒ Ⓓ　　26. Ⓐ Ⓑ Ⓒ Ⓓ　　46. Ⓐ Ⓑ Ⓒ Ⓓ　　66. Ⓐ Ⓑ Ⓒ Ⓓ　　86. Ⓐ Ⓑ Ⓒ Ⓓ

7. Ⓐ Ⓑ Ⓒ Ⓓ　　27. Ⓐ Ⓑ Ⓒ Ⓓ　　47. Ⓐ Ⓑ Ⓒ Ⓓ　　67. Ⓐ Ⓑ Ⓒ Ⓓ　　87. Ⓐ Ⓑ Ⓒ Ⓓ

8. Ⓐ Ⓑ Ⓒ Ⓓ　　28. Ⓐ Ⓑ Ⓒ Ⓓ　　48. Ⓐ Ⓑ Ⓒ Ⓓ　　68. Ⓐ Ⓑ Ⓒ Ⓓ　　88. Ⓐ Ⓑ Ⓒ Ⓓ

9. Ⓐ Ⓑ Ⓒ Ⓓ　　29. Ⓐ Ⓑ Ⓒ Ⓓ　　49. Ⓐ Ⓑ Ⓒ Ⓓ　　69. Ⓐ Ⓑ Ⓒ Ⓓ　　89. Ⓐ Ⓑ Ⓒ Ⓓ

10. Ⓐ Ⓑ Ⓒ Ⓓ　　30. Ⓐ Ⓑ Ⓒ Ⓓ　　50. Ⓐ Ⓑ Ⓒ Ⓓ　　70. Ⓐ Ⓑ Ⓒ Ⓓ　　90. Ⓐ Ⓑ Ⓒ Ⓓ

11. Ⓐ Ⓑ Ⓒ Ⓓ　　31. Ⓐ Ⓑ Ⓒ Ⓓ　　51. Ⓐ Ⓑ Ⓒ Ⓓ　　71. Ⓐ Ⓑ Ⓒ Ⓓ　　91. Ⓐ Ⓑ Ⓒ Ⓓ

12. Ⓐ Ⓑ Ⓒ Ⓓ　　32. Ⓐ Ⓑ Ⓒ Ⓓ　　52. Ⓐ Ⓑ Ⓒ Ⓓ　　72. Ⓐ Ⓑ Ⓒ Ⓓ　　92. Ⓐ Ⓑ Ⓒ Ⓓ

13. Ⓐ Ⓑ Ⓒ Ⓓ　　33. Ⓐ Ⓑ Ⓒ Ⓓ　　53. Ⓐ Ⓑ Ⓒ Ⓓ　　73. Ⓐ Ⓑ Ⓒ Ⓓ　　93. Ⓐ Ⓑ Ⓒ Ⓓ

14. Ⓐ Ⓑ Ⓒ Ⓓ　　34. Ⓐ Ⓑ Ⓒ Ⓓ　　54. Ⓐ Ⓑ Ⓒ Ⓓ　　74. Ⓐ Ⓑ Ⓒ Ⓓ　　94. Ⓐ Ⓑ Ⓒ Ⓓ

15. Ⓐ Ⓑ Ⓒ Ⓓ　　35. Ⓐ Ⓑ Ⓒ Ⓓ　　55. Ⓐ Ⓑ Ⓒ Ⓓ　　75. Ⓐ Ⓑ Ⓒ Ⓓ　　95. Ⓐ Ⓑ Ⓒ Ⓓ

16. Ⓐ Ⓑ Ⓒ Ⓓ　　36. Ⓐ Ⓑ Ⓒ Ⓓ　　56. Ⓐ Ⓑ Ⓒ Ⓓ　　76. Ⓐ Ⓑ Ⓒ Ⓓ　　96. Ⓐ Ⓑ Ⓒ Ⓓ

17. Ⓐ Ⓑ Ⓒ Ⓓ　　37. Ⓐ Ⓑ Ⓒ Ⓓ　　57. Ⓐ Ⓑ Ⓒ Ⓓ　　77. Ⓐ Ⓑ Ⓒ Ⓓ　　97. Ⓐ Ⓑ Ⓒ Ⓓ

18. Ⓐ Ⓑ Ⓒ Ⓓ　　38. Ⓐ Ⓑ Ⓒ Ⓓ　　58. Ⓐ Ⓑ Ⓒ Ⓓ　　78. Ⓐ Ⓑ Ⓒ Ⓓ　　98. Ⓐ Ⓑ Ⓒ Ⓓ

19. Ⓐ Ⓑ Ⓒ Ⓓ　　39. Ⓐ Ⓑ Ⓒ Ⓓ　　59. Ⓐ Ⓑ Ⓒ Ⓓ　　79. Ⓐ Ⓑ Ⓒ Ⓓ　　99. Ⓐ Ⓑ Ⓒ Ⓓ

20. Ⓐ Ⓑ Ⓒ Ⓓ　　40. Ⓐ Ⓑ Ⓒ Ⓓ　　60. Ⓐ Ⓑ Ⓒ Ⓓ　　80. Ⓐ Ⓑ Ⓒ Ⓓ　　100. Ⓐ Ⓑ Ⓒ Ⓓ

Practice Exam

In each case, write the letter of the word or expression that best completes the statement or answers the question.

1. Acid rain, the kind that causes environmental problems, is caused primarily by
   (a) carbonic acid
   (b) hydrochloric acid
   (c) acetic acid
   (d) sulfuric acid

2. On June 21, insolation and polar elevation are greatest at
   (a) the Equator
   (b) 0° latitude
   (c) 23.5° S
   (d) 23.5° N

3. Because of the seasonal lag, the warmest temperatures across most of the nation are observed
   (a) several weeks after the summer solstice
   (b) at the same time as the summer solstice
   (c) around June 21
   (d) a few weeks before the summer solstice

4. The climate across most of the United States is best described as
   (a) tropical
   (b) moist
   (c) temperate, with warm summers and cold winters
   (d) polar

5. When satellites are used as remote sensing tools to study sea-surface temperatures, the techniques employed may include measuring
   (a) X-ray emissions from the ocean surface
   (b) infrared emissions from the ocean surface
   (c) visible emissions from the ocean surface
   (d) radio-wave emissions from the ocean surface

6. During winter in the United States, what season is observed in Australia?
   (a) winter
   (b) spring
   (c) summer
   (d) autumn

7. The ozone responsible for blocking incoming ultraviolet energy from reaching Earth's surface is located in the
   (a) stratosphere
   (b) troposphere
   (c) lithosphere
   (d) mesosphere

8. Thunderstorms most often occur ahead of or along with
   (a) old fronts
   (b) warm fronts
   (c) stationary fronts
   (d) occluded fronts

9. Upwelling usually
   (a) produces high concentrations of phytoplankton
   (b) produces moderate concentrations of phytoplankton
   (c) produces low concentrations of phytoplankton
   (d) has no effect upon the concentration of phytoplankton

10. If you lived on the Arctic Circle, you would expect
    (a) 2 days per year of complete darkness
    (b) 6 months of darkness
    (c) 6 months of daylight
    (d) 1 day of complete darkness and 1 day of complete daylight

11. Lake effect snow most often affects
    (a) western New York and western Pennsylvania
    (b) the mountains of Vermont and New Hampshire
    (c) the Poconos and western New Jersey
    (d) the New York City metropolitan area

12. When Appalachian damming occurs, the U.S. northeastern seaboard often experiences
    (a) cold, dry weather
    (b) cool, damp, rainy weather
    (c) warm, dry weather
    (d) very significant thunderstorm activity

13. When you look up at the base of clouds in the sky, the atmosphere is
    (a) not saturated below the base of the clouds
    (b) saturated at the height where the clouds are located
    (c) saturated immediately above Earth's surface
    (d) filled with water vapor

14. Global warming may
    (a) raise sea levels significantly, thereby flooding coastal locations
    (b) increase Earth's cloud cover
    (c) cause deserts where crops are now grown
    (d) have all of the above effects

15. Scientists are in agreement that
    (a) greenhouse-gas concentrations are increasing in Earth's atmosphere
    (b) global warming has begun and mankind is responsible
    (c) carbon dioxide and other greenhouse gases absorb large amounts of heat
    (d) both a and c

16. El Niño is an unusual
    (a) cooling of the eastern Pacific
    (b) warming of the eastern Pacific
    (c) cooling of the Atlantic
    (d) warming of the Atlantic

17. Climate change is
    (a) a phenomenon caused by humans
    (b) a natural phenomenon that has occurred continuously in the past
    (c) a rare event
    (d) both a natural and a human-caused phenomenon

18. The mountain rain shadow effect causes
    (a) much rainfall west of the Cascades
    (b) a high desert (semiarid) climate east of the Cascades
    (c) large shadows in the Cascades
    (d) a desert to the west of the Cascades

19. When the temperature and dew point are very close, the relative humidity is
    (a) very low
    (b) around 50 percent
    (c) very high
    (d) 100 percent

20. Cyclonic circulation in the Northern Hemisphere is
    (a) clockwise
    (b) counterclockwise
    (c) rising
    (d) falling

21. As a cyclone approaches a region, the air pressure there will
    (a) fall, then rise
    (b) fall
    (c) rise
    (d) remain constant

22. When a cold front passes by, to what direction do the winds often shift and what is the effect on the temperature?
    (a) southwest; it rises
    (b) northwest; it rises
    (c) southwest; it falls
    (d) northwest; it falls

23. Tornadoes are most often accompanied by
    (a) severe thunderstorms
    (b) hail
    (c) frequent lightning
    (d) all of the above

24. Acid rain poses a danger to fish because
    (a) fish can live only within a certain pH range
    (b) acid rain dissolves fish
    (c) fish are poisoned by the heavy metals released into acidified waters
    (d) both a and c

25. When a cold front travels from southern Canada to the eastern seaboard, the air mass most likely to follow it is
    (a) continental Polar
    (b) maritime Polar
    (c) maritime Tropical
    (d) continental Tropical

26. Warm fronts are often preceded by
    (a) stratus clouds, then cirrus clouds
    (b) stratus clouds, then cumulus clouds
    (c) cirrus clouds, then stratus clouds
    (d) cirrus clouds, then cumulus clouds

27. Cold fronts are often preceded by
    (a) stratus clouds, then cirrus clouds
    (b) stratus clouds, then cumulus clouds
    (c) cirrus clouds, then stratus clouds
    (d) cumulus clouds and cumulonimbus clouds

28. The acid primarily responsible for the natural acidity of rain is
    (a) carbonic acid
    (b) hydrochloric acid
    (c) acetic acid
    (d) sulfuric acid

29. Sleet often occurs when the temperature in the clouds is
  (a) above freezing, but the temperature below the clouds is below freezing
  (b) above freezing, and the temperature below the clouds is above freezing
  (c) below freezing, but the temperature below the clouds is above freezing
  (d) below freezing, and the temperature below the clouds is below freezing

30. El Niño can cause
  (a) changes in the trade winds
  (b) shifts in precipitation patterns throughout the tropics
  (c) droughts and floods throughout the United States
  (d) all of the above

31. Earth's atmosphere is composed primarily of
  (a) carbon dioxide
  (b) methane
  (c) oxygen
  (d) nitrogen

32. Which city has the greatest annual temperature range?
  (a) San Francisco
  (b) Chicago
  (c) New York
  (d) London

33. An example of convection is the fact that
  (a) a hot plate heats a pot of water
  (b) you tan while enjoying the afternoon sun
  (c) you experience a sea breeze during an afternoon at the shore
  (d) you generate heat when you rub your hands together

34. The Sun's energy reaches Earth's surface by
  (a) radiation
  (b) conduction
  (c) convection currents
  (d) osmosis

35. The term *albedo* describes Earth's
  (a) convection currents
  (b) degree of reflectivity
  (c) tectonic activity
  (d) energy received from the Sun

36. The day with the longest daylight of the year in much of the Northern Hemisphere is
    (a) March 21
    (b) June 21
    (c) July 21
    (d) August 21

37. The warmest day of the year throughout much of the Northern Hemisphere is usually closest to
    (a) March 21
    (b) June 21
    (c) July 21
    (d) August 21

38. The belt of global winds over the midlatitudes is known as the
    (a) easterlies
    (b) doldrums
    (c) ITCZ
    (d) westerlies

39. Convection cells are present in
    (a) the atmosphere
    (b) the oceans
    (c) Earth's mantle
    (d) all of the above

40. A front is
    (a) a low-pressure system
    (b) a boundary between two differing air masses
    (c) a region where it snows
    (d) an air mass

41. Air masses tend to produce
    (a) large regions of rain or snow
    (b) high winds
    (c) large regions of fair weather
    (d) warm temperatures

42. Cyclones tend to form
    (a) in air masses
    (b) over the ocean
    (c) along the coasts
    (d) along fronts

43. The term *dew point* describes
    (a) how much dew is in the atmosphere
    (b) at what time dew will form on the ground
    (c) the temperature at which the atmosphere is saturated
    (d) the amount of moisture in the air

44. The jet stream is located in the
    (a) oceans
    (b) troposphere
    (c) stratosphere
    (d) mesosphere

45. An inversion occurs when
    (a) the temperature is colder on a mountain than in a valley
    (b) the temperature is colder in a valley than on a mountain
    (c) a front is passing by
    (d) the temperature is extremely cold

46. The diagram below shows Earth, the Moon, and the Sun's rays as viewed
    from space.

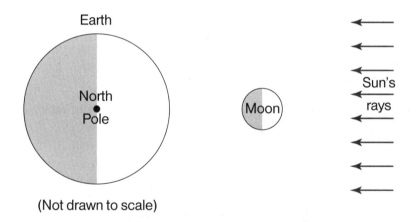

(Not drawn to scale)

For observers on Earth, which phase of the Moon is represented by the diagram?

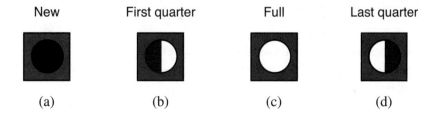

47. In which list are celestial features placed in the correct order from smallest to largest?
     (a) galaxy, solar system, universe, planet
     (b) planet, solar system, galaxy, universe
     (c) solar system, galaxy, planet, universe
     (d) universe, galaxy, solar system, planet

48. The diagram below shows the stump of a tree whose root grew into a small crack in bedrock and split the rock apart.

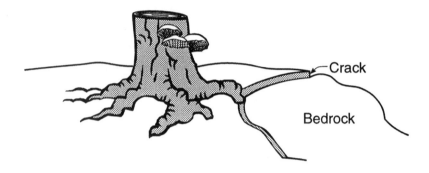

The splitting of the bedrock by the root is an example of
     (a) chemical weathering
     (b) deposition
     (c) erosion
     (d) physical weathering

49. The cross sections of crust below represent two regions of sedimentary rock layers that have been altered.

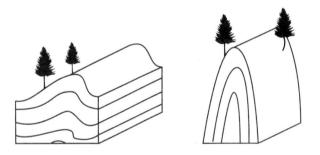

The sedimentary bedrock in both regions originally formed as
     (a) horizontal layers
     (b) recrystallized layers
     (c) faulted layers
     (d) folded layers

50. The map below shows the present-day locations of South America and Africa. Remains of *Mesosaurus*, an extinct freshwater reptile, have been found in similarly aged bedrock formed from lake sediments at locations *X* and *Y*.

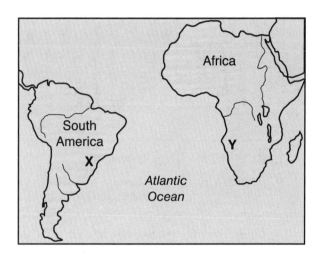

Which statement represents the most logical conclusion to draw from this evidence?

(a) *Mesosaurus* migrated across the ocean from location *X* to location *Y*.
(b) *Mesosaurus* came into existence on several widely separated continents at different times.
(c) The continents of South America and Africa were joined when *Mesosaurus* lived.
(d) The present climates at locations *X* and *Y* are similar.

51. At the Aleutian Trench and the Peru-Chile Trench, tectonic plates are generally

(a) moving along a transform boundary
(b) moving over a mantle hot spot
(c) diverging
(d) converging

52. The diagrams below represent two different geologic cross sections in which an igneous formation is found in sedimentary bedrock layers. The layers have not been overturned.

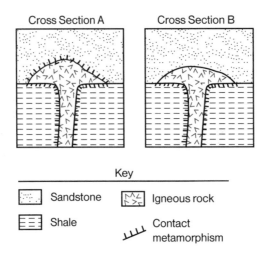

Cross Section A   Cross Section B

Key

Sandstone   Igneous rock

Shale   Contact metamorphism

Which statement best describes the relative age of each igneous formation compared to the age of the overlying sandstone bedrock?

(a) In A the igneous rock is younger than the sandstone, and in B the igneous rock is older than the sandstone.

(b) In A the igneous rock is older than the sandstone, and in B the igneous rock is younger than the sandstone.

(c) In both A and B, the igneous rock is younger than the sandstone.

(d) In both A and B, the igneous rock is older than the sandstone.

53. The profile below shows four regions of the ocean bottom.

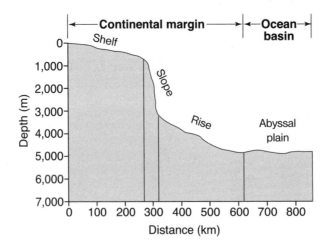

In which list are these regions arranged in order of gradient from least steep to most steep?

(a) rise, abyssal plain, shelf, slope

(b) slope, rise, shelf, abyssal plain

(c) abyssal plain, shelf, rise, slope

(d) shelf, abyssal plain, rise, slope

54. The photograph below shows the igneous rock obsidian.

The obsidian's glassy texture indicates that it formed from a magma that cooled

(a) slowly, deep below Earth's surface

(b) slowly, on Earth's surface

(c) quickly, deep below Earth's surface

(d) quickly, on Earth's surface

Base your answers to questions 55 and 56 on the map below. Dots on the map show the distribution of major earthquake epicenters. The shaded circle labeled *A* represents a location on Earth's surface.

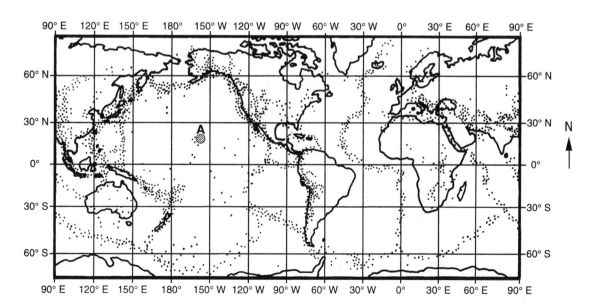

55. Which conclusion can best be inferred from the data shown on this map?
(a) Earthquakes generally are evenly distributed over Earth's surface.
(b) Most earthquakes occur west of the Prime Meridian and north of the Equator.
(c) Most earthquakes are concentrated in zones along plate boundaries.
(d) Most earthquakes occur on continents.

56. Location *A* is best described as an area that is
(a) within a rift valley at a mid-ocean ridge
(b) at a boundary between two diverging plates
(c) within a deep-sea trench between two converging plates
(d) above a mantle hot spot near the center of a crustal plate

57. The formation and subsequent development of our solar system is a topic most likely to be researched by
(a) a meteorologist
(b) an oceanographer
(c) an astronomer
(d) a geologist

58. Changes taking place in the atmosphere can also impact the
(a) hydrosphere
(b) biosphere
(c) geosphere
(d) all of the above

59. Redshifts provide evidence to support the
(a) big bang theory
(b) nebular hypothesis
(c) planetary accretion theory
(d) plate tectonics theory

60. The Milky Way is
(a) a spiral galaxy
(b) an elliptical galaxy
(c) a dwarf galaxy
(d) an irregular galaxy

61. The primary source of energy for all stars is
(a) nuclear fusion
(b) nuclear fission
(c) combustion of gasoline
(d) burning of oxygen

62. Most stars spend the greater part of their life cycles as
    (a) supernovas
    (b) brown dwarfs
    (c) main-sequence stars
    (d) red giants

63. The most abundant element in stars and in the universe overall is
    (a) hydrogen
    (b) helium
    (c) oxygen
    (d) nitrogen

64. The Sun is estimated to be
    (a) about 5 billion years old
    (b) about 15 billion years old
    (c) about as old as all the planets
    (d) older than Earth

65. The distances to nearby stars are determined by using
    (a) the Hertzsprung-Russell diagram
    (b) estimates of the absolute and apparent magnitudes of the stars
    (c) the parallax method
    (d) Distances cannot be determined.

66. The planets in the solar system are often organized into two groups, the inner and outer planets. The planets in each group have similar characteristics. A planet that fails to fit well into either category is
    (a) Mercury
    (b) Earth
    (c) Jupiter
    (d) Pluto

67. Venus's atmosphere is rich in
    (a) oxygen
    (b) nitrogen
    (c) hydrogen
    (d) carbon dioxide

68. Other than Earth, which planet is the most likely candidate for finding some form of life that either exists now or once existed?
    (a) Mercury
    (b) Venus
    (c) Mars
    (d) Jupiter

69. One of the most tectonically active celestial bodies in the entire solar system is
    (a) Europa
    (b) Io
    (c) Titan
    (d) the Moon

70. Which naturally occurring satellite has a frozen surface and may contain an ocean of water below that surface?
    (a) Europa
    (b) Io
    (c) Titan
    (d) the Moon

71. The heliocentric model of the universe was first proposed by
    (a) Ptolemy
    (b) Copernicus
    (c) Galileo
    (d) Newton

72. Early proof that Earth rotates on an axis was provided by the
    (a) pendulum
    (b) telescope
    (c) sextant
    (d) Hubble space telescope

73. The entire lunar surface is covered with
    (a) maria
    (b) regolith
    (c) highlands
    (d) water

74. Minerals rich in silicon and oxygen are classified as
    (a) oxides
    (b) sulfides
    (c) silicates
    (d) quartz

75. Igneous rock is classified
    (a) as intrusive or extrusive
    (b) as foliated or nonfoliated
    (c) by particle size
    (d) by hardness

76. Marble is an example of a
    (a) nonfoliated sedimentary rock
    (b) nonfoliated metamorphic rock
    (c) foliated sedimentary rock
    (d) foliated metamorphic rock

77. Most exposed rocks on the surface are
    (a) sedimentary
    (b) igneous
    (c) metamorphic
    (d) none of the above

78. Limestone is an example of
    (a) an intrusive igneous rock
    (b) an extrusive igneous rock
    (c) a chemical sedimentary rock
    (d) a nonfoliated metamorphic rock

79. Rocks provide
    (a) clues to ancient climates
    (b) clues to ancient life-forms
    (c) evidence of plate tectonics
    (d) all of the above

80. The rock cycle indicates that magma may cool beneath the surface to form
    (a) intrusive igneous rock
    (b) extrusive igneous rock
    (c) sedimentary rock
    (d) metamorphic rock

81. In which of the following are the geologic times listed in correct chronological order?
    (a) Precambrian, Paleozoic, Cenozoic, Mesozoic
    (b) Precambrian, Cenozoic, Paleozoic, Mesozoic
    (c) Precambrian, Paleozoic, Mesozoic, Cenozoic
    (d) Cenozoic, Mesozoic, Paleozoic, Precambrian

82. The first mammals appeared during the
    (a) Precambrian era
    (b) Paleozoic era
    (c) Mesozoic era
    (d) Cenozoic era

83. A great extinction of life-forms, including the dinosaurs, marked the end of the
    (a) Precambrian era
    (b) Paleozoic era
    (c) Mesozoic era
    (d) Cenozoic era

84. The law of superposition predicts that, if rock layers have not been overturned, the oldest layer will be
    (a) at the top of a sequence of rock strata
    (b) at the base of a sequence of rock strata
    (c) in the middle of a sequence of rock strata
    (d) No prediction can be made using the law of superposition.

85. Rock layers from one site may be correlated to those from another site by
    (a) similar color
    (b) the presence of index fossils
    (c) similar minerals
    (d) all of the above

86. If a period of deposition is followed by a period of erosion and then more deposition occurs, the likely result is
    (a) an unconformity
    (b) a dike
    (c) a sill
    (d) a laccolith

87. An alpha particle is
    (a) an electron
    (b) a positron
    (c) a helium nucleus
    (d) a daughter element

88. If one-eighth of the original radioactive parent element remains in a rock sample, through how many half-lives has the element gone?
    (a) 1
    (b) 2
    (c) 3
    (d) 4

89. Carbon-14 is not useful as an isotope to determine Earth's age because
    (a) it is too rare
    (b) it is dangerous to handle
    (c) its half-life is too short
    (d) its half-life is too long

90. Convection cells that drive movements in the plates at Earth's surface are located in the
    (a) crust
    (b) mantle
    (c) lithosphere
    (d) asthenosphere

91. The continents have been drifting apart since Pangaea split approximately
    (a) 100 million years ago
    (b) 200 million years ago
    (c) 400 million years ago
    (d) 4 billion years ago

92. New material is being produced at or along
    (a) mid-ocean ridges
    (b) deep ocean trenches
    (c) subduction zones
    (d) transform faults

93. The San Andreas Fault is the site of
    (a) a subduction zone
    (b) a converging plate boundary
    (c) a transform fault
    (d) a deep ocean trench

94. The earthquake of greatest magnitude in U.S. history occurred in
    (a) California
    (b) Alaska
    (c) Oregon
    (d) Missouri

95. Damage to buildings generally occurs when an earthquake exceeds a Richter-scale magnitude of
    (a) 4
    (b) 5
    (c) 6
    (d) 7

96. Low-viscosity magma tends to produce gentle, flowing lava eruptions. A classic location to illustrate this action is
    (a) the Hawaiian Islands
    (b) Mount Saint Helens
    (c) Mount Rainier
    (d) Mount Vesuvius

97. A large region of intrusive igneous rock located deep within Earth's crust is called a
    (a) laccolith
    (b) batholith
    (c) sill
    (d) dike

98. Moraines are deposits left behind by
    (a) wind
    (b) rivers
    (c) glaciers
    (d) mass wasting

99. Mountains that form as a result of convergence along plate boundaries are known as
    (a) fault block mountains
    (b) folded mountains
    (c) domed mountains
    (d) volcanic mountains

100. A layer where permeable rock (rock that can transmit water through pore spaces) or sediment allows groundwater to travel freely is known as a
    (a) cavern
    (b) aquifer
    (c) well
    (d) sill

Answers to Practice Exam

1. (d)	21. (b)	41. (c)	61. (a)	81. (c)
2. (d)	22. (d)	42. (d)	62. (c)	82. (c)
3. (a)	23. (c)	43. (c)	63. (a)	83. (c)
4. (c)	24. (d)	44. (b)	64. (a)	84. (b)
5. (b)	25. (a)	45. (b)	65. (c)	85. (b)
6. (c)	26. (c)	46. (a)	66. (d)	86. (a)
7. (a)	27. (d)	47. (b)	67. (d)	87. (c)
8. (a)	28. (a)	48. (d)	68. (c)	88. (c)
9. (a)	29. (a)	49. (a)	69. (b)	89. (c)
10. (d)	30. (d)	50. (c)	70. (a)	90. (b)
11. (a)	31. (d)	51. (d)	71. (b)	91. (b)
12. (b)	32. (b)	52. (a)	72. (a)	92. (a)
13. (a)	33. (c)	53. (c)	73. (b)	93. (c)
14. (d)	34. (a)	54. (d)	74. (c)	94. (d)
15. (d)	35. (b)	55. (c)	75. (a)	95. (c)
16. (b)	36. (b)	56. (d)	76. (b)	96. (a)
17. (b)	37. (c)	57. (c)	77. (a)	97. (b)
18. (b)	38. (d)	58. (d)	78. (c)	98. (c)
19. (c)	39. (d)	59. (a)	79. (d)	99. (b)
20. (b)	40. (b)	60. (a)	80. (a)	100. (b)

INDEX

A

Abrasion, 208
Absolute magnitude, 45–46
Abyssal plains, 255–256
Accretion, 68
Acid rain deposition, 393
Acidified soil, 393
Adiabatic cooling, 378
Advection, 351
Aerosols, 293
Aftershocks, 179
Air masses, 337, 350–351
Air pressure
    altitude-based variations, 334
    measuring of, 334–335
    wind and, relationship between, 339–340
Albedo, 298–299
Allotrope, 293
Alpha particles, 154
Aluminum, 119
Animals, 209–210
Anions, 115
Anticlines, 215
Anticyclones, 337, 350
Apollo, 11, 82
Appalachian Mountains, 175, 351, 390–391
Apparent brightness, 45–46
Aquifers, 128
Arctic air masses, 350
Arêtes, 211
Artesian formation/wells, 231
Asteroid belt, 64–65
Asthenosphere, 169
Astronomical high tide, 273
Astronomy, 4
Atmosphere
    characteristics of, 9, 68
    circulation of, 307–312
    clouds. see Clouds
    components of, 292–293
    convection of, 300–301
    definition of, 9
    early, 293–294
    free oxygen in, 293–294

    global circulation, 379
    greenhouse gases in, 80
    mesosphere, 296
    moisture. see Moisture
    stability of, 330
    stratosphere, 295–296
    structure of, 294–296
    thermosphere, 296
    trace gases in, 292–293
    troposphere, 294–295, 377
    warming of, 300–301

B

Background radiation, 24–25
Banks, 234
Barometer, 334
Barred spiral galaxies, 33
Basalt, 124
Batholiths, 198–199
Bauxite, 119, 121
Beach restoration, 278–281
Benthic zone, 250
Bergeron process, 330
Big bang theory, 4, 24
Binary stars, 44
Biosphere, 11
Black dwarf, 38–39, 42
Black hole, 42–43
Boiling point, 228
Brahe, Tycho, 70, 72
Breakers, 275–276

C

Calcite, 117
Cambrian period, 144–145
Carbon dioxide, 292
Carbonic acid, 393
Cascade Mountain Range, 173–174
Catastrophic collision hypothesis, 96
Cation, 115
Caverns, 233–234
Caves, 233–234
Cenozoic era, 147–148
Cepheid variable stars, 46

Challenger, 82–83
Channel, 234
Charge-coupled devices, 29
Charon, 62
Chemosynthetic organisms, 143
Chlorofluorocarbons, 300
Cinder cone volcanoes, 197–198
Cirrus clouds, 329
Cleavage, 119
Climate
    changes in, 392–394
    definition of, 376
    factors that affect, 377
    global patterns, 376–377
    global zones, 379–380
    horse latitudes, 308, 311, 380
    human impact on, 392–394
    humid continental, 390
    humid oceanic, 385
    Internet resources, 394
    of Moon, 104
    mountain rain shadow effect, 386–388
    natural changes in, 391
    polar, 376
    polar storm track, 380
    subtropical, 376, 385
    temperate, 376
    tropical, 376
    U.S. patterns
        Desert Southwest, 388–389
        Great Lakes effect, 390
        Midwest, 390
        mountain rain shadow effect, 386–388
        mountains, 389
        Northeast, 390–391
        semiarid regions, 388–389
        study of, 381–383
        western seaboard, 384–386
    vertical motion, 377–379
Climatology, 5
Clouds
    cirrus, 329
    composition of, 328
    cumulonimbus, 329

Clouds *(continued)*
cumulus, 329
formation of, 328–329
nimbostratus, 330
stratus, 329
types of, 329–330
Coal, 129
Cold front, 354
Collision-coalescence, 330
Columbia, 83
Columnar jointing, 216
Comet Shoemaker-Levy, 81
Comets, 66
Composite cone volcanoes, 198
Condensation, 378
Condensation nuclei, 328, 330
Conglomerate, 128
Constellations, 46–47
Continental Divide, 236
Continental drift hypothesis,
166–169, 172
Continental margins, 257–258
Continental polar air masses, 350
Continental tropical air masses, 351
Convection, 300–301
Convection cells, 307
Convection currents, 301
Convergent plate boundaries, 168,
171, 173–175
Copernicus, Nicolaus, 70–71
Coriolis effect, 266, 268, 311, 336
Corrasion, 208
Correlation, 149–150
Cosmic Background Explorer, 25,
27
Crab nebula, 41
Crater Lake, 185
Cratons, 156
Cretaceous period, 146
Cross cuts, 150–151
Crust, 169
Cryosphere, 10
Cumulonimbus clouds, 329
Cumulus clouds, 329
Cyclones
characteristics of, 336–337,
354–355
development of, 355
dissipation of, 357
formation of, 354–355

precipitation patterns associated
with, 355–357
tropical, 361–362

D
Daughter element, 154–155
Dawn Mission, 64–65
Desalination, 246
Desert Southwest, 388–389
Devonian period, 145
Dew point, 325–327
Diamond, 117–118
Dikes, 199
Dinosaurs, 143–147
Diurnal tides, 273–274
Divergent plate boundaries, 168,
171–173
Divide, 236
Dolomite, 117
Domed mountains, 214
Doppler effect, 26
Doppler radar, 360
Drainage basin, 234–236
Drizzle, 331–332
Drought, 209
Drumlins, 213
Dunes, 280

E
Earth
age of, 155–157
atmosphere of. *see* Atmosphere
characteristics of, 59
description of, 56, 59
gravitational field of, 97
layers of, 11
oceans of, 98–99
orbit of, 45
rotation of, 301–307
size of, 96–98
sun and, relationship between,
296–299, 302
surface of
deformations, 215–216
factors that affect, 208–213
features of, 94–96
heating of, 294
mountains. *see* Mountain(s)
systems of
atmosphere. *see* Atmosphere

biosphere, 11
cryosphere, 10
description of, 8–9, 94
geosphere, 11
hydrosphere, 9–10
Earth systems science
definition of, 2
focus of, 2
scientists involved in, 2
studying of, 12–13
Earthquakes
aftershocks, 179
definition of, 177
epicenter of, 179–181
focus of, 179–180
foreshocks, 179
intensity of, 181–183
Internet resources, 185–186
L-waves, 178–179
magnitude of, 181–182
Mercalli intensity scale of,
182–183
plate boundaries and, 176–178
predicting of, 184
P-waves, 178–179, 183
Richter magnitude scale for, 181
seismograph of, 179
S-waves, 178–179, 183
understanding of, 183–184
Echo sounder, 251
Eclipse
lunar, 103
solar, 103–104
El Niño, 5, 247–248, 266, 269–271
Elastic limit, 215
Electromagnetic waves, 296–297
Elliptical galaxies, 33
Environmental science, 8
Epicenter, 179–181
Epicycles, 69
Equator, 308
Equinoxes, 306
Erosion
Earth's surface affected by, 208
groundwater-related, 232
wave-related, 276–277
Erratics, 213
Eskers, 213
Euphotic zone, 249
Europa, 62

European Space Agency, 84
Evaporation, 327–328
Evaporational cooling, 328
Evaporites, 127
Extra-tropical storms. *see* Cyclones
Extrusive rock, 122–124
Eye, 361
Eye wall, 361

F
Fault(s), 215–216
Fault boundaries, 175–176
Fault-block mountains, 213–214
Fetch, 275
Focus, 179–180
Fog, 331
Folded mountains, 214
Folds, 215
Forecasting of weather
    description of, 364
    modern methods of, 365
Foreshocks, 179
Fossils
    definition of, 151
    description of, 4–5
    index, 150
    Internet resources, 157–158
    rock dating using, 150–152
Foucault pendulum, 76
Fracture, 119
Free oxygen, 293–294
Freezing drizzle, 332
Freezing rain, 332
Front
    cold, 354
    definition of, 328, 351
    occluded, 354
    stationary, 351
    warm, 351–353
Frontolysis, 354
Frost wedging, 210
Frost-thaw cycle, 210

G
Galaxy
    definition of, 32
    description of, 26
    formation of, 32–33
    Milky Way, 32–33
    types of, 33

Galena, 118
GALEX, 31
Galileo, Galilei, 74
Gamma ray spectrometer, 80
Gas giants, 56, 61
Gems, 121
Geologic time scale
    Cenozoic era, 147–148
    Mesozoic era, 145–147
    need for, 142
    organization of, 142–148
    Paleozoic era, 144–145
    Precambrian era, 142–144
Geologic timescale, 4, 6
Geology
    description of, 4–7
    historical, 4
    physical, 4
Geosphere, 11
Geysers, 232
Glaciation, 211
Glaciers, 210–213
Global convection, 328
Global convection cells, 307–308
Global interconnected ocean ridges
    system, 257
Global Precipitation Measurement,
    253–254
Global warming, 3, 230, 392
GLOBE Project, 12–13
Gneiss, 129–130
Granite, 123
Graptolites, 5
Gravitational field, 97
Gravity, 67, 75
Great Lakes effect, 390
Great Plains region, 389
Great Smoky Mountains, 185
Greenhouse effect, natural, 299–300
Greenhouse gases, 80, 300, 393
Groins, 278–279
Groundwater
    description of, 231
    erosion caused by, 232
    geysers, 232
    hot springs, 232
    wells, 232–233
Gulf of Mexico, 311, 359
Guyot, 257
Gypsum, 117, 128

H
Hadley cell, 308
Hail, 331–332
Half-life, 155
Halite, 117, 128
Halley's comet, 66
Hawaiian Islands, 176–177
Heat, latent, 328
Heliocentric theory, 70–71
Hertzsprung-Russell diagram,
    37–38, 43–44
Highland Physiographic Province,
    123, 145
High-pressure systems, 350–351
Historical geology, 4
Holocene epoch, 148
Hook, 251
Horse latitudes, 308, 311, 380
Hot plumes, 176
Hot spots, 176–177
Hot springs, 232
Hourglass nebula, 39–40
Hubble, Edwin, 26
Hubble space telescope, 29–30,
    79–80
Humid continental climate, 390
Humid oceanic climate, 385
Humidity, relative, 322–327
Hurricane
    classification of, 362–363
    destruction caused by, 363–364
    storm surge, 364
    structure of, 361
Hutton, James, 142
Hydrogen, 25
Hydrogen bonding, 226
Hydrosphere, 9–10, 59
Hydrothermal vent, 250
Hygrometer, 326

I
Ice crystals, 328
Ice pellets, 332
Igneous rocks
    classification of, 123–124
    colors of, 124
    extrusive, 122–124
    identification of, 125
    intrusive, 122–124
    lava, 121–123

Igneous rocks (*continued*)
  magma, 121–123
  societal uses of, 125
Index fossils, 150
Insolation, 299, 305, 377
Intermolecular attraction, 226
International Panel on Climate
      Changes, 392
International Space Station, 84
Internet resources, 13
Intertidal zone, 250
Intertropical convergence zone,
      308, 310, 328, 379–380
Intrusive rock, 122–124
Io, 62
Ionic bond, 116
Ionic compounds, 115–117
Irregular galaxies, 33
Isotopes
  definition of, 154
  half-life of, 155
  radioactive, 155–156
ITCZ. *see* Intertropical
      convergence zone

J
Jetties, 278
Joints, 216
Jovian planets, 56, 60–61
Jupiter, 60–61, 81
Jurassic period, 146

K
Kennedy, John F., 81
Kepler, Johannes, 72–74
Kilauea, 200
K-T boundary, 147
Kuiper belt, 62

L
La Niña, 5, 269–271
Laccoliths, 199
Latent heat, 328
Latitudes
  horse, 308, 311, 380
  midlatitudes, 312
Lava, 121–123, 253
Lava flows, 196
Lava plateaus, 198
Law of superposition, 148–149

Laws of planetary motion, 72–74
Lightning, 357
Light-year, 44–45
Limestone, 127
Lithification, 125
Lithosphere, 11, 169
Loess deposits, 209
Longshore current, 276
Long-wave radiation, 299
Low-pressure systems, 336–337,
      341
Lunar eclipse, 103
Lunar maria, 95
Luster, 118
L-waves, 178–179

M
Magma, 121–123, 195
Magnetite, 119
Main-sequence star, 37
Mantle
  convergent plate boundaries,
      171, 173–175
  crust's relationship to, 169
Marble, 131
Marianas Trench, 175
Marine-life zones, 249–251
Mariner, 4, 60
Maritime polar air masses, 350
Maritime tropical air masses, 351
Mars, 59–60, 80–81
Martian Equator, 80
Mass wasting, 210
MAVEN, 59
Mercalli intensity scale, 182–183
Mercury, 58
Mercury barometer, 334
Mesoscale weather systems, 336
Mesosphere, 296
Mesozoic era, 145–147
Messenger, 58
Metamorphic rocks, 129–131
Metamorphism, 128
Meteor Crater, 65
Meteoroids, 65, 156
Meteorology, 5, 8
Microscale weather systems, 336
Mid-Atlantic Ridge, 171, 257
Midlatitudes, 312
Mid-ocean ridges, 257

Milky Way galaxy, 32–33
Millibar, 334, 341
Mineralogy, 114
Minerals
  chemistry of, 114–117
  cleavage of, 119
  definition of, 121
  fracture of, 119
  grouping of, 119–120
  Internet resources, 132
  Moh's hardness scale, 118
  properties of, 117–119
  rock-forming, 120
  in seawater, 244, 246
  societal uses of, 121
  types of, 114–115
Mir, 83–84
"Mission to Planet Earth," 84
Modified continental climate,
      391
Moho, 183
Mohorovic discontinuity, 183
Moh's hardness scale, 118
Moisture
  dew point, 325–327
  latent heat, 328
  relative humidity, 322–327
  saturation, 323–324
Moon
  characteristics of, 60
  climate of, 104
  description of, 9
  Earth's oceans affected by,
      98–99, 272–273
  eclipse of, 103
  formation hypotheses regarding,
      96–98
  full, 100
  future missions to, 84
  Internet resources, 105
  motions of, 99
  new phase of, 101
  orbit of, 99
  phases of, 100–101
  regolith of, 95
  revolution of, 77, 100–101
  rotation of, 100–101
  scientific study of, 98
  size of, 96–98
  surface features of, 94–96, 98

Moon (continued)
    waxing gibbous phase of, 101
    waxing phases of, 101
Moonrise, 101
Moonset, 101
Moraines, 212–213
Mount Rainier, 185
Mount Saint Helens, 195–196,
        199–200
Mountain(s)
    domed, 214
    fault-block, 213–214
    folded, 214
    upwarped, 214
    volcanic, 215
Mountain rain shadow effect,
        386–388
Mouth, 234

N
National Aeronautics and Space
        Administration (NASA),
        2, 84
National Center for Environmental
        Prediction, 365
National parks, 185
Natural greenhouse effect, 299–300
Natural sciences
    astronomy. see Astronomy
    geology. see Geology
    meteorology, 5, 8
    oceanography, 8
Nazca Plate, 173
Neap tides, 273
Nebula, 36–37, 39–40, 66–67
Nebular hypothesis of solar
        system, 66–67
Negatively charged anions, 115–116
Neptune, 60
Neritic zone, 250
Neutron, 154
Neutron star, 39, 42
New Horizons, 62
New Madrid earthquakes, 182
Newton, Isaac, 75
Nimbostratus clouds, 330
Nitrogen, 9
Nor'easter, 339
North Atlantic Oscillation, 249
North Pole, 303, 380

Northeast climate, 390–391
Northern Hemisphere, 312
Nuclear chemistry, 153–154
Nuclear fusion, 34–36, 56
Nucleosynthesis, 35

O
Occluded front, 354
Occlusion, 357
Ocean(s). See also Seawater; Water
    aphotic zone of, 249
    climactic patterns, 267–269
    deep, benthic zone of, 246
    euphotic zone of, 249
    Internet resources, 281
    layered structure of, 246–249
    mapping of, 251–254
    marine-life zones, 249–251
    moon's effect on, 98–99,
        272–273
    neritic zone of, 250
    photic zone of, 250
    sea surface temperature, 269
    surface currents of, 266–271
    surface mixed zone of, 246
    thermocline of, 246–247
    tides. see Tides
    upwelling, 247–248, 266–268
    warming of, 230
    waves. see Wave(s)
Ocean circulation
    deep, 271
    surface patterns of, 266
Ocean currents, 231
Ocean trenches, 254–255
Oceanic zone, 251
Oceanographers, 8
Oceanography, 8
Olivine, 117–118
Orbit
    Earth's, 45, 72
    Moon's, 99
Ordovician period, 145
Ore, 121
Original horizontality, principle of,
        149
Original remains, 152
Orographic feature, 328
Oxygen, 9
Ozone, 293, 298, 393–394

P
Paleomagnetism, 251
Paleozoic era, 144–145
Pangaea, 166, 171
Parallax, 45, 72
Parallax shift, 45
Parent element, 154
Peneplain, 214
Penumbra, 103
Periodic table, 115
Photosynthetic organisms, 143
Physical geology, 4
Phytoplankton, 249
Piedmont Physiographic Province,
        123
Planets
    classification of, 56
    Earth. see Earth
    inner vs. outer, 68
    Internet resources, 84–85
    Jovian, 56, 60–61
    Jupiter, 60–61, 81
    Mars, 59–60
    Mercury, 58
    motions of, 76–79
    Neptune, 60
    Pluto, 61–62
    precession of, 77–79
    revolution of, 77
    rocky core, 56–58
    rotation of, 76–77
    Saturn, 60–61
    terrestrial, 56
    Venus, 59
Plants, 209–210
Plate boundaries
    convergent, 168, 171, 173–175
    divergent, 168, 171–173
    earthquake activity near,
        176–177
    volcanic activity near, 176–177
Plate tectonics
    continental drift hypothesis,
        166–169, 172
    schematic diagram of, 171
    tectonic plates, 169–170
Pleistocene epoch, 148
Pluto, 61–62
Plutons, 198
Polar climates, 376

Polar molecule, 227
Polar storm track, 380
Polyatomic ion, 116
Positively charged anions, 115–116
Potassium feldspar, 118
Precambrian era, 142–144
Precession, 77–79
Precipitation
    Bergeron process, 330
    condensation nuclei, 328, 330
    cyclones and, 355–357
    formation of, 330–331
    forms of, 331–333
    measuring of, 333
    rain, 331
    snow, 331–333
Pressure gradient, 341
Principle of original horizontality,
        149
Project Mercury, 82
Protostar, 36
Ptolemaic system, 68–69
Pulsar, 41
P-waves, 178–179, 183
Pyrite, 118
Pyroclastic materials, 196

Q
Quartz, 118
Quaternary period, 148

R
Radiation
    background, 24–25
    long-wave, 299
    short-wave, 299
    solar, 296–298
Radio telescope, 30–31
Radioactive decay, 152–153
Radioactive isotopes, 155–156
Radiometric dating of rocks,
        153–155
Radiosonde, 365
Rain
    acid, 393
    characteristics of, 331
Red giant, 38–39, 44
Red supergiant, 41
Redshift, 25–26
Reflecting telescope, 28

Refracting telescope, 28
Regional faults, 216
Regolith, 95
Relative humidity, 322–327
Remote sensing instruments
    description of, 27–28
    radio telescope, 30–31
    spectroscope, 30
    telescopes, 28–30
Replaced remains, 152
Revolution
    of Earth, 302
    of Moon, 77
    of planets, 77
Rhyolite, 124
Richter magnitude scale, 181
Ring of Fire, 194
Rivers, 234–236
Rock(s)
    absolute dating of, 152–155
    animal actions on, 209–210
    dating of, 148–155
    definition of, 121
    igneous. *see* Igneous rocks
    Internet resources, 132
    metamorphic, 129–131
    plant actions on, 209–210
    radiometric dating of, 153–155
    relative dating of, 148–152
    sedimentary. *see* Sedimentary
        rocks
    weathering of, 208–209
Rock cycle, 131
Rock layers
    correlation of, 149–150
    cross cutting, 150–151
    fossils in, 150–152
    unconformities in, 149, 151
Rocky core planets, 56–58
Rocky Mountains, 175
Rotation
    of Earth, 301–307
    of Moon, 100–101
    of planets, 76–77
Running water, 208

S
Salinity, 244–245, 271
Salyut 1, 83
San Andreas Fault, 176, 184, 216

Sand dunes, 280
Satellites
    man-made, 2–3
    natural, 62–64
Saturation, 323–324
Saturn, 60–61
Seafloor
    abyssal plains, 255–256
    features of, 254–258
    mapping of, 251
    mid-ocean ridges, 257
    ocean trenches, 254–255
    seamounts, 256–257
Seamounts, 256–257
Seasonal lag, 306
Sea-viewing Wide-Field-of-view
        Sensor, 253
Seawall, 279–280
Seawater
    composition of, 244–245
    Internet resources, 244
    minerals in, 244, 246
    percentage of Earth's water,
        226
    resources from, 246
    salinity of, 244–245, 271
Sedimentary rocks
    chemical, 125–126
    chemical origin, 126–127
    classification of, 126–128
    detrital origin, 126–127
    formation of, 125–126
    identification of, 127
    importance of, 128–129
    layers of, 126
    organic, 126
    value of, 128–129
    Wentworth scale for, 128
Seismogram, 167
Seismograph, 167, 179
Semiarid regions, 388–389
Semidiurnal tides, 273–274
Shale, 129
Shenandoah Mountains, 185
Shield volcanoes, 197
Short-wave radiation, 299
Sidereal month, 101–103
Silicon, 119–120
Sills, 123, 199
Silurian period, 145

Sling psychrometer, 326
Snow, 331–333
Sodium, 25
Solar eclipse, 103–104
Solar radiation, 294–295, 377
Solar system
    age of, 157
    asteroid belt, 64–65
    comets, 66
    description of, 79–80
    formation of, 66–68
    human exploration of, 81–84
    Internet resources, 84–85
    meteoroids, 65
    natural satellites, 62–64
    nebular hypothesis of, 66–67
    planets. *see* Planets
    Ptolemaic system, 68–69
    schematic diagram of, 56
    Soviet/Russian contributions,
        83–84
    sun. *see* Sun
    U.S. exploration of, 81–83
Solstices, 301–303
South Pole, 380
Specific gravity, 119
Specific heat, 229
Spectroscope, 30
Spiral galaxies, 33
Spit, 251, 277
Spring tides, 273
Sputnik, 83
Squall line, 356–357
Standard rain gauge, 333
Star
    absolute magnitude of, 45–46
    apparent brightness of, 45–46
    binary, 44
    brightness of, 45–46
    cepheid variable, 46
    constellations of, 46–47
    definition of, 34
    energy source for, 34–36
    evolution of, 36–40
    formation of, 36
    life cycle of, 36
    main-sequence, 37
    medium-mass, 38
    nebulae, 36–37
    neutron, 39, 42

remnants of, 42–43
supernova, 40–42
Stationary front, 351
Stellar parallax, 70, 72
Storm surge, 364
Strata, 126
Stratosphere, 295–296, 393–394
Stratovolcanoes, 198
Stratus clouds, 329
Streak color, 118
Streambed, 234
Streams, 234–236
Strike-slip faults, 216
Stromatolites, 143
Subduction, 214
Subducts, 173
Submarine canyons, 256
Subtropical climates, 376
Subtropical dry climate, 385
Subtropical Southeast, 391–392
Summer solstice, 303
Sun
    characteristics of, 44, 56
    distance to, 45
    Earth and, relationship between,
        296–299, 302
    eclipse of, 103–104
    life span of, 44
    nuclear fusion by, 56
Superclusters, 33
Supercooled, 328
Supernova, 40–42, 67
S-waves, 178–179, 183
Synclines, 215
Synodic month, 101–103
Synoptic meteorology, 5
Synoptic scale, 336, 338
Systems
    Earth's. *see* Earth, systems of
    interactions and interdepen-
        dencies among, 3
    solar. *see* Solar system

T
Talc, 118
Talus slopes, 210
Tectonic plates, 169–170
Telescopes
    Galileo's contributions, 74
    types of, 28–30

Temperate climates, 376
Terra, 2–3
Terrestrial planets, 56
Tertiary period, 148
Thermocline, 246
Thermohaline circulation, 247,
    271
Thermosphere, 296
Thunderstorms, 336, 357–359
Tidal motion, 98–99
Tidal range, 272
Tides
    causes of, 272–273
    diurnal, 273–274
    mixed, 273–274
    neap, 273
    semidiurnal, 273–274
    spring, 273
Titan, 62
Tornadoes, 336–337, 359–361
Trade winds, 309
Transform fault boundaries, 175
Transmutation, 154
Triangulation, 180
Triassic period, 146
Trilobites, 5, 144
Tropic of Cancer, 303
Tropical climates, 376
Tropical depression, 361
Tropical Rainfall Measuring
    Mission, 253
Tropical storm, 361–362
Tropopause, 294–295, 308
Troposphere, 294–295, 377
Tsunamis, 184–185

U
Ultraviolet radiation, 296–298
Unconformity, 149, 151
United States climate patterns
    Desert Southwest, 388–389
    Great Lakes effect, 390
    Midwest, 390
    mountain rain shadow effect,
        386–388
    mountains, 389
    Northeast, 390–391
    semiarid regions, 388–389
    study of, 381–383
    western seaboard, 384–386

Universe
    background radiation, 24–25
    big bang theory of, 4, 24
    description of, 24
    redshift, 25–26
    remote sensing instruments for
        studying. *see* Remote
        sensing instruments
    theories regarding origin of, 27
Upwarped mountains, 214
Upwelling, 247–248, 266–268
Uranium-235, 155

V

Van Allen radiation belts, 298
Venus, 59, 80
Vernal equinox, 306
Vertical air currents, 377–378
Very large array radio telescope,
    80
Viking 1, 60
Virga, 325
Volcanic mountains, 215
Volcanoes
    active, 194
    batholiths, 198–199
    cinder cone, 197–198
    composite cone, 198
    description of, 194
    dikes, 199
    eruption of, 195–196
    formation of, 177
    Internet resources, 196

    intrusive features of, 198–199
    Kilauea, 200
    laccoliths, 199
    lava, 123
    lava flows, 196
    lava plateaus, 198
    magma, 121–123, 195
    Mount Saint Helens, 195–196,
        199–200
    pyroclastic materials, 196
    Ring of Fire, 194
    shield, 197
    sills, 199
    stratovolcanoes, 198
    structure of, 197
    types of, 196–198
Voyager 1, 64
Voyager 2, 64

W

Warm front, 351–353
Wasting, mass, 210
Watch, 358
Water. *See also* Ocean
    boiling point of, 228
    density of, 271
    dissolved minerals in, 229
    distribution of, 226
    freezing of, 228
    groundwater. *see* Groundwater
    hydrogen bonding, 226
    polarity of, 229
    properties of, 226–230

    running, 208
    seawater. *see* Seawater
    shortage of, 226
    solvent uses, 229
    specific heat of, 229
    surface tension of, 228
    warming trend in, 230
Water molecule, 226–227
Watershed, 234–236
Wave(s)
    breaking, 275–276
    causes of, 275
    characteristics of, 275
    erosion caused by, 276–277
Wave height, 275
Wave period, 275
Wavelength, 275
Weather forecasting
    description of, 364
    modern methods of, 365
Weathering, 208–209
Wells, 232–233
Wentworth scale, 128
Westerlies, 336, 363, 385
Western seaboard, 384–386
White caps, 275
White dwarf, 37, 39
Wind
    air pressure and, relationship
        between, 339–340
    description of, 209
    factors that affect, 340–341
    thunderstorm, 357–359